Advances in Industrial Control

Other titles published in this series:

Digital Controller Implementation and Fragility
Robert S.H. Istepanian and James F. Whidborne (Eds.)

Optimisation of Industrial Processes at Supervisory Level
Doris Sáez, Aldo Cipriano and Andrzej W. Ordys

Robust Control of Diesel Ship Propulsion
Nikolaos Xiros

Hydraulic Servo-systems
Mohieddine Jelali and Andreas Kroll

Model-based Fault Diagnosis in Dynamic Systems Using Identification Techniques
Silvio Simani, Cesare Fantuzzi and Ron J. Patton

Strategies for Feedback Linearisation
Freddy Garces, Victor M. Becerra, Chandrasekhar Kambhampati and Kevin Warwick

Robust Autonomous Guidance
Alberto Isidori, Lorenzo Marconi and Andrea Serrani

Dynamic Modelling of Gas Turbines
Gennady G. Kulikov and Haydn A. Thompson (Eds.)

Control of Fuel Cell Power Systems
Jay T. Pukrushpan, Anna G. Stefanopoulou and Huei Peng

Fuzzy Logic, Identification and Predictive Control
Jairo Espinosa, Joos Vandewalle and Vincent Wertz

Optimal Real-time Control of Sewer Networks
Magdalene Marinaki and Markos Papageorgiou

Process Modelling for Control
Benoît Codrons

Computational Intelligence in Time Series Forecasting
Ajoy K. Palit and Dobrivoje Popovic

Modelling and Control of Mini-Flying Machines
Pedro Castillo, Rogelio Lozano and Alejandro Dzul

Ship Motion Control
Tristan Perez

Hard Disk Drive Servo Systems (2nd Ed.)
Ben M. Chen, Tong H. Lee, Kemao Peng and Venkatakrishnan Venkataramanan

Measurement, Control, and Communication Using IEEE 1588
John C. Eidson

Piezoelectric Transducers for Vibration Control and Damping
S.O. Reza Moheimani and Andrew J. Fleming

Manufacturing Systems Control Design
Stjepan Bogdan, Frank L. Lewis, Zdenko Kovačić and José Mireles Jr.

Windup in Control
Peter Hippe

Nonlinear H_2/H_∞ Constrained Feedback Control
Murad Abu-Khalaf, Jie Huang and Frank L. Lewis

Practical Grey-box Process Identification
Torsten Bohlin

Control of Traffic Systems in Buildings
Sandor Markon, Hajime Kita, Hiroshi Kise and Thomas Bartz-Beielstein

Wind Turbine Control Systems
Fernando D. Bianchi, Hernán De Battista and Ricardo J. Mantz

Advanced Fuzzy Logic Technologies in Industrial Applications
Ying Bai, Hanqi Zhuang and Dali Wang (Eds.)

Practical PID Control
Antonio Visioli

(continued after Index)

Iulian Munteanu • Antoneta Iuliana Bratcu
Nicolaos-Antonio Cutululis • Emil Ceangă

Optimal Control of Wind Energy Systems

Towards a Global Approach

Iulian Munteanu, Dr.-Eng.
"Dunărea de Jos" University of Galaţi
Faculty of Electrical Engineering and
  Electronics
Department of Electronics and
  Telecommunications
800008-Galaţi
Romania

Antoneta Iuliana Bratcu, Dr.-Eng.
"Dunărea de Jos" University of Galaţi
Faculty of Electrical Engineering and
  Electronics
Department of Electrical Energy
  Conversion Systems
800008-Galaţi
Romania

Nicolaos-Antonio Cutululis, Dr.-Eng.
Wind Energy Department
Risø National Laboratory
Technical University of Denmark (DTU)
DK-4000 Roskilde
Denmark

Emil Ceangă, Dr.-Eng.
"Dunărea de Jos" University of Galaţi
Faculty of Electrical Engineering and
  Electronics
Department of Electrical Energy
  Conversion Systems
800008-Galaţi
Romania

ISBN 978-1-84800-079-7   e-ISBN 978-1-84800-080-3

DOI 10.1007/978-1-84800-080-3

Advances in Industrial Control series ISSN 1430-9491

British Library Cataloguing in Publication Data
A catalogue record for this book is available from the British Library

Library of Congress Control Number: 2007942442

© 2008 Springer-Verlag London Limited

MATLAB® and Simulink® are registered trademarks of The MathWorks, Inc., 3 Apple Hill Drive, Natick, MA 01760-2098, USA. http://www.mathworks.com

Apart from any fair dealing for the purposes of research or private study, or criticism or review, as permitted under the Copyright, Designs and Patents Act 1988, this publication may only be reproduced, stored or transmitted, in any form or by any means, with the prior permission in writing of the publishers, or in the case of reprographic reproduction in accordance with the terms of licences issued by the Copyright Licensing Agency. Enquiries concerning reproduction outside those terms should be sent to the publishers.

The use of registered names, trademarks, etc. in this publication does not imply, even in the absence of a specific statement, that such names are exempt from the relevant laws and regulations and therefore free for general use.

The publisher makes no representation, express or implied, with regard to the accuracy of the information contained in this book and cannot accept any legal responsibility or liability for any errors or omissions that may be made.

Cover design: eStudio Calamar S.L., Girona, Spain

Printed on acid-free paper

9 8 7 6 5 4 3 2 1

springer.com

Advances in Industrial Control

Series Editors

Professor Michael J. Grimble, Professor of Industrial Systems and Director
Professor Michael A. Johnson, Professor (Emeritus) of Control Systems and Deputy Director

Industrial Control Centre
Department of Electronic and Electrical Engineering
University of Strathclyde
Graham Hills Building
50 George Street
Glasgow G1 1QE
United Kingdom

Series Advisory Board

Professor E.F. Camacho
Escuela Superior de Ingenieros
Universidad de Sevilla
Camino de los Descubrimientos s/n
41092 Sevilla
Spain

Professor S. Engell
Lehrstuhl für Anlagensteuerungstechnik
Fachbereich Chemietechnik
Universität Dortmund
44221 Dortmund
Germany

Professor G. Goodwin
Department of Electrical and Computer Engineering
The University of Newcastle
Callaghan
NSW 2308
Australia

Professor T.J. Harris
Department of Chemical Engineering
Queen's University
Kingston, Ontario
K7L 3N6
Canada

Professor T.H. Lee
Department of Electrical Engineering
National University of Singapore
4 Engineering Drive 3
Singapore 117576

Professor Emeritus O.P. Malik
Department of Electrical and Computer Engineering
University of Calgary
2500, University Drive, NW
Calgary
Alberta
T2N 1N4
Canada

Professor K.-F. Man
Electronic Engineering Department
City University of Hong Kong
Tat Chee Avenue
Kowloon
Hong Kong

Professor G. Olsson
Department of Industrial Electrical Engineering and Automation
Lund Institute of Technology
Box 118
S-221 00 Lund
Sweden

Professor A. Ray
Pennsylvania State University
Department of Mechanical Engineering
0329 Reber Building
University Park
PA 16802
USA

Professor D.E. Seborg
Chemical Engineering
3335 Engineering II
University of California Santa Barbara
Santa Barbara
CA 93106
USA

Doctor K.K. Tan
Department of Electrical Engineering
National University of Singapore
4 Engineering Drive 3
Singapore 117576

Professor Ikuo Yamamoto
The University of Kitakyushu
Department of Mechanical Systems and Environmental Engineering
Faculty of Environmental Engineering
1-1, Hibikino,Wakamatsu-ku, Kitakyushu, Fukuoka, 808-0135
Japan

To our families

Series Editors' Foreword

The series *Advances in Industrial Control* aims to report and encourage technology transfer in control engineering. The rapid development of control technology has an impact on all areas of the control discipline. New theory, new controllers, actuators, sensors, new industrial processes, computer methods, new applications, new philosophies…, new challenges. Much of this development work resides in industrial reports, feasibility study papers and the reports of advanced collaborative projects. The series offers an opportunity for researchers to present an extended exposition of such new work in all aspects of industrial control for wider and rapid dissemination.

Electrical power generation from wind energy conversion systems is a growth industry in the European Union, as it is globally. Targets within the countries of the EU are set at 12% market share by 2020 but, as the authors of this *Advances in Industrial Control* monograph observe: wind energy conversion at the parameter and technical standards imposed by the energy markets is not possible without the essential contribution of automatic control. In keeping with this assertion, authors Iulian Munteanu, Antoneta Iuliana Bratcu, Nicolaos-Antonio Cutululis and Emil Ceangă proceed to outline their vision of how control engineering techniques can contribute to the control of various types of wind turbine power systems. The result is a wide-ranging monograph that begins from the basic characteristics of wind as a renewable energy resource and finishes at hardware-in-the-loop concepts and test-rigs for the assessment of prototype controller solutions.

The research journey passes through those phases that are common to any in-depth investigation into the control of a complex nonlinear industrial system. Understanding the wind energy process and deriving models and performance specifications occupies the first three chapters of the monograph. The next three then concentrate on control designs as they evolve to meet more complex sets of performance objectives. The monograph concludes with an assessment of the value that can be obtained from hardware-in-the-loop performance tests.

Thus, *Optimal Control of Wind Energy Systems* with its full assessment of a variety of optimal control strategies makes a welcome contribution to the wind power control literature. The volume nicely complements the *Advances in Industrial Control* monograph *Wind Turbine Control Systems: Principles,*

Modelling and Gain Scheduling Design by Fernando Bianchi and his colleagues that was published in July 2006. Together these volumes provide a thorough research framework for the study of the control of wind energy conversion systems.

Industrial Control Centre	*M.J. Grimble*
Glasgow	*M.A. Johnson*
Scotland, UK	
2007	

Preface

Actual strategies for sustainable energy development have as prior objective the gradual replacement of fossil-fuel-based energy sources by renewable energy ones. Among the clean energy sources, wind energy conversion systems currently carry significant weight in many developed countries. Following continual efforts of the international research community, a mature wind energy conversion technology is now available to sustain the rapid dynamics of concerned investment programs.

The main problem regarding wind power systems is the major discrepancy between the irregular character of the primary source (wind speed is a random, strongly non-stationary process, with turbulence and extreme variations) and the exigent demands regarding the electrical energy quality: reactive power, harmonics, flicker, *etc.* Thus, wind energy conversion within the parameters imposed by the energy market and by technical standards is not possible without the essential contribution of automatic control.

The stochastic nature of the primary energy source represents a risk factor for the viability of the mechanical structure. The literature concerned emphasises the importance of the reliability criterion, sometimes more important than energy conversion efficiency (*e.g.*, in the case of off-shore farms), in assessing global economic efficiency. This aspect must be taken into account in control strategies.

Many research works deal with wind power systems control, aiming at optimising the energetic conversion, interfacing wind turbines to the grid and reducing the fatigue load of the mechanical structure. Meanwhile, the gap between the development of advanced control algorithms and their effective use in most of the practical engineering domain is widely recognized. Much work has been and continues to be done, especially by the research community, in order to bridge this gap and ease the technology transfer in control engineering.

This book is aimed at presenting a point of view on the wind power generation optimal control issues, covering a large segment of industrial wind power applications. Its main idea is to propose the use of a set of optimization criteria which comply with a comprehensive set of requirements, including the energy conversion efficiency, mechanical reliability, as well as quality of the energy provided. This idea opens the perspective toward a multi-purpose global control approach.

A series of control techniques are analyzed, assessed and compared, starting from the classical ones, like PI control, maximum power point strategies, LQG optimal control techniques, and continuing with some modern ones: sliding-mode techniques, feedback linearization control and robust control. The discussion is aimed at identifying the benefits of dynamic optimization approaches to wind power systems. The main results are presented along with illustration by case studies and MATLAB®/Simulink® simulation assessment. The corresponding software programmes and block diagrams are included on the back-of-book software material. For some of the case studies presented real-time simulation results are also available.

The discourse of this book concludes by stressing the point on the possibility of designing WECS control laws based upon the frequency separation principle. The idea behind this is simple. First, one must define the set of quality demands the control law must comply with. Then one seeks to split this set into contradictory pairs, for each of them a component of the control law being separately synthesized. Finally, these components are summed to yield the total control input. This approach is possible because the different WECS dynamic properties usually involved in the imposed quality requirements are exhibited in disjointed frequency ranges.

Offering a thorough description of wind energy conversion systems – principles, functionality, operation modes, control goals and modelling – this book is mainly addressed to researchers with a control background wishing either to approach or to go deeper in their study of wind energy systems. It is also intended to be a guide for control engineers, researchers and graduate students working in the field in learning and applying systematic optimization procedures to wind power systems.

The book is organised in seven chapters preceded by a glossary and followed by a concluding chapter, three appendices, a list of pertinent references and an index.

Chapter 1 realises an introduction about the wind energy resource and systems. Chapter 2 presents a systemic analysis of the main parts of a wind energy conversion system and introduces the associated control objectives. The modelling development needed for control purposes is presented in the Chapter 3. Chapter 4 is dedicated to explaining the fundamentals of the wind turbine control systems. In Chapter 5 some powerful control methods for energy conversion maximization are presented, each of which is illustrated by a case study. Chapter 6 deals with mixed optimization criteria and introduces the frequency separation principle in the optimal control of the wind energy systems, whose effectiveness is suggested by two case studies. Chapter 7 is focused on using the hardware-in-the-loop simulation philosophy for building development systems that experimentally validate the wind energy systems control laws. A case study is presented to illustrate the proposed methodology. Chapter 8 discusses general conclusions and suggestions for future development of WECS control laws.

Appendix A offers detailed information about the features of systems used in the case studies. Appendix B resumes the main theoretical results supporting the sliding-mode, feedback linearization and QFT robust control methods. Finally,

Appendix C presents some illustrations accompanying the implementation of the reported case studies.

We would like to acknowledge the Romanian National Authority for Scientific Research (ANCS – CEEX Research Programme) and the Romanian National University Research Council (CNCSIS) for their partial financial support during the period in which this manuscript was written.

Galați and Roskilde,
August 2007

Iulian Munteanu
Antoneta Iuliana Bratcu
Nicolaos-Antonio Cutululis
Emil Ceangă

Contents

Notation xix

1 Wind Energy 1
   1.1   Introduction 1
   1.2   State of the Art and Trends in Wind Energy Conversion Systems 1
         1.2.1   Issues in WECS Technology 2
         1.2.2   Wind Turbines 3
         1.2.3   Low-power WECS 5
         1.2.4   Issues in WECS Control 5
   1.3   Outline of the Book 6

2 Wind Energy Conversion Systems 9
   2.1   Wind Energy Resource 9
   2.2   WECS Technology 13
   2.3   Wind Turbine Aerodynamics 15
         2.3.1   Actuator Disc Concept 15
         2.3.2   Wind Turbine Performance 16
   2.4   Drive Train 19
   2.5   Power Generation System 19
         2.5.1   Fixed-speed WECS 20
         2.5.2   Variable-speed WECS 21
   2.6   Wind Turbine Generators in Hybrid Power Systems 23
   2.7   Control Objectives 25

3 WECS Modelling 29
   3.1   Introduction and Problem Statement 29
   3.2   Wind Turbine Aerodynamics Modelling 30
         3.2.1   Fixed-point Wind Speed Modelling 30
         3.2.2   Wind Turbine Characteristics 37
         3.2.3   Wind Torque Computation Based on the Wind Speed Experienced by the Rotor 42

3.3 Electrical Generator Modelling 46
    3.3.1 Induction Generators 47
    3.3.2 Synchronous Generators 51
3.4 Drive Train Modelling 54
    3.4.1 Rigid Drive Train 55
    3.4.2 Flexible Drive Train 56
3.5 Power Electronics Converters and Grid Modelling 57
3.6 Linearization and Eigenvalue Analysis 60
    3.6.1 Induction-generator-based WECS 60
    3.6.2 Synchronous-generator-based WECS 66
3.7 Case Study (1): Reduced-order Linear Modelling
    of a SCIG-based WECS 69

4 Basics of the Wind Turbine Control Systems 71
4.1 Control Objectives 71
4.2 Physical Fundamentals of Primary Control Objectives 72
    4.2.1 Active-pitch Control 73
    4.2.2 Active-stall Control 73
    4.2.3 Passive-pitch Control 74
    4.2.4 Passive-stall Control 74
4.3 Principles of WECS Optimal Control 75
    4.3.1 Case of Variable-speed Fixed-pitch WECS 75
    4.3.2 Case of Fixed-speed Variable-pitch WECS 78
4.4 Main Operation Strategies of WECS 80
    4.4.1 Control of Variable-speed Fixed-pitch WECS 80
    4.4.2 Control of Variable-pitch WECS 86
4.5 Optimal Control with a Mixed Criterion: Energy Efficiency – Fatigue
    Loading 90
4.6 Gain-scheduling Control for Overall Operation 92
4.7 Control of Generators in WECS 95
    4.7.1 Vector Control of Induction Generators 95
    4.7.2 Control of Permanent-magnet Synchronous Generators 100
4.8 Control Systems for Grid-connected Operation and Energy Quality
    Assessment 101
    4.8.1 Power System Stability 101
    4.8.2 Power Quality 106

5 Design Methods for WECS Optimal Control with Energy Efficiency
Criterion 109
5.1 General Statement of the Problem and State of the Art 109
    5.1.1 Optimal Control Methods Using the Nonlinear Model 110
    5.1.2 Optimal Control Methods Using the Linearized Model 113
    5.1.3 Concluding Remarks 115
5.2 Maximum Power Point Tracking (MPPT) Strategies 116
    5.2.1 Problem Statement and Literature Review 116
    5.2.2 Wind Turbulence Used for MPPT 119

		5.2.3	Case Study (2): Classical MPPT *vs.* MPPT with Wind Turbulence as Searching Signal 124

- 5.2.3 Case Study (2): Classical MPPT *vs.* MPPT with Wind Turbulence as Searching Signal 124
- 5.2.4 Conclusion 128
- 5.3 PI Control 129
  - 5.3.1 Problem Statement 129
  - 5.3.2 Controller Design 130
  - 5.3.3 Case Study (3): 2 MW WECS Optimal Control by PI Speed Control 132
  - 5.3.4 Case Study (4): 6 kW WECS Optimal Control by PI Power Control 134
- 5.4 On–Off Control 135
  - 5.4.1 Controller Design 135
  - 5.4.2 Case Study (5) 140
- 5.5 Sliding-mode Control 142
  - 5.5.1 Modelling 143
  - 5.5.2 Energy Optimization with Mechanical Loads Alleviation 143
  - 5.5.3 Case Study (6) 146
  - 5.5.4 Real-time Simulation Results 147
  - 5.5.5 Conclusion 150
- 5.6 Feedback Linearization Control 150
  - 5.6.1 WECS Modelling 151
  - 5.6.2 Controller Design 152
  - 5.6.3 Case Study (7) 156
- 5.7 QFT Robust Control 158
  - 5.7.1 WECS Modelling 158
  - 5.7.2 QFT-based Control Design 158
  - 5.7.3 Case Study (8) 160
- 5.8 Conclusion 166

6 WECS Optimal Control with Mixed Criteria 169
- 6.1 Introduction 169
- 6.2 LQ Control of WECS 170
  - 6.2.1 Problem Statement 170
  - 6.2.2 Input–Output Approach 170
  - 6.2.3 Case Study (9): LQ Control of WECS with Flexibly-coupled Generator Using R-S-T Controller 173
- 6.3 Frequency Separation Principle in the Optimal Control of WECS 176
  - 6.3.1 Frequency Separation of the WECS Dynamics 176
  - 6.3.2 Optimal Control Structure and Design Procedure (2LFSP) 177
  - 6.3.3 Filtering and Prediction Algorithms for Wind Speed Estimation 180
- 6.4 2LFSP Applied to WECS with Rigidly-coupled Generator 182
  - 6.4.1 Modelling 182
  - 6.4.2 Steady-state Optimization Within the Low-frequency Loop 185
  - 6.4.3 LQG Dynamic Optimization Within the High-frequency Loop. 185
  - 6.4.4 LQ Dynamic Optimization Within the High-frequency Loop.... 187
  - 6.4.5 Case Study (10) 190
  - 6.4.6 Global Real-time Simulation Results 193

xviii Contents

    6.5   2LFSP Applied to WECS with Flexibly-coupled Generator 197
            6.5.1   Modelling 197
            6.5.2   Steady-state Optimization Within the Low-frequency Loop 199
            6.5.3   Dynamic Optimization Within the High-frequency Loop 199
            6.5.4   Case Study (11) 201
    6.6   Concluding Remarks on the Effectiveness of 2LFSP 204
    6.7   Towards a Multi-purpose Global Control Approach 205
            6.7.1   Control Objectives in Large Wind Power Plants 205
            6.7.2   Global Optimization *vs.* Frequency Separation Principle
                    for a Multi-objective Control 206
            6.7.3   Frequency-domain Models of WECS 208
            6.7.4   Spectral Characteristics of the Wind Speed Fluctuations 209
            6.7.5   Open-loop Bandwidth Limitations of WECS Control Systems . 211
            6.7.6   Frequency Separation Control of WECS 214

**7   Development Systems for Experimental Investigation
of WECS Control Structures 219**
    7.1   Introduction 219
    7.2   Electromechanical Simulators for WECS 220
            7.2.1   Principles of Hardware-in-the-loop (HIL) Systems 220
            7.2.2   Systematic Procedure of Designing HIL Systems 223
            7.2.3   Building of Physical Simulators for WECS 223
            7.2.4   Error Assessment in WECS HIL Simulators 225
    7.3   Case Study (12): Building of a HIL Simulator for a DFIG-based
          WECS 229
            7.3.1   Requirements Imposed to the WECS Simulator 230
            7.3.2   Building of the Real-time Physical Simulator (RTPS) 230
            7.3.3   Building of the Investigated Physical System (IPS) and
                    Electrical Generator Control 233
            7.3.4   Global Operation of the Simulated WECS 236
    7.4   Conclusion 237

8   General Conclusion 239

A   Features of WECS Used in Case Studies 243

B   Elements of Theoretical Background and Development 247
    B.1   Sliding-mode Control 247
    B.2   Feedback Linearization Control 249
    B.3   QFT Robust Control 255

C   Photos, Diagrams and Real-time Captures 261

References 269

Index 281

Notation

Wind Power System

Aerodynamic Subsystem and Drive Train

v, v_s, v_t	Total, steady-state and turbulence wind speed [m/s]
w	Relative wind speed to the blades [m/s]
ρ	Air density [kg/m³]
I_t, L_t	Turbulence intensity [–] and length [m]
Ω_l, Ω_h	Rotational speed of a wind turbine rotor (low-speed shaft) and of the high-speed shaft respectively [rad/s]
R	Blade length of a wind turbine [m]
β	Pitch angle
N_b	Number of blades of a wind turbine
$A = \pi \cdot R^2$	Area swept by the rotor blades [m²]
σ_P	Prandtl's coefficient [–]
λ, λ_{opt}	Tip speed ratio of a wind turbine and its optimal value [–]
v_S, v_n, v_M	Cut-in, rated and respectively cut-out wind speed of a wind turbine [m/s]
P_{air}	Total power of a delimited moving mass of air [W]
P, P_{em}	Generated active power and generator mechanical power respectively [W]
P_{wt}, P_n	Harvested and respectively rated power of a wind turbine [W]

C_p, $C_{p_{max}} \equiv C_{p_{opt}}$	Power coefficient of a wind turbine and its maximum value [–]
Γ_{wt}, Γ_h	Torque of a wind turbine rotor (low-speed shaft) and of the high-speed shaft respectively [N·m]
P_{opt}	Maximum captured power from wind [W]
$\Gamma_{opt} = \Gamma_{wt}(\lambda_{opt}, v)$	Wind torque corresponding to the optimal tip speed ratio [N·m]
C_Γ, $C_{\Gamma_{max}}$	Torque coefficient of a wind turbine and its maximum value [–]
$C_{\Gamma_{opt}} = C_\Gamma(\lambda_{opt})$	Torque coefficient corresponding to λ_{opt} [–]
J_{wt}	Inertia of a wind turbine rotor [Kg·m²]
i	Gear box (drive train) ratio [–]
K_s, B_s	Drive train stiffness [N·m] and damping respectively [N·m·s]
η	Efficiency [–]
Γ	Drive train torsional torque [N·m]

Generators

J_g	Generator shaft inertia [kg·m²]
Γ_G	Generator torque [N·m]
p	Number of pole pairs
ω_S, ω_R	Stator (synchronous) and respectively rotor frequency [rad/s]

Induction Generator

V_S, V_R	Stator and rotor RMS voltage respectively [V]
i_S, i_R	Stator and rotor current respectively [A]
Φ_S, Φ_R	Stator and rotor flux respectively [Wb]
R_S, R_R	Stator and rotor winding resistance respectively [Ω]
L_S, L_R, L_m	Stator, rotor and respectively mutual winding inductance [H]
ω	Rotational speed in electrical rad/s
θ_S, θ_R	Stator and rotor flux vector position respectively [°]
σ	Leakage factor [–]

Synchronous Generator
R_l Load resistance [Ω]
u, i Stator voltage [V] and current [A]
L Stator inductance

Modelling

s	Laplace operator
j	$\sqrt{-1}$
$\dot{x}$	Time derivative of x [units of x / s]
x^*	Reference value for x variable [units of x]
$\bar{x}$	Value of a variable in a given steady-state operating point [units of x]
$\Delta x = \bar{x} - x$	Variation around the steady-state operating point $\bar{x}$ [units of x]
$\overline{\Delta x} = \Delta x / \bar{x}$	Normalized variation around the steady-state operating point $\bar{x}$ [–]
$\sigma(x)$	Standard deviation of x
$S_{xx}(\omega)$	Spectral power density of x
$e(t)$	White noise
$E\{x\}$	Expectation of x
T_w	Time constant of a low-pass shaping filter [s]
γ	Torque parameter of a wind turbine [–]
J_l	Equivalent inertia rendered at the low-speed shaft [kg·m²]
J_h	Equivalent inertia rendered at the high-speed shaft [kg·m²]
J_T, J_G	Time constants of the wind turbine's linearized model [s]
x_d, x_q	Generator x variable in (d,q) axis [units of x]
x_a, x_b, x_c	Generator x variable in (a,b,c) 3-phase system [units of x]
T_G	Time constant describing the equivalent dynamics of the electromagnetic subsystem [s]

Acronyms and Abbreviations

2LFSP	Two-loop control structure based on the frequency separation principle
AS	Aerodynamic subsystem
BET	Blade element theory
BPS	Basic physical system
CS	Control subsystem
DFIG	Doubly-fed induction generator
DFT	Discrete Fourier Transform
DT	Drive train
EFT	Effector (part of the HIL simulator)
EMS	Electromagnetic subsystem
EPS	Emulated physical system
EPSM	Model of the EPS
FFT	Fast Fourier Transform
H/V AWT	Horizontal-/vertical-axis wind turbine
HFL	High-frequency loop
HIL	Hardware-in-the-loop
HILS	Hardware-in-the-loop simulation
HPF	High-pass filter
HSS	High-speed shaft
IPS	Investigated physical system
LFL	Low-frequency loop
LPF	Low-pass filter
LSS	Low-speed shaft
OP	Operating point
OOP	Optimal operating point
ORC	Optimal regimes characteristic
PMSG	Permanent-magnet synchronous generator
PWM	Pulse-width modulation
RTPS	Real-time physical simulator
RTSS	Real-time software simulator
SCIG	Squirrel-cage induction generator
TSC	Tip speed controller
TSR	Tip speed ratio
(V/C S) WECS / WPS	(Variable-/constant-speed) wind energy conversion system/wind power system
WRIG	Wound-rotor induction generator
WRSG	Wound-rotor synchronous generator

1
Wind Energy

1.1 Introduction

The use of the wind has a history of thousands of years. Since ancient times wind power has been used for different purposes, varying from agricultural activities, like grain milling and water pumping to, nowadays, electricity production. Since the early 1970s oil crisis, wind power technology has experienced an important development, moving – in just two decades – from a low level, experimental technology used mainly for batteries charging to a mainstream power technology. Today, wind power is by far the fastest-growing renewable energy source.

Wind power is free, clean and endless. Furthermore, the cost of the electricity produced by wind turbines is fixed once the plant has been built (EWEA 2005) and it has already reached the point where the cost of the electricity produced by wind is comparable with that of electricity produced by some of the conventional, fossil-based power plants (Parfit and Leen 2005).

The power produced by wind worldwide reached, at the end of 2004, 48 GW, representing 0.57% of the total world electricity supply. The figure might not seem impressive, but when compared to other renewable energy technologies, it becomes clear that wind power is the most promising one. As an example, wind power is still a small electricity player on the European market, producing 2.4% of its total electricity production. This will change as the European Union has decided to make wind power a major electricity source, with a 12% market share in 2020 and 20% in 2030 (EWEA 2005).

1.2 State of the Art and Trends in Wind Energy Conversion Systems

Wind energy conversion systems (WECS) constitute a mainstream power technology that is largely underexploited. Wind technology has made major progression from the prototypes of just 25 years ago. Two decades of technological progress has resulted in today's wind turbines looking and being much more like power stations, in addition to being modular and rapid to install. A single wind turbine can produce 200 times more power than its equivalent two decades ago (EWEA 2005). The low-power WECS have not however lost their importance, being nowadays of great interest in islanding generation, hybrid microgrid systems, distributed energy production, *etc*. Today, WECS represent a mature technology still with important development potential.

1.2.1 Issues in WECS Technology

The development of various wind turbine concepts in the last decade has been very dynamic. The main differences in wind turbine concepts are in the electrical design and control. Thus, WECS can be classified according to *speed control* and *power control ability*, leading to wind turbine classes differentiated by the generating system (speed control) and the method employed for limiting the aerodynamic efficiency above the rated power (power control).

The speed-control criterion leads to two types of WECS: *fixed-speed* and *variable-speed* wind turbines, while the power control ability divides WECS into three categories: stall-controlled, pitch-controlled and active-stall-controlled wind turbines.

Fixed-speed WECS
Fixed-speed wind turbines are the pioneers of the wind turbine industry. They are simple, reliable and use low-cost electrical parts. They use induction generators and they are connected directly to the grid, giving them an almost constant rotor speed stuck to the grid frequency, regardless of the wind speed.

Variable-speed WECS
Variable-speed wind turbines are currently the most used WECS. Their advantages, compared to fixed-speed wind turbines, are numerous. First of all and most important, the decoupling between the generating system and the grid frequency makes them more flexible in terms of control and optimal operation. Of course, this comes at a price, namely the use of power electronic converters, which are the interfaces between the electrical generator and the grid and thus they actually make the variable-speed operation possible. But still, the high controllability offered by the variable-speed operation is a powerful advantage in achieving higher and higher wind energy penetration levels (Sørensen *et al.* 2005; Hansen and Hansen 2007).

The variable-speed operation allows the rotational speed of the wind turbine to be continuously adapted (accelerated or decelerated) in such a manner that the wind turbine operates constantly at its highest level of aerodynamic efficiency. While fixed-speed wind turbines are designed to achieve maximum aerodynamic efficiency at one wind speed, variable-speed wind turbines achieve maximum aerodynamic efficiency over a wide range of wind speeds. Furthermore, variable-speed operation allows the use of advanced control methods, with different objectives: reduced mechanical stress, reduced acoustical noise, increased power capture, *etc.* (Ackermann 2005; Burton *et al.* 2001).

Power control ability refers to the aerodynamic performances of wind turbines, especially in the power limiting operation range. All wind turbines have some sort of power control.

Stall-controlled WECS

The simplest form of power control is reducing the aerodynamic efficiency by using the stall effect in high winds without changes in blade geometry. As the wind velocity increases, the rotor aerodynamics drives the rotor in the stall regime "naturally". The key factor in this method is a special design of blade profile, providing accentuated stall effect around rated power without undesired collateral aerodynamic behaviour.

The drawbacks of this power control method are: high mechanical stress caused by wind gusts, no assisted start and variations in the maximum steady-state power due to variations in air density and grid frequency (Hansen and Hansen 2007).

Pitch- and Active-stall-controlled WECS

Another method to control power is modifying the pitch angle, thus modifying the blade geometry. This method, widely used today, implies modifying the so-called pitch angle, thus modifying the way the wind speed is seen by the blade, turning it away or into the wind. Depending on the direction that the blade is turned (upwind or downwind), this method is further split into pitch control and active-stall control. The details of the phenomena and differences between these two methods will be presented later in the book.

The main advantages of controlling the pitch angle are good power control performance, assisted start-up and emergency-stop power reduction. On the other hand, they add cost and complexity due to the pitch mechanism and the control system.

1.2.2 Wind Turbines

Wind turbine design objectives have changed a lot in the past 10 years. Modern wind turbines have become larger and they have moved from being fixed-speed, stall-controlled and with rudimentary control systems to variable-speed, pitch-controlled, drive-train with or without gearboxes and highly controllable. Thus, they have moved from being convention-driven to being optimization-driven.

The market share of the different wind turbine concepts, for the European market, is presented next. The classification is made upon the speed control abilities, leading to four WECS concepts (EWEA 2005): fixed-speed (one or two speeds), limited variable-speed, improved variable-speed and variable-speed with full-scale frequency converter (see Table 1.1).

Table 1.1. WECS concept European market share (EWEA 2005)

WECS Concept	European market share (cum.) %
Fixed-speed	30
Limited variable-speed	10
Improved variable-speed	45
Variable-speed with full-scale frequency converter	15

The characteristics of the wind turbines installed today are given in Table 1.2 (EWEA 2005).

In Hansen and Hansen (2007) a thorough market share trend analysis is presented. The analysis shows that the fixed-speed WECS concept are maintaining a rather stable market share, especially in the United States, due to the Kenetech variable-speed operation patent (Richardson and Erdman 1992), while the limited variable-speed WECS concepts are being phased out of the market. On the other hand, the improved variable-speed WECS concept is clearly the dominant one on the market today. Together with the variable-speed with full-scale frequency converter concept, they seem to represent the future of WECS.

Table 1.2. Wind turbine characteristic (EWEA 2005)

Wind turbine characteristic	<Range>, Typical value
Rated power [MW]	<0.850 – 6.0>, 3.0
Rotor diameter [m]	<58 – 103>, 90
Specific rated power [W/m^2]	<300 – 500>, 470
Capacity factor (=load factor)[a] (%)	<18 – 40>
Full load equivalent[b] (h)	<1800 – 4000>
Specific annual energy output[c] (kW/m^2 year)	<600 – 1500>
Technical availability[d]	<95 – 99>, 97.5

[a] Depends largely on site average wind speed and on matching of specific power and site average wind speed
[b] The same as above
[c] Normalised to rotor swept area, depending on site average wind speed
[d] Values are valid on-shore, including planned outages for regular maintenance

The development of high-power WECS technology is more and more influenced by the grid connection requirements and thus by the power system operators. The important growth of the installed WECS and, more important, the planned increase of the wind power penetration level, bring more and more focus on wind turbine control capabilities to act as conventional power plants.

The development of the different components of a WECS (aerodynamic efficiency, generators, power electronics, *etc.*) depends on the control capabilities of the individual components, as well as on the WECS as a whole.

In conclusion, some generic trends in wind turbines development can be summarized as follows. The interest in fixed-speed wind turbines will keep decreasing, especially since the grid connection requirements are becoming stricter. The current fixed-speed WECS technology cannot meet those requirements. On the other hand, using high-voltage direct-current (HVDC) technology with fixed-speed WECS wind farms could be a solution in meeting the grid code requirements (Hansen *et al.* 2001). Variable-speed wind turbines will probably dominate the market in the future. The focus is on developing very large wind turbines (8–10 MW), both on-shore and off-shore. From a control perspective, the focus is – besides optimal operation – on load reduction, on grid integration and on developing the conventional power plant capabilities of both wind turbines and wind farms (UpWind 2006).

1.2.3 Low-power WECS

Interest has also grown in low-power wind turbines, due to their application in insulated grids and distributed energy production, from which the microgrid concept has emerged (Kanellos and Hatziargyriou 2002).

Low-power WECS are being incorporated both in stand-alone generation systems, as well as elements of hybrid power systems. Related to the latter, some typical applications are the hybrid wind-photovoltaic generation systems or wind turbines in conjunction with fuel-cell/diesel, all of them using accumulator batteries for energy storage.

Because of the very high penetration level, the control problems are here somehow different from those related to wind farms, being strongly dependent on the current application. For example, for water pumping or house heating the control objectives are obviously different from ensuring power quality standards of an insulated utility grid.

Therefore, the main problems in insulated grids relate to the wind energy sources scheduling depending on the instantaneous consumption and on the power reserve from other generators (taking account of energy storage). Besides the captured power maximisation and the reliability-related issues, control focuses on the local power system stability and the delivered power conditioning (fluctuations, harmonics, *etc.*).

In some cases the generators contained in hybrid systems (*e.g.*, wind-photovoltaic-accumulator) feed a common DC-bus. Here, the problem is the global control of the system in order to ensure continuity of power supply, while complying with operation requirements. The latter can relate to the life-time of the system components which must not be affected by the control action (*e.g.*, regularity of accumulator charge/discharge cycles, the diesel-generator on/off regimes, *etc.*).

1.2.4 Issues in WECS Control

The challenge in WECS control is to ensure good quality electrical energy delivery from a profoundly irregular primary source, the wind.

Modern wind generation systems are equipped with control and supervision subsystems implementing the supervisory control and data acquisition (SCADA) concept. Generally, there are three low-level control systems, which are briefly reviewed in the following.

Aerodynamic power control acting on the blades is based upon well-established and widely-used techniques. Industrial applications already benefit from the classical PI or optimal control structure. As regards *generator control* ensuring variable-speed operation, the literature offers a multitude of control techniques waiting for field testing; however, none of them has become classical such as to be widely used by wind turbine integrators. A unitary variable-speed strategy has not yet been established and the real-world applications actually implement only the basic control laws. Finally, *grid interface control and output power conditioning* are intensively researched because the grid connection standards are continuously changing. The control objectives, problem formulations and their methods of solution depend greatly on the current generation structure, local utility grid, operating regime (*i.e.*, islanding or grid-connected), *etc*.

Many research works deal with WECS control, aiming at optimising the energy conversion, interfacing wind turbines to the grid and even reducing the fatigue load of the mechanical structure. The idea of building unitary approaches based on optimization criteria, complying with a comprehensive set of requirements that depend on the actual application, opens the perspective toward a multi-criteria global control approach.

1.3 Outline of the Book

The book is organised in eight chapters preceded by a glossary and followed by three appendices, a list of references and an index.

After this first, introductory chapter, in the second chapter the wind energy resource is presented and the main parts of a wind energy conversion system are analysed from a functional point of view: the turbine (rotor), the drive train and the electrical subsystem. The associated control objectives are stated at the end of this chapter.

The modelling development needed for control purposes is presented in the third chapter. The analysis starts with the exogenous variable, namely the wind, and provides fixed-point wind speed models and also models of the wind speed experienced by the turbine rotor. Then the models of the subsystems described in the previous chapter are detailed. This chapter ends with a case study illustrating the dynamic properties analysis of a class of wind power systems.

The fourth chapter is dedicated to explaining the fundamentals of the wind turbine control systems. Here are included the closed-loop systems to fulfil the so-called primary objectives – stall and pitch control – as well as more advanced control systems generally derived from mixed optimization criteria. Controllers for the reactive power and for the energy quality when operating under grid conditions are also presented.

In the fifth chapter some powerful control methods for energy conversion maximization in the partial-load regime are presented, which can be classified depending on how rich the knowledge is that they use about the system. Each such method is illustrated by a case study in order to allow the assessment of their performances and drawbacks. The conclusion of this chapter suggests the idea of expressing the various WECS control requirements by mixed criteria.

When mixed optimization criteria are formulated, for example, if, apart the energy conversion maximization, a mechanical reliability constraint is imposed, then more complex control structures are needed. In the sixth chapter the frequency separation principle in the optimal control of the wind energy systems is formulated, which is fundamental for the intended design methodology. Two case studies are presented here to illustrate the application of this principle to rigidly- and flexibly-coupled-generator-based wind power systems.

The seventh chapter deals with development systems used for experimentally validating the control laws associated with wind power systems. These experimental simulators are based on the hardware-in-the-loop (HIL) philosophy, consisting of closed-loop connecting hardware and software elements, in order to replicate the real-world systems and their operating conditions. A case study is presented to illustrate the closed-loop optimised functioning of an induction-generator-based variable-speed wind energy conversion system.

The last chapter of the book presents some general conclusions and suggests future directions in developing WECS control laws.

Appendix A provides extensive information about the parameters of WECS used in the case studies. Both low- and high-power, rigid- and flexible-drive-train, induction- or permanent-magnet-synchronous-generator-based WECS have been chosen as illustrative examples. Appendix B resumes the main theoretical results supporting the sliding-mode, feedback linearization and QFT robust control methods. Appendix C groups together some photos, diagrams and real-time captures that accompany the implementation of the reported case studies.

2

Wind Energy Conversion Systems

2.1 Wind Energy Resource

The characteristics of the wind energy resource are important in different aspects regarding wind energy exploitation. The first step in every wind energy project is the identification of suitable sites and prediction of the economic viability of the wind project.

The energy available in the wind varies as the cube of wind speed. Wind is highly variable, both in space and in time. The importance of this variability becomes critical since it is amplified by the cubic relation of the available energy.

The variability in time of the wind can be divided into three distinct time scales (Burton *et al.* 2001). First, the large time scale variability describes the variations of the amount of wind from one year to another, or even over periods of decades or more. The second is the medium time scale, covering periods up to a year. These seasonal variations of the wind are much more predictable. Therefore, the suitability of a given site, in terms of wind variability, is usually assessed in terms of monthly variations, covering one year. The assessment is done by statistical analysis of long time (several years) measurements of wind speed. Finally, the short term time scale variability, covering time scales of minutes to seconds, called turbulence, is also well known and it presents interest in the wind turbine design process.

The medium time scale wind variability, further called monthly variation, is typically characterized in terms of probability distribution over one year. The Weibull distribution is commonly used to fit the wind speed frequency distribution (Burton *et al.* 2001).

The Weibull distribution is a two-parameter function widely used in statistical analysis and is given by Seguro and Lambert (2000)

$$P(v < v_i < v + dv) = P(v > 0)\left(\frac{k}{c}\right)\left(\frac{v_i}{c}\right)^{k-1} \exp\left[-\left(\frac{v_i}{c}\right)^k\right], \qquad (2.1)$$

where c is the Weibull *scale parameter*, with units equal to the wind speed units, k is the unitless Weibull *shape parameter*, v is the wind speed, v_i is a particular wind speed, dv is the wind speed increment, $P(v < v_i < v + dv)$ is the probability that the wind speed is between v and $v + dv$ and $P(v > 0)$ is the probability that the wind speed exceeds zero.

The cumulative distribution function is given by

$$P(v < v_i) = P(v \geq 0)\left\{1 - \exp\left[-\left(\frac{v_i}{c}\right)^k\right]\right\} \tag{2.2}$$

The two Weibull parameters and the average wind speed are related by

$$\bar{v} = c \cdot \Gamma\left(1 + \frac{1}{k}\right), \tag{2.3}$$

where $\bar{v}$ is the average wind speed and $\Gamma(\cdot)$ is the complete gamma function.

A special case is when $k = 2$, the Weibull distribution becoming a Rayleigh distribution. In this case, the factor $\Gamma(1+1/k)$ has the value $\sqrt{\pi}/2 = 0.8862$. The influence of the k parameter on the probability density function is presented in Figure 2.1, with the scale factor kept constant. Simply speaking, the variation of the hourly mean speed around the annual mean is small as k is higher, as depicted in Figure 2.1.

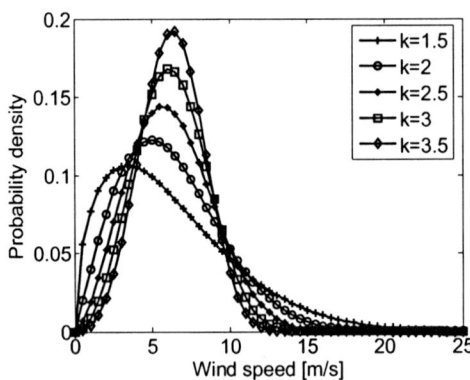

Figure 2.1. Weibull distributions as a function of k (constant c)

The scale factor c shows how "windy" a location is or, in other words, how high the annual mean speed is. The influence of the scale factor on the probability density function is presented in Figure 2.2, with the shape factor kept constant.

The estimation of the Weibull distribution parameters – c and k – is usually done with two methods.

One method for calculating parameters c and k is starting from Equation 2.2 and taking the natural logarithm of both sides:

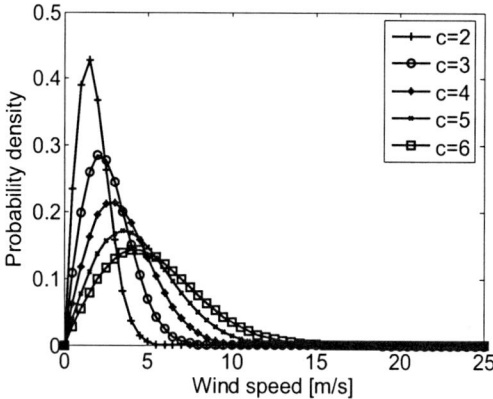

Figure 2.2. Weibull distributions as a function of c (constant k)

$$\ln\left[-\ln\left(1-P(v_i)\right)\right] = k\left(\ln v_i - \ln c\right) \quad (2.4)$$

Using the notation

$$y = \ln\left[-\ln\left(1-P(v_i)\right)\right]; \quad u = \ln v_i; \quad k_0 - \ln c, \quad (2.5)$$

Equation 2.4 becomes:

$$y = ku + k_0, \quad (2.6)$$

where k and k_0 are calculated using linear regression of the cumulative distribution function.

Finally, the scale factor c is calculated:

$$c = \exp\left(-\frac{k}{k_0}\right) \quad (2.7)$$

The second method, the maximum likelihood method (Seguro and Lambert 2000), uses the time-series wind data instead of the cumulative distribution function. The two parameters are calculated respectively with

$$k = \left(\frac{\sum_{i=1}^{n} v_i^k \ln(v_i)}{\sum_{i=1}^{n} v_i^k} - \frac{\sum_{i=1}^{n} \ln(v_i)}{n}\right)^{-1}, \quad (2.8)$$

$$c = \left(\frac{1}{n}\sum_{i=1}^{n} v_i^k\right)^{1/k}, \quad (2.9)$$

where v_i is the wind speed in time step i and n is the number of nonzero wind speed data points. Since Equation 2.8 must be solved using an iterative procedure, it is suitable to start with the initial guess $k = 2$.

Knowing the annual variation of the wind on a given site is important, but it is not sufficient for assessing the economic viability of the wind turbine installation. For that purpose, the level of wind resource is often defined in terms of the wind-power-density value, expressed in watts per square meter (W/m²). This value incorporates the combined effects of the wind speed frequency distribution and the dependence on the air density and the cube of the wind speed.

The power of the wind over an area A is given by

$$P_t = \frac{1}{2}\rho A v^3, \qquad (2.10)$$

where ρ is the air density.

Thus, the *mean wind power density*, over an area A, can be calculated with

$$\frac{P_{t_{mean}}}{A} = \frac{1}{2}\rho \langle v^3 \rangle, \qquad (2.11)$$

where the mean value of the cube of the wind speed is

$$\langle v^3 \rangle = \int_0^\infty v^3 p(v) dv, \qquad (2.12)$$

with $p(v)$ being the probability density function.

After integrating Equation 2.12 and using the Weibull function

$$\langle v^3 \rangle = \frac{\Gamma(1+3/k)}{\Gamma^3(1+1/k)} \langle v^3 \rangle, \qquad (2.13)$$

where

$$e(k) = \frac{\Gamma(1+3/k)}{\Gamma^3(1+1/k)} \qquad (2.14)$$

is called *energy pattern factor* (EFP – Jamil 1990). Using Equations 2.13 and 2.14, Equation 2.11 becomes

$$\frac{P_{t_{mean}}}{A} = \frac{1}{2}\rho \cdot e(k) \cdot \langle v^3 \rangle \qquad (2.15)$$

Thus, the mean wind power density is proportional to the EFP and the cube of the wind speed.

The evolution of the mean wind power density as a function of the Weibull distribution function parameters, c and k, is presented in Figure 2.3.

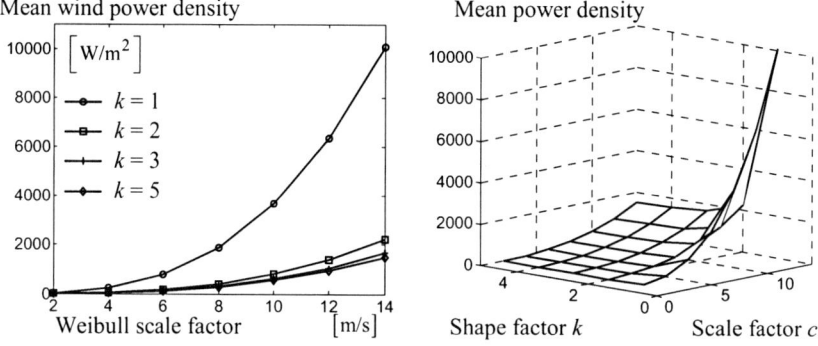

Figure 2.3. Mean wind power density

When the power conversion efficiency (power coefficient; see Section 2.3.1) is constant, the wind speed, denoted v_{opt}, for which the maximum energy is obtained from the condition

$$\frac{d}{dv}\left[v^3 p(v)\right] = 0, \qquad (2.16)$$

which, using the Weibull distribution, results in (Jamil 1990)

$$v_{opt} = c \cdot (1 + 2/k)^{1/k} \qquad (2.17)$$

On the other hand, the *most probable wind speed*, denoted v^* for the given site, can be deduced from the condition

$$\frac{d}{dv}[v \cdot p(v)] = 0 \qquad (2.18)$$

as

$$v^* = c \cdot (1 - 1/k)^{1/k} \qquad (2.19)$$

The pair (v_{opt}, v^*) calculated for a given site by using the Weibull distribution, can offer useful *qualitative* information about the wind energy resource.

In conclusion, the Weibull distribution gives information about the annual variation of the wind speed as well as on the mean power density of a given site. A good case scenario is having a site which can be characterized by a Weibull distribution with a high scale factor (c) and a reduced shape factor (k).

2.2 WECS Technology

A WECS is a structure that transforms the kinetic energy of the incoming air stream into electrical energy. This conversion takes place in two steps, as follows.

The extraction device, named *wind turbine rotor* turns under the wind stream action, thus harvesting a mechanical power. The rotor drives a rotating electrical machine, the generator, which outputs electrical power.

Several wind turbine concepts have been proposed over the years. A historical survey of wind turbine technology is beyond the scope here, but someone interested can find that in Ackermann (2005). There are two basic configurations, namely *vertical axis wind turbines* (VAWT) and, *horizontal axis wind turbines* (HAWT). Today, the vast majority of manufactured wind turbines are horizontal axis, with either two or three blades.

HAWT is comprised of the tower and the nacelle, mounted on the top of the tower (Figure 2.4). Except for the energy conversion chain elements, the nacelle contains some control subsystems and some auxiliary elements (*e.g.*, cooling and braking systems, *etc.*).

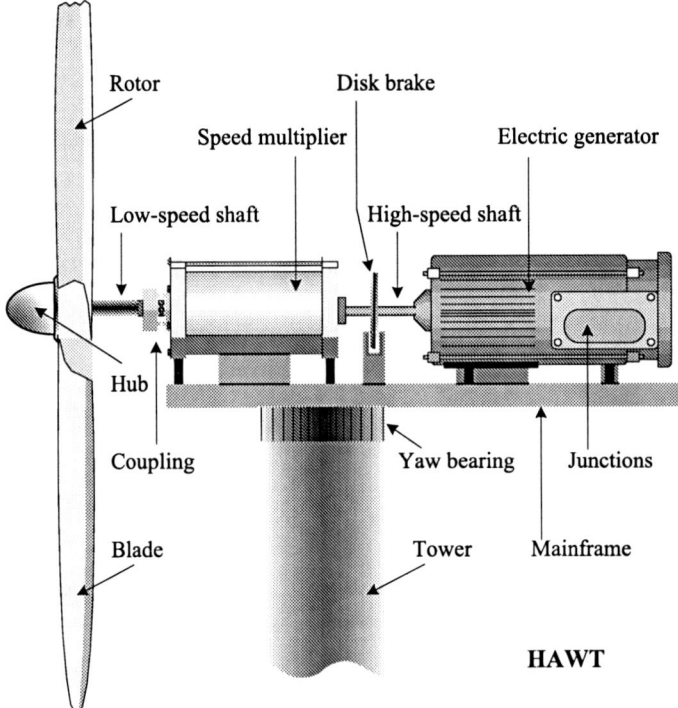

Figure 2.4. Main elements of a two-bladed HAWT

The energy conversion chain is organised into four subsystems:
- aerodynamic subsystem, consisting mainly of the turbine rotor, which is composed of blades, and turbine hub, which is the support for blades;
- drive train, generally composed of: low-speed shaft – coupled with the turbine hub, speed multiplier and high-speed shaft – driving the electrical generator;
- electromagnetic subsystem, consisting mainly of the electric generator;
- electric subsystem, including the elements for grid connection and local grid.

All wind turbines have a mechanism that moves the nacelle such that the blades are perpendicular to the wind direction. This mechanism could be a tail vane (small wind turbines) or an electric yaw device (medium and large wind turbines).

Concerning the power conversion chain, it involves naturally some loss of power. Because of the nonzero wind velocity behind the wind turbine rotor one can easily understand that its efficiency is less than unity. Also, depending on the operating regime, both the motion transmission and the electrical power generation involve losses by friction and by Joule effect respectively. Being directly coupled one with the other, the energy conversion chain elements dynamically interact, mutually influencing their operation.

2.3 Wind Turbine Aerodynamics

The wind turbine rotor interacts with the wind stream, resulting in a behaviour named aerodynamics, which greatly depends on the blade profile.

2.3.1 Actuator Disc Concept

The analysis of the aerodynamic behaviour of a wind turbine can be done, in a generic manner, by considering the extraction process (Burton *et al.* 2001).

Consider an actuator disc (Figure 2.5) and an air mass passing across, creating a stream-tube.

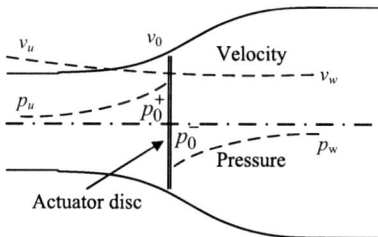

Figure 2.5. Energy extracting actuator disc

The conditions (velocity and pressure) in front of the actuator disc are denoted with subscript u, the ones at the disc are denoted with 0 and, finally, the conditions behind the disc are denoted with w.

The momentum $H = m(v_u - v_w)$ transmitted to the disc by the air mass m passing through the disc with cross-section A produces a force, expressed as

$$T = \frac{\Delta H}{\Delta t} = \frac{\Delta m(v_u - v_w)}{\Delta t} = \frac{\rho A v_0 \Delta t (v_u - v_w)}{\Delta t} = \rho A v_0 (v_u - v_w) \qquad (2.20)$$

or

$$T = A\left(p_0^+ - p_0^-\right) \qquad (2.21)$$

Using Bernoulli's equation, the pressure difference is

$$p_0^+ - p_0^- = \frac{1}{2}\rho\left(v_u^2 - v_w^2\right) \qquad (2.22)$$

and, replacing Equation 2.12 in Equation 2.21, results

$$T = \frac{1}{2}\rho A\left(v_u^2 - v_w^2\right) \qquad (2.23)$$

From Equations 2.20 and 2.23 one gets

$$v_0 = \frac{1}{2}(v_u + v_w) \Rightarrow v_u - v_w = 2(v_u - v_0) \qquad (2.24)$$

The kinetic energy of an air mass travelling with a speed v is

$$E_k = \frac{1}{2}mv^2, \qquad (2.25)$$

where m is the air mass that passes the disc in a unit length of time, e.g., $m = \rho A v_0$; then the power extracted by the disc is

$$P = \frac{1}{2}\rho A v_0 \left(v_u^2 - v_w^2\right) \qquad (2.26)$$

or

$$P = \frac{1}{2}\rho A v^3 4a(1-a)^2, \qquad (2.27)$$

with $a = 1 - v_0/v_u$.

The *power coefficient*, denoting the power extraction efficiency, is defined as

$$C_p = \frac{P}{P_t} = \frac{0.5 \cdot \rho A v^3 \cdot 4a(1-a)^2}{0.5 \cdot \rho A v^3} \qquad (2.28)$$

Therefore

$$C_p = 4a(1-a)^2 \qquad (2.29)$$

The maximum value of C_p occurs for $a = 1/3$ and is $C_{p\max} = 0.59$, known as the Betz limit (Betz 1926) and represents the maximum power extraction efficiency of a wind turbine.

2.3.2 Wind Turbine Performance

A wind turbine is a power extracting device. Thus, the performance of a wind turbine is primarily characterized by the manner in which the main indicator – power – varies with wind speed. Besides that, other indicators like torque and

2.3 Wind Turbine Aerodynamics

thrust are important when the performances of a wind turbine are assessed.

The generally accepted way to characterize the performances of a wind turbine is by expressing them by means of non-dimensional characteristic performance curves (Burton *et al.* 2001).

The *tip speed ratio* of a wind turbine is a variable expressing the ratio between the peripheral blade speed and the wind speed. It is denoted by λ and computed as

$$\lambda = \frac{R \cdot \Omega_l}{v}, \qquad (2.30)$$

where R is the blade length, Ω_l is the rotor speed (the low-speed shaft rotational speed) and v is the wind speed. The tip speed ratio is a key variable in wind turbine control and will be extensively used in the rest of the book. It characterizes the power conversion efficiency and it is also used to define the acoustic noise levels.

The power coefficient, C_p, describes the power extraction efficiency of a wind turbine. The aerodynamic performance of a wind turbine is usually characterized by the variation of the non-dimensional C_p vs. λ curve. Based upon Equation 2.28, the power extracted by a wind turbine whose blade length is R is expressed as

$$P_{wt} = \frac{1}{2} \cdot \rho \cdot \pi R^2 \cdot v^3 \cdot C_p(\lambda) \qquad (2.31)$$

Therefore, the $C_p(\lambda)$ performance curve gives information about the power efficiency of a wind turbine. Figure 2.6 presents this curve for a typical two-bladed wind turbine. One can see that the conversion efficiency is lower than the Betz limit (0.59), which is normal since the Betz limit assumes perfect blade design. The theoretical reasons for such an allure of the $C_p - \lambda$ curve lie in the aerodynamic blade theory; some justifications are given in Chapter 3.

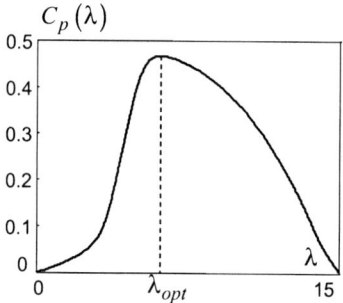

Figure 2.6. $C_p(\lambda)$ performance curve

For control purposes, useful information arising from the $C_p - \lambda$ performance curve is the fact that the power conversion efficiency has a well determined maximum for a specific tip speed ratio, denoted by λ_{opt}.

From Equation 2.31, one finds that the captured power characteristic, $P_{wt} - \Omega_l$, at constant wind velocity, has the same allure as in Figure 2.6. This means that the turbine rotor outputs non-negligible mechanical power if rotating in an intermediary speed range, which depends on the wind speed.

The *torque coefficient*, denoted by C_Γ, characterizes the rotor output (wind) torque, Γ_{wt}. It is derived from the power coefficient simply by dividing it by the tip speed ratio:

$$C_\Gamma(\lambda) = \frac{C_p(\lambda)}{\lambda} \quad (2.32)$$

The torque coefficient *vs.* the tip speed ratio curve, compared to the power coefficient curve, does not give any additional information about the wind turbine performance but it is useful for torque assessment and for control purposes (*e.g.*, assisted start-up process). $C_\Gamma(\lambda)$ gives the rotor mechanical characteristic allure, $\Gamma_{wt} - \Omega_l$, for a fixed wind velocity.

The rotor has a finite number of blades, usually two or three. This number has an impact on the supporting structure; thus, two-bladed wind turbines have a lighter tower top and can be built with a lighter support structure, reducing costs (Gasch and Twelve 2002). On the other hand, three-bladed wind turbines have a balanced rotor inertia and are therefore easier to handle (Thresher *et al.* 1998). Their speed range is smaller than that of the two-bladed wind turbines but the peak output torque is larger.

The wind turbine operates, with different dynamics, from the *cut-in wind speed* (usually 3–4 m/s, for modern wind turbines) to the *cut-out wind speed* (around 25 m/s), as shown in Figure 2.7. The output power evolves according to Equation 2.31 (proportionally with the wind speed cubed), until it reaches the wind turbine rated power. This happens at *rated wind velocity*, which splits the wind turbine operation range in two: below rated (also called partial load region) and full load region, where the captured power must be limited to rated.

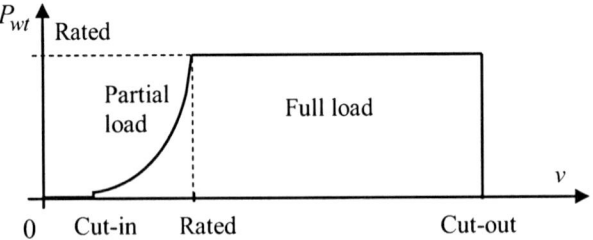

Figure 2.7. Output power *vs.* wind speed characteristic

Therefore, for safety reasons, above the rated wind speed the captured power is prevented from increasing further by using an aerodynamic power control subsystem. This modifies the aerodynamic properties of the rotor by severely decreasing its power coefficient, C_p. To this end, multiple power control solutions are usually employed in WECS. Some of them are passive (*e.g.*, stall control), using blade profile properties; others are active (*e.g.*, pitch control), changing blades position relative to the rotating plane. As further detailed in Section 4.2, the blades can be turned into the wind (upwind) or away from the wind (downwind). Some control solutions aim at turning the entire rotor away from the wind in order to diminish the aerodynamic efficiency.

2.4 Drive Train

The rotational motion of the turbine rotor is transmitted to the electrical generator by means of a mechanical transmission called drive train. Its structure strongly depends on each particular WECS technology. For example, the turbines employing multipole synchronous generators use the direct drive transmission (the generator and the rotor are coupled on the same shaft). But most of the systems (*e.g.*, those employing induction machines) employ speed multipliers (*i.e.*, gearboxes with a certain multiplying ratio) for the mechanical power transmission. Therefore, the electrical machine will experience an increased rotational speed and a reduced electromagnetic torque.

The speed multiplier dissociates the transmission in two parts: the low-speed shaft (LSS) on which the rotor is coupled and the high-speed shaft (HSS) relied on by the electrical generator.

The coupling between the two shafts can be either rigid or flexible. In the second case, the LSS and HSS have different instantaneous rotational speeds. This kind of decoupling is used for damping the mechanical efforts generated either by wind speed or by electromagnetic torque variations. The result is a "compliant" and more reliable transmission, which is less affected by load transients and therefore by mechanical fatigue.

The technology used for speed multiplier construction is out of interest in this book, but one must note that the multiplying ratio depends mostly on the rated power and can generally involve more than one stage (*e.g.*, based on spur or helical gears). The speed multiplier is critical equipment, severely affecting the WECS in terms of weight and reliability, and therefore overall efficiency.

2.5 Power Generation System

The electrical power generation structure contains both electromagnetic and electrical subsystems. Besides the electrical generator and power electronics converter it generally contains an electrical transformer to ensure the grid voltage compatibility. However, its configuration depends on the electrical machine type and on its grid interface (Heier 2006).

2.5.1 Fixed-speed WECS

Fixed-speed WECS operate at constant speed. That means that, regardless of the wind speed, the wind turbine rotor speed is fixed and determined by the grid frequency. Fixed-speed WECS are typically equipped with squirrel-cage induction generators (SCIG), softstarter and capacitor bank and they are connected directly to the grid, as shown in Figure 2.8. This WECS configuration is also known as the "*Danish concept*" because it was developed and widely used in Denmark (Hansen and Hansen 2007).

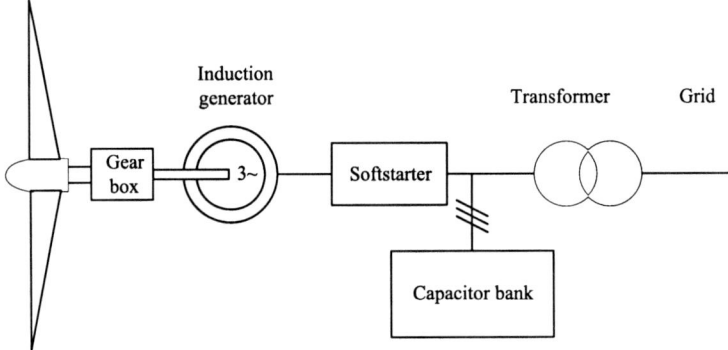

Figure 2.8. General structure of a fixed-speed WECS

Initially, the induction machine is connected in motoring regime such that it generates electromagnetic torque in the same direction as the wind torque. In steady-state, the rotational speed exceeds the synchronous speed and the electromagnetic torque is negative. This corresponds to the squirrel-cage induction machine operation in generation mode (or in the over-synchronous regime – Bose 2001). As it is directly connected to the grid, the SCIG works on its natural mechanical characteristic having an accentuated slope (corresponding to a small slip) given by the rotor resistance. Therefore, the SCIG rotational speed is very close to the synchronous speed imposed by the grid frequency. Furthermore, the wind velocity variations will induce only small variations in the generator speed. As the power varies proportionally with the wind speed cubed, the associated electromagnetic variations are important.

SCIG are preferred because they are mechanically simple, have high efficiency and low maintenance cost. Furthermore, they are very robust and stable. One of the major drawbacks of the SCIG is the fact that there is a unique relation between active power, reactive power, terminal voltage and rotor speed (Ackermann 2005). That means that an increase in the active power production is possible only with an increase in the reactive power consumption, leading to a relatively low full-load power factor. In order to limit the reactive power absorption from the grid, SCIG-based WECS are equipped with capacitor banks. The softstarter's role is to smooth the inrush currents during the grid connection (Iov 2003).

SCIG-based WECS are designed to achieve maximum power efficiency at a unique wind speed. In order to increase the power efficiency, the generator of some fixed-speed WECS has two winding sets, and thus two speeds. The first set is used at low wind speed (typically eight poles) and the other at medium and large wind speeds (typically four to six poles).

Fixed-speed WECS have the advantage of being simple, robust and reliable, with simple and inexpensive electric systems and well proven operation. On the other hand, due to the fixed-speed operation, the mechanical stress is important. All fluctuations in wind speed are transmitted into the mechanical torque and further, as electrical fluctuations, into the grid. Furthermore, fixed-speed WECS have very limited controllability (in terms of rotational speed), since the rotor speed is fixed, almost constant, stuck to the grid frequency.

An evolution of the fixed-speed SCIG-based WECS are the limited variable-speed WECS. They are equipped with a wound-rotor induction generator (WRIG) with variable external rotor resistance; see Figure 2.9. The unique feature of this WECS is that it has a variable additional rotor resistance, controlled by power electronics. Thus, the total (internal plus external) rotor resistance is adjustable, further controlling the slip of the generator and therefore the slope of the mechanical characteristic. Obviously, the range of the dynamic speed control is determined by how big the additional resistance is. Usually the control range is up to 10% over the synchronous speed.

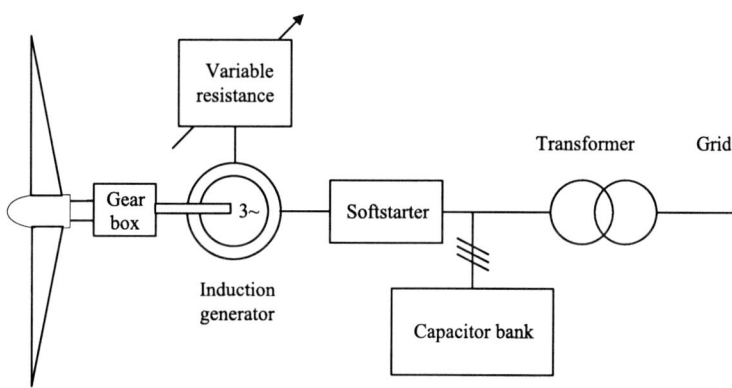

Figure 2.9. General structure of a limited variable-speed WECS

2.5.2 Variable-speed WECS

Variable-speed wind turbines are currently the most used WECS. The variable-speed operation is possible due to the power electronic converters interface, allowing a full (or partial) decoupling from the grid.

The doubly-fed-induction-generator (DFIG)-based WECS (Figure 2.10), also known as improved variable-speed WECS, is presently the most used by the wind turbine industry.

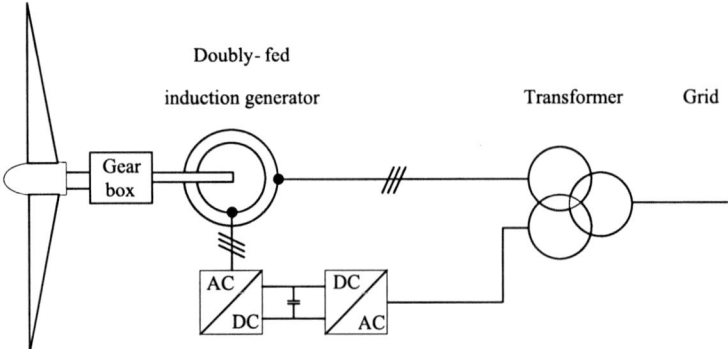

Figure 2.10. General structure of an improved variable-speed WECS

The DFIG is a WRIG with the stator windings connected directly to the three-phase, constant-frequency grid and the rotor windings connected to a back-to-back (AC–AC) voltage source converter (Akhmatov 2003; Ackermann 2005). Thus, the term "doubly-fed" comes from the fact that the stator voltage is applied from the grid and the rotor voltage is impressed by the power converter. This system allows variable-speed operation over a large, but still restricted, range, with the generator behaviour being governed by the power electronics converter and its controllers.

The power electronics converter comprises of two IGBT converters, namely the rotor side and the grid side converter, connected with a direct current (DC) link. Without going into details about the converters, the main idea is that the rotor side converter controls the generator in terms of active and reactive power, while the grid side converter controls the DC-link voltage and ensures operation at a large power factor.

The stator outputs power into the grid all the time. The rotor, depending on the operation point, is feeding power into the grid when the slip is negative (over-synchronous operation) and it absorbs power from the grid when the slip is positive (sub-synchronous operation). In both cases, the power flow in the rotor is approximately proportional to the slip (Lund *et al.* 2007).

The size of the converter is not related to the total generator power but to the selected speed variation range. Typically a range of ±40% around the synchronous speed is used (Akhmatov 2003).

DFIG-based WECS are highly controllable, allowing maximum power extraction over a large range of wind speeds. Furthermore, the active and reactive power control is fully decoupled by independently controlling the rotor currents. Finally, the DFIG-based WECS can either inject or absorb power from the grid, hence actively participating at voltage control.

Full variable-speed WECS are very flexible in terms of which type of generator is used. As presented in Figure 2.11, it can be equipped with either an induction (SCIG) or a synchronous generator. The synchronous generator can be either a wound-rotor synchronous generator (WRSG) or a permanent-magnet synchronous generator (PMSG), the latter being the one mostly used by the wind turbine industry. The back-to-back power inverter is rated to the generator power and its operation is similar to that in DFIG-based WECS. Its rotor-side ensures the rotational speed being adjusted within a large range, whereas its grid-side transfers the active power to the grid and attempts to cancel the reactive power consumption. This latter feature is important especially in the case of SCIG-equipped WECS.

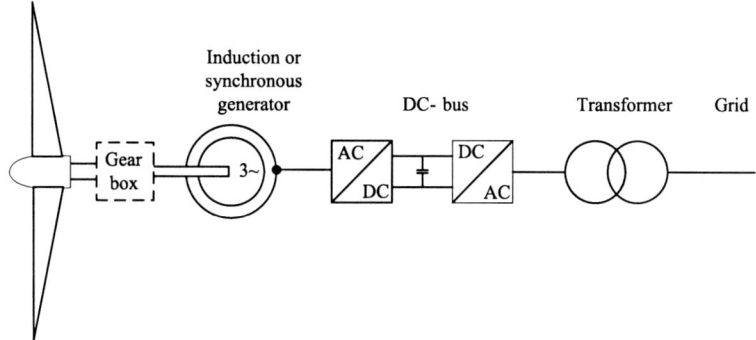

Figure 2.11. General structure of a full variable-speed WECS

The PMSG is considered, in many research articles, a good option to be used in WECS, due to its self-excitation property, which allows operation at high power factor and efficiency (Alatalo 1996).

PMSG does not require energy supply for excitation, as it is supplied by the permanent magnets. The stator of a PMSG is wound and the rotor has a permanent-magnet pole system. The salient pole of PMSG operates at low speeds, and thus the gearbox (Figure 2.11) can be removed. This is a big advantage of PMSG-based WECS as the gearbox is a sensitive device in wind power systems. The same thing can be achieved using direct driven multipole PMSG with large diameter.

The synchronous nature of PMSG may cause problems during start-up, synchronization and voltage regulation and they need a cooling system, since the magnetic materials are sensitive to temperature and they can loose their magnetic properties if exposed to high temperatures (Ackermann 2005).

2.6 Wind Turbine Generators in Hybrid Power Systems

Two main types of hybrid generation structures embedding WECS can be found in insulated grid utilities (Cutululis 2005). Typically, the generators feed energy into a common AC bus, as Figure 2.12 depicts (Manwell *et al.* 1993; Jeffries *et al.* 1996; Tomilson *et al.* 1998; Papathanassiou and Papadopoulos 2001; Bialasiewicz and Muljadi 2002; Baring-Gould *et al.* 2004; Cutululis *et al.* 2006a). An alternative solution envisages the power sources coupling on a common DC-bus, the electrical power being further transformed by an inverter in order to feed AC loads (Figure 2.13 – Borowy and Salameh 1997; De Broe *et al.* 1999; Ruin and Carlson 2000; Vechiu *et al.* 2004; Cutululis *et al.* 2006b; El Mokadem *et al.* 2003).

In Figure 2.12 the example of two sources is taken, namely a wind turbine and a diesel generator feed common AC bus. This structure presents two kinds of energy storage. The first, envisaging the long-term electro-chemical energy storage, is realised by the accumulator battery. The second, acting in the short-term, stores kinetic energy by means of a high-speed flywheel. These two storage elements allow bidirectional power flow.

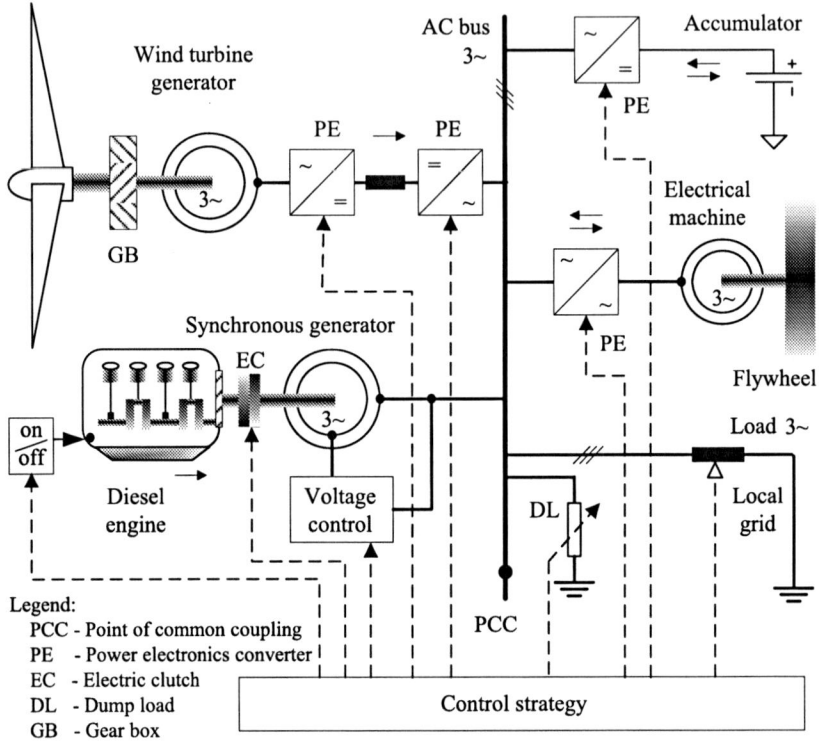

Figure 2.12. AC-coupled hybrid generation system

Here are the guidelines of how the system in Figure 2.12 operates. When the wind speed is low, WECS generates less power than the load (local utilities) needs, and the diesel engine is turned on, compensating the active power imbalance. When the wind velocity is sufficiently high, the diesel engine is shut down, and the synchronous generator acts as reactive power compensator. Concerning the flywheel, it accumulates kinetic energy when the high winds induce energy in excess. In case of short wind gaps the flywheel delivers power to the AC bus. The accumulator represents the emergency solution.

Concerning the DC-connected generation system, an example is given in Figure 2.13. In this configuration some short-term energy storage units can be supplementary introduced (*e.g.*, the flywheel device driven by a switched-reluctance machine used by El Mokadem *et al.* 2006).

Figure 2.13 is self-explanatory; each moment the power fed into the DC-bus is the result of the wind turbine, diesel-generator, photovoltaic array and accumulator power contributions, depending on their operating regimes.

The control objective aims at the AC load being continuously supplied with energy. Depending on the operating regime the turbine can be controlled either for maximum power point tracking or for power limiting. The reliability requirements are important in these structures. The control of hybrid generation structure envisages the entire system; its management and supervision are beyond the scope of this book.

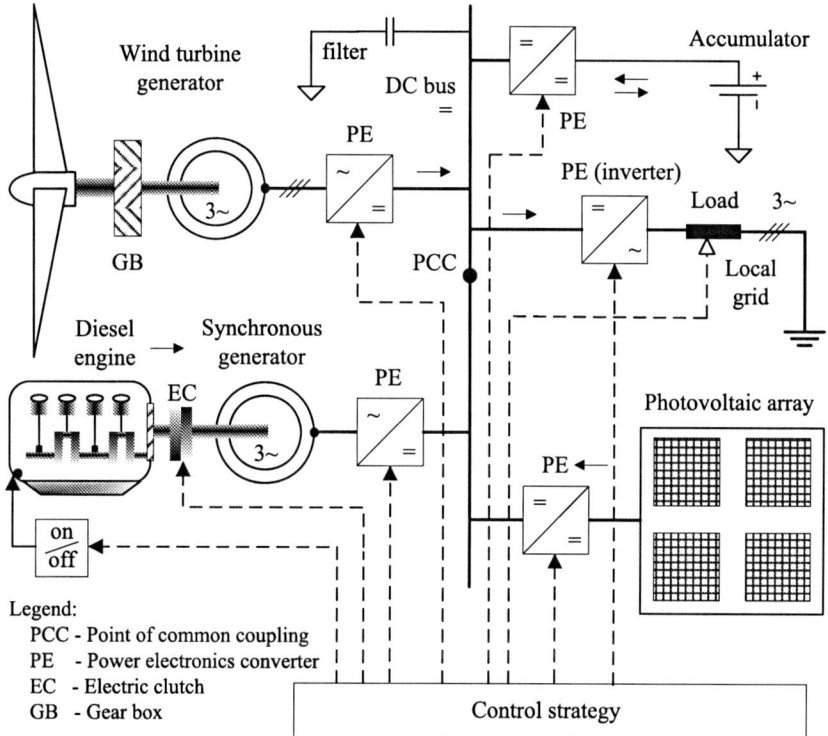

Figure 2.13. DC-coupled hybrid generation system

2.7 Control Objectives

Control plays an ever increasing role in modern WECS. There are numerous research articles dedicated to WECS control, all of them having starting from the idea that control can and does significantly improve all aspects of WECS.

In any process, control has two main objectives: protection and optimization of operation. Furthermore, when applied to WECS, control becomes more important, in all aspects, as the main characteristic of WECS is that they have to cope with the highly variable, intermittent and unpredictable nature of the wind.

To this end, as previously mentioned, all WECS have some sort of power control. The passive-stall wind turbines manage to limit the aerodynamic power, for protection reasons, without any active controllers. This approach is simple and offers hardware robustness, but can lead to unacceptable levels of mechanical loads

(Burton *et al.* 2001). Thus, control in that sense has as its only objective the protection of wind turbines.

Active stall implies that WECS are equipped with several additional hardware components: electromechanical or hydraulic actuators used to move the blades (or parts of them), sensors and controllers. All of these add complexity and increase the operation and maintenance costs but they also allow one to extend the control objectives to increase the power capture, thus optimizing the WECS operation.

Fixed-speed WECS, with either passive or active stall, dominated the wind power industry for a long time. Their main drawback is their rigidity, as the fixed generator speed does not offer any control flexibility. This disappears with the use of DFIG-based WECS and, later, with the use of full scale power converter WECS. Variable-speed operation became possible by incorporating power electronics converters.

Variable-speed WECS control system generally includes three main control subsystems:
− aerodynamic power control, through pitch control;
− variable-speed operation and energy capture maximization, by means of generator control;
− grid power transfer control, through the power electronics converter.

Furthermore, the specific objectives of each control subsystem vary in accordance to the operating regime (see Figure 2.7).

When the wind speed is between the cut-in and the rated speed (partial load regime), the pitch control system is typically inactive, with two exceptions: when the pitch system is used to assist the start-up process, as the two- or three-bladed wind turbines have a relatively low starting torque, and when the rotational speed is limited by pitch control as the wind speed approaches the rated value. The pitch control system is active when the wind speed exceeds the rated wind speed. Its objective is to limit the aerodynamic power to the rated one and, when the wind speed reaches the cut-out value, to stop the wind turbine. Thus, the pitch control system deals mainly with alleviating the mechanical loads on the wind turbine structure.

During the partial load regime, the generator control is the only active control and aims at maximizing the energy captured from the wind and/or at limiting the rotational speed at rated. This is possible by continuously accelerating or decelerating the generator speed in such a way that the optimum tip speed ratio is tracked. At rated wind speed, the generator control limits the generator speed. Thus, the generator control deals mainly with the power conversion efficiency optimization. Sometimes this means that the generator torque varies along with the wind speed and, in some conditions, can induce supplementary mechanical stress to the drive train. Consequently, maximizing the power conversion efficiency through generator control should be done, bearing in mind the possibility that supplementary loads are induced to the mechanical structure.

Finally, the power electronics converter control ensures that the strict power quality standards (frequency, power factor, harmonics, flicker, *etc.*) are met. Recently, the increasing requirements for WECS to remain connected and to provide active grid support have added control objectives for the power electronics converters. In the case of a grid fault, the WECS should remain connected; thus they should cope with sudden and important loads, and even assist the grid in voltage or frequency control (Akhmatov 2003; Ackermann 2005; Sørensen *et al.* 2005). Thus, the power electronics converter control deals mainly with power quality standards.

The role and objectives of WECS control, as presented above, can be summarized as follows (De La Salle *et al.* 1990; Leithead *et al.* 1991):
- starting on the WECS at the cut-in speed, stopping it at the cut-out speed and switching controllers corresponding to the specific operating conditions;
- controlling the aerodynamic power and the rotational speed above rated wind speed;
- maximising the wind harvested power in partial load zone, with respect to the speed and captured power constraints;
- alleviating the variable loads, in order to guarantee a certain level of resilience of the mechanical parts, in all operating regimes;
- guaranteeing a desired response to isolated wind gusts;
- transferring the electrical power to the grid at an imposed level, for wide range of wind velocities;
- meeting strict power quality standards (power factor, harmonics, flicker, *etc.*);
- protecting the WECS and, at the same time, offer active grid support during grid faults.

The list is not exhaustive; several other control objectives, deriving from those listed above, can be formulated. Variable-speed WECS is a highly nonlinear time-variant system excited by stochastic inputs which significantly affect its reliability and leads to non-negligible variations in the dynamic behaviour of the system over its operating range. This is the reason why the control of variable-speed wind turbines is still in the phase of searching technical solutions suitable to be widely implemented in the wind turbine industry.

3

WECS Modelling

3.1 Introduction and Problem Statement

At present, there are several variable-speed WECS configurations being widely used. In this book, some typical configurations are studied, which have the following specific properties:
- they are based upon variable-speed horizontal-axis wind turbines – with two or three blades – and fixed- or variable-pitch;
- the mechanical transmission is single (fixed) multiplying ratio, either rigid or flexible;
- the electrical generator is either synchronous (permanent-magnet) or induction generator (squirrel-cage/doubly-fed), equipped with front-end AC–AC power electronic converter, thus ensuring wide range variable-speed operation; the power transferred to the grid is controlled through the power electronics.

Figure 3.1 presents two wind power conversion systems approached in this book. The essential differences between them concern the power flow control. In the first case (Figure 3.1a), which uses either squirrel-cage induction generator (SCIG) or permanent-magnet synchronous generator (PMSG), the AC–AC converter is *stator* grid-connected and rated at the generator's power level. As regards the second case (Figure 3.1b), using doubly-fed induction generator (DFIG), the power electronics converter interfaces the *rotor* with the grid, thus transferring only part of the generated power to the grid.

From a system viewpoint, the conversion chain can be divided into four interacting main components which will be separately modelled (Figure 3.1a,b): the aerodynamic subsystem (AS) – S_1 and the electromagnetic subsystem (EMS) – S_2 interact by means of the drive train (mechanical transmission, DT) – S_3, whereas S_4 denotes the grid interface. The action of two control subsystems is suggested on these figures: ensuring the variable-speed operation (exerted on $S_1 \div S_3$) and respectively controlling the power transfer at grid imposed parameters (exerted on S_4). LSS and HSS denote low- and high-speed shafts respectively.

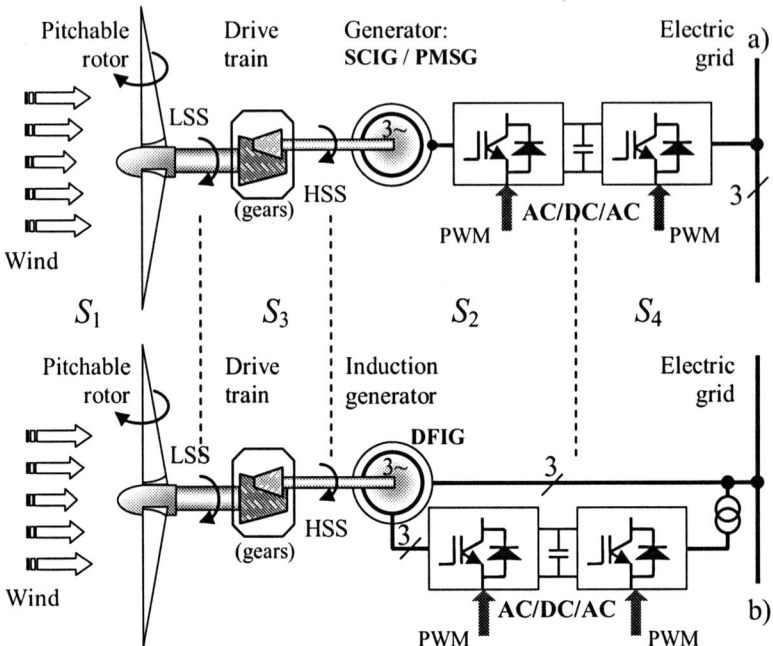

Figure 3.1. General configuration of a variable-speed-controlled WECS from a system point of view: **a** SCIG/PMSG-based; **b** DFIG-based

The WECS subsystems are usually treated individually; a global WECS model suitable for control structure design is obtained by adding models of their interactions. The modelling of WECS components, presented in the following, relies upon some assumptions, which are specified when they appear. These modelling assumptions depend on the desired level of detail, the operating regime and on the control goal. Thus, since the variable-speed operation regime is effective in partial load, some wind turbine simplified models – *e.g.*, fixed aerodynamic characteristics – can be used for generator control purposes, irrespective of the turbine size. Furthermore, since the issue of getting accurate measure information is beyond the scope here, all necessary measurements, such as the wind velocity at the hub level, the blades' position, the HSS rotational speed and all necessary electrical variables, are assumed available for control/supervision.

3.2 Wind Turbine Aerodynamics Modelling

3.2.1 Fixed-point Wind Speed Model

From a system point of view, the wind speed represents the main exogenous signal applied to the WECS and determines its behaviour. Its erratic variation, highly dependent on the given site and on the atmospheric conditions, makes the wind speed quite difficult to model. Usually the thermic equilibrium of the atmosphere nearby Earth is assumed (*neutral atmosphere* – Burton et al. 2001). Therefore, turbulence results mainly from the friction between air and ground, due to the ground roughness. When designing WECS, the history of the wind speed extreme values (gusts) is considered for the mechanical structure design and also for control purposes.

Wind near the Earth's surface is generally modelled by a spatial (3D) speed distribution. Assuming that the turbine is equipped with a vane (or yawing equipment) and that changes in wind direction are sufficiently slow, then the turbine rotor is maintained normal to the wind and WECS analysis requires only the longitudinal wind speed being synthesized/modelled. Thus, in the present book only *scalar* (1D) wind speed models will be used. As the interest here is focused on WECS behaviour in normal operating regimes, the developed models will not include extreme operating conditions like wind gusts.

Wind dynamics result from combining meteorological conditions with particular features of a given site. Thus, wind speed is modelled in the literature as a non-stationary random process, yielded by superposing two components (Burton et al. 2001; Nichita et al. 2002; Vihriälä 2002; Bianchi et al. 2006):

$$v(t) = v_s(t) + v_t(t), \qquad (3.1)$$

where $v_s(t)$ is the *low-frequency component* (describing long term, low-frequency variations) and $v_t(t)$ is the *turbulence component* (corresponding to fast, high-frequency variations).

These components can be identified in Van der Hoven's large band (six decades) model (Figure 3.2). The spectral gap of around 0.5 mHz suggests that the turbulence component can be modelled as a zero average random process (there is little energy in the spectral range between 2 h and 10 min). v_s is considered constant (equal to the average wind speed) when viewed at the turbulence time scale. Averaging is usually performed on a 10-min time window (Burton et al. 2001).

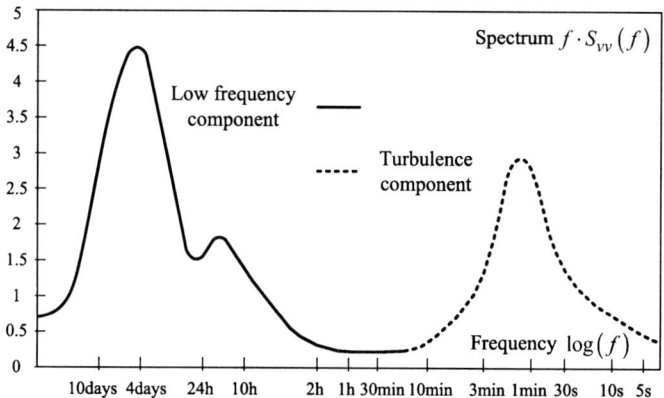

Figure 3.2. Van der Hoven's spectral model of the wind speed

The *low-frequency component* corresponds to the very slow wind speed variations and characterizes the site from the energy viewpoint. It can be modelled as a Weibull's distribution (see Chapter 2) or a Rayleigh's distribution (Leithead et al. 1991):

$$v_s = a \cdot v \cdot e^{-1/2 a \bar{v}^2},$$

where $\bar{v}$ is the wind speed's hourly average and a is a parameter depending on the wind speed's very long term average. The value of this component influences the turbulence amplitude, but its evolution is not crucial for short and medium term dynamic behaviour of WECS.

Fast wind speed variations (typically occurring within 10 min) are modelled by the *turbulence component*. This is mathematically described as a zero average normal distribution, whose standard deviation, σ, depends on the current value of the hourly average, v_s. The *turbulence intensity* is a measure of the global level of turbulence, depends on the ground surface roughness and is defined as

$$I_t = \frac{\sigma}{v_s} \tag{3.2}$$

The mathematical description of the turbulence's dynamical properties, $v_t(t)$, can be obtained by using two kinds of spectra: von Karman's and Kaimal's respectively. According to Burton et al. (2001), Kaimal's spectrum reflects better the correspondence to experimental data, when turbulence is present. But von Karman's spectrum is more consistently theoretically founded (an analytical connection with the correlation function is provided) and allows a realistic representation of turbulence data in wind tunnels. The von Karman's model for the longitudinal component of the turbulence is

$$\frac{f \cdot S_{vv}(f)}{\sigma^2} = \frac{4f \cdot L_t / v_s}{\left(1 + 70.8 (f \cdot L_t / v_s)^2\right)^{5/6}}, \tag{3.3}$$

where $S_{vv}(f)$ is the power spectral density, L_t is the *length of turbulence*, specific to the site (ground roughness), and f is the frequency in Hz.

Kaimal's spectral model has the form

$$\frac{f \cdot S_{vv}(f)}{\sigma^2} = \frac{4f \cdot L_t / v_s}{\left(1 + 6 f \cdot L_t / v_s\right)^{5/3}} \tag{3.4}$$

One can note that in both models the power spectral density is influenced by the turbulence intensity, I_t, which determines the turbulence "level" (*i.e.*, its variance, σ^2) and the turbulence length, L_t, which impresses the turbulence dynamic properties (the spectral function bandwidth). Both these parameters are adopted according to various standards. For example, in the Danish standard (DS 742 2007), the following relations are used to compute these parameters:

$$I_t = \frac{1}{\ln(z/z_0)} \tag{3.5}$$

and respectively

$$L_t = \begin{cases} 150 \text{ m}, & \text{if } z \geq 30 \text{ m} \\ 5 \cdot z \text{ m}, & \text{if } z < 30 \text{ m} \end{cases}, \tag{3.6}$$

where z is the height from ground where the wind speed is computed and z_0 is the roughness length.

Figure 3.3 comparatively presents the spectral functions at Equations 3.3 and 3.4 for the same values of parameters z, z_0 and v_s. For an easier analysis of von Karman's and Kaimal's spectra, in Figure 3.4a one can see the corresponding power spectral densities, $S_{vv}(f)$, whereas Figure 3.4b shows the Bode diagrams of the non-integer-order shaping filter outputting the turbulence component when fed with a white noise.

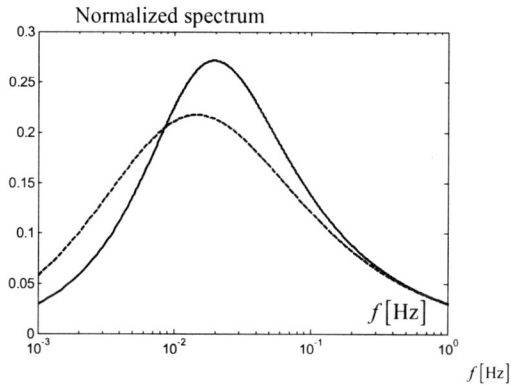

Figure 3.3. Comparison between the von Karman's (Equation 3.3 – *solid line*) and Kaimal's (Equation 3.4 – *dashed line*) normalized spectra (z=30 m, z_0=0.01 m, $v_s = 10$ m/s, Danish standard DS472)

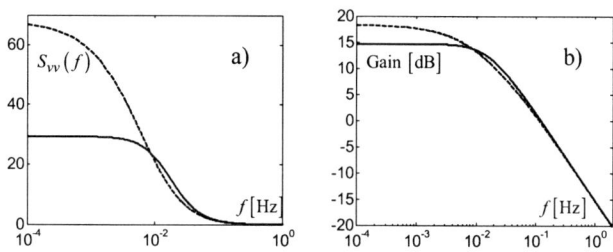

Figure 3.4. Von Karman's (*solid line*) vs. Kaimal's (*dashed line*) spectral models (z=30 m, z_0=0.01 m, $v_s = 10$ m/s, Danish standard DS472): **a** power spectral densities, $S_{vv}(f)$; **b** shaping filter gains [dB]

The following remarks can be formulated:
- the asymptotical properties (at high frequency) of the two spectra are identical, since the shaping filters gain diagrams have the same slope, equal to $-20\frac{5}{6}$ dB/dec;
- in Kaimal's model the spectral function's slope varies within a larger range, especially at low frequency, thus allowing a more accurate description of turbulence; this further ensures a better fit to atmospheric turbulence measurements;
- the shaping filter function of von Karman's model is close to that of a first-order filter; this remark enables a simplified model of turbulence to be used, which is based upon a first-order shaping filter.

The nonstationary character of the turbulence component is illustrated in Figure 3.5a, where the von Karman's spectral characteristic, $f \cdot S_{vv}(f)$, for two different values of the wind speed, 10 m/s and 15 m/s, and for the same values of z and z_0 is shown. Figure 3.5b displays the Bode diagrams of the corresponding shaping filters. One can note that the wind speed influences both the gain and the cut-off frequency of the shaping filter.

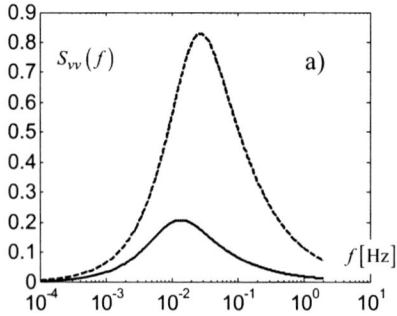
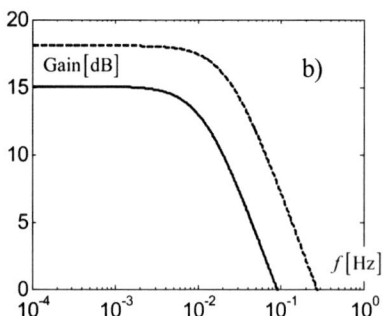

Figure 3.5. Von Karman's spectrum for two different values of the average wind speed: $v_s = 7$ m/s – solid line; $v_s = 14$ m/s – dashed line ($z=30$ m, $z_0=0.01$ m): **a** power spectral densities, $f \cdot S_{vv}(f)$; **b** shaping filter gains [dB]

The turbulence component, $v_t(t)$, is usually synthesized by feeding a suitable shaping filter, $H_t(j\omega)$, with a white noise. This procedure is based on the fundamental relation from the linear systems statistical dynamics (Damper 1995):

$$S_{vv}(\omega) = |H_t(j\omega)|^2 \cdot S_{wn}(\omega), \qquad (3.7)$$

where $S_{wn}(\omega) = $ constant is the power spectral density of the white noise. By using von Karman's model, the transfer function of the shaping filter has the following form:

$$H_t(j\omega) = \frac{K_F}{(1+j\omega T_F)^{5/6}}, \qquad (3.8)$$

where parameters K_F and T_F depend on the low-frequency wind speed, v_s.

Welfonder et al. (1997) propose a procedure for obtaining the turbulence component, which is based upon experimentally identifying the shaping filter's parameters from Equation 3.8. The time constant results from $T_F = L_t/v_s$, with L_t being found empirically. The filter is fed with a normally distributed white noise having one-unit variance, whose sampling time, T_s, can be configured. By computing the static gain, K_F:

$$K_F = \sqrt{\frac{2\pi}{B(1/2,1/3)} \cdot \frac{T_F}{T_s}}, \qquad (3.9)$$

where B is the beta function, one gets a one-unit-variance coloured noise at the filter's output. In order to obtain variance σ^2 corresponding to the average wind speed, v_s, the coloured noise is multiplied by the product $I \cdot v_s$, with the turbulence intensity, I, being empirically determined.

The procedure proposed by Welfonder et al. (1997) is adapted by Nichita et al. (2002) for obtaining the non-stationary wind speed, within a large time window, by using the block diagram from Figure 3.6. Here, the time scales of the two components, $v_s(t)$ and $v_t(t)$, are different. Usually the sampling time are T_{ss} = 10 min for v_s and T_{st} = 1 s for the turbulence component.

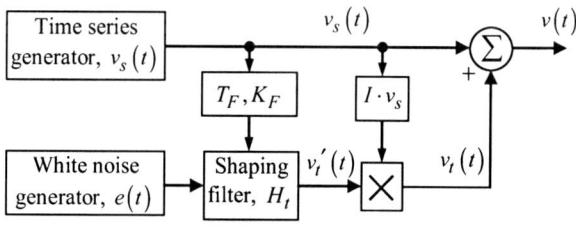

Figure 3.6. Nonstationary wind speed generation

The low-frequency component is obtained either based upon a model fitted to measured data, or by using a generic model, e.g., the van der Hoven's spectrum. In the latter case, the low-frequency component must be sampled. Let ω_i, $i = \overline{1,N}$, be the discrete angular frequency and $S_{v_s v_s}(\omega_i)$ the corresponding values of the power spectral density. The ω_i harmonic has the amplitude

$$A_i = \frac{2}{\pi}\sqrt{\frac{1}{2}\left(S_{v_s v_s}(\omega_i) + S_{v_s v_s}(\omega_{i+1})\right) \cdot (\omega_{i+1} - \omega_i)} \qquad (3.10)$$

and component v_s is thus computed:

$$v_s(t) = \sum_{i=0}^{N} A_i \cos(\omega_i t + \varphi_i), \qquad (3.11)$$

where phase φ_i is generated randomly in the $[-\pi,\pi]$ range. For $\omega_0=0$ it is set $\varphi_0=0$ and $A_0 = \bar{v}$, where $\bar{v}$ is the average wind speed, calculated on a time horizon greater than the largest period in Van der Hoven's characteristic (*i.e.*, $T = 2\pi/\omega_1$).

For each new computed value of v_s, the current time constant of the shaping filter, $T_F(t) = L_t/v_s(t)$, is computed. The static gain K_F is computed based upon Equation 3.9, ensuring unitary variance at the filter's output, and then the turbulence's variance is adjusted by means of factor $I \cdot v_s$, as shown in Figure 3.6. Parameters T_F, K_F and turbulence's variance remain constant along the time interval T_{ss} as long as v_s is constant; they are re-calculated as soon as a new v_s value is obtained.

The turbulence component of the wind speed is simulated at T_{st} sampling time and involves numerical difficulties if the non-integer order filter at Equation 3.8 is employed. This filter can be approximated by a two pole and one zero transfer function (Nichita *et al.* 2002):

$$H_t(s) = K_F \cdot \frac{m_1 T_F s + 1}{(T_F s + 1)(m_2 T_F s + 1)}, \quad m_1 = 0.4, \; m_2 = 0.25 \qquad (3.12)$$

The non-stationary wind speed, computed with the procedure presented above, can be seen in Figure 3.7. It covers a 5-h time range. The evolution of the low-frequency wind speed component is highlighted. Figure 3.8 presents details of the wind speed profile when v_s varies in two distinct ranges.

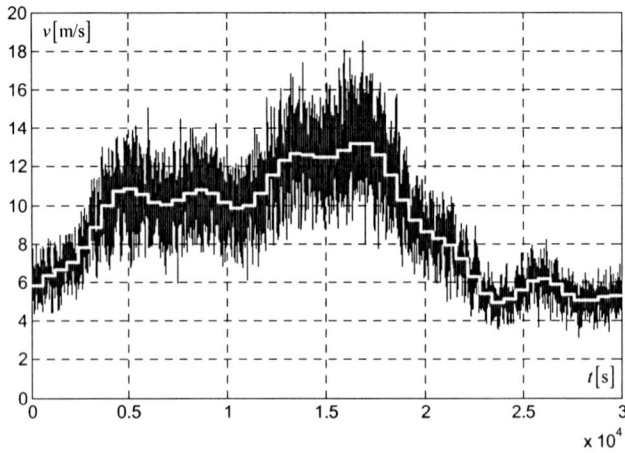

Figure 3.7. Non-stationary wind speed: total wind speed, $v(t)$ (*black*), and its low-frequency component, $v_s(t)$ (*white*) (z=50 m, z_0=0.005 m, Danish standard DS472)

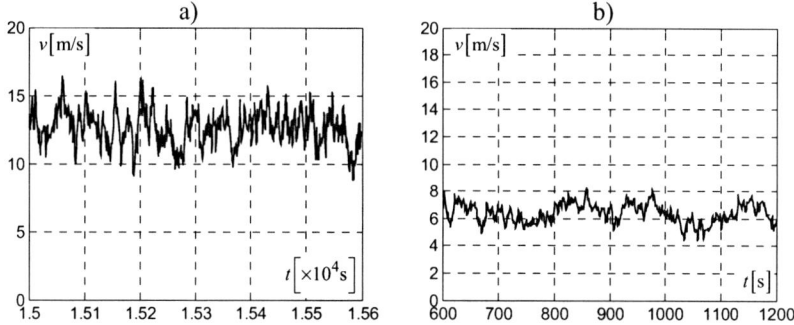

Figure 3.8. Details from Figure 3.7: **a** v_s varies around 12 m/s; **b** v_s varies around 7 m/s

3.2.2 Wind Turbine Characteristics

Variable-pitch Case
The main modelling purpose is to provide the wind torque developed by the turbine rotor in the form

$$\Gamma_{wt} = \Gamma_{wt}(\beta, v, \Omega_l) \qquad (3.13)$$

as the wind velocity, v, rotational speed, Ω_l, and blade pitch, β, are given. For this purpose, the *blade element theory* (BET) is used. Some very interesting aspects concerning the turbine operation and interaction between the rotor and the air stream can also be revealed.

According to BET, the blade is divided into a number of transversal elements, placed along the blade. A blade element j (Figure 3.9) is obtained by sectioning the blade with two parallel planes, situated at distances r and $r+dr$ from the hub, and normally disposed to the blade.

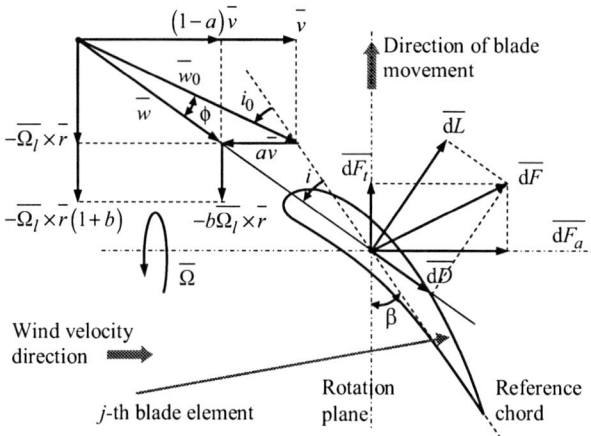

Figure 3.9. Aerodynamic loads along the blade profile

Consider that the turbine is situated in an air flow of v velocity, normal to the rotor. The blade element profile and the undertaken aerodynamic loads are given in Figure 3.9, where notations have the following meaning:

- v wind speed,
- Ω_l rotational speed of the turbine,
- w relative wind speed to the blades (vortex motion is considered),
- w_0 relative wind speed to the blades when the vortex motion is not considered,
- a axial flow interference factor,
- b tangential flow interference factor,
- $\overline{dF}$ total force acting on the blade element,
- $\overline{dD}$ the elementary drag force,
- $\overline{dL}$ the elementary lift force,
- $\overline{dF_t}$ the elementary tangential force in the direction of rotation,
- $\overline{dF_a}$ the elementary axial thrust force.

The main modelling assumptions within BET are:
- neglecting the interactions within the adjacent elements of the same blade,
- neglecting the radial component of the speed,
- the aerodynamic coefficients are functions of incidence angle and blade profile; any effects related to the Reynolds number influence are neglected,
- infinite number of blades is assumed.

Each element j is characterized by the following variables:
- the distance to the hub axis, $r(j)$,
- the corresponding chord, $c(j)$ – the chord variation is known for a given profile,
- the corresponding pitch angle with respect to the hub, $\beta(j)$ – the blade longitudinal torsion is known,
- the elementary tip speed ratio, $\lambda_r(j) = \dfrac{r(j) \cdot \Omega_l}{v}$.

Without detailing the involved phenomenology (for details see Burton et al. 2001; Freris 1990), the essential aspect in turbine operation is that the wind flow has a relative motion to the rotating blade, described by the relative wind speed to the blades, w. This variable has an angle of incidence, i, with the blade element reference chord. This is the key variable, as it ultimately determines the aerodynamic behaviour of the turbine. The incidence can be affected by wind velocity, rotational speed and pitch angle variations, as further detailed.

The elementary drag force in the direction of rotation, $\overline{dF_a}$, depends proportionally on the so-called *drag coefficient*, C_x, and the elementary lift force, $\overline{dF_t}$, depends proportionally on the *lift coefficient*, C_z. The ratio C_z/C_x is a measure of the turbine aerodynamic efficiency.

The elementary axial and tangential interference factors, $a(j)$ and $b(j)$

respectively, are computed based on the $\lambda_r(j)$ and on the Lagrange coefficient, K (Dumitrescu et al. 1990):

$$a(j) = \frac{K}{(1-K)^2} \cdot \frac{\lambda_r^2(j)}{1+\lambda_r^2(j)/(1-K)^2} \quad (3.14)$$

$$b(j) = \frac{K}{(1-K)} \cdot \frac{1}{1+\lambda_r^2(j)/(1-K)^2} \quad (3.15)$$

The incidence angle of the j-th element, $a(j)$, is given by

$$i(j) = \operatorname{atan}\left(\frac{1}{\lambda_r^2(j)} \cdot \frac{1-a(j)}{1+b(j)}\right) - \beta(j) \quad (3.16)$$

The wind speed relative to the blade, $w(j)$, is deduced from the speed diagram (see Figure 3.9):

$$w(j) = v \cdot \sqrt{(1-a(j))^2 + \lambda_r^2(j) \cdot (1+b(j))^2} \quad (3.17)$$

The elementary lift force in the direction of rotation computes as

$$dF_t(j) = 0.5\rho c(j) w^2(j) C_Z(i) \cdot \sin(\beta(j)+i(j)) \cdot \left[1 - \varepsilon \cdot \operatorname{ctan}(\beta(j)+i(j))\right] \cdot dr, \quad (3.18)$$

where $C_z(i)$ is the lift coefficient, $C_x(i)$ is the drag coefficient and $\varepsilon(i)$ is a ratio reflecting the turbine aerodynamic efficiency:

$$\varepsilon(i) = C_x(i)/C_z(i) \quad (3.19)$$

Dependence of coefficients $C_z(i)$ and $C_x(i)$ vs. the incidence angle are known for a given blade profile (Le Gourières 1982 – Figure 3.10).

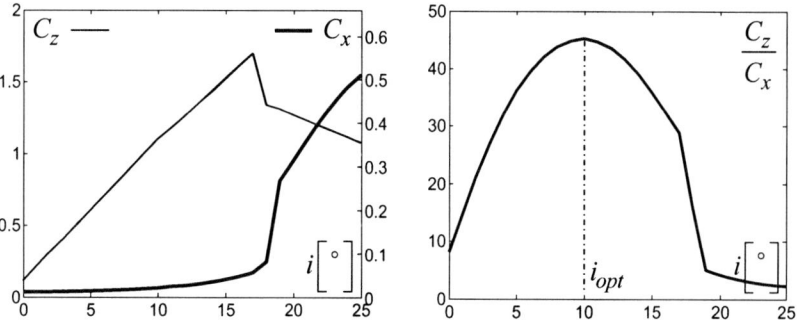

Figure 3.10. C_x, C_z and $1/\varepsilon$ profiles vs. the incidence angle, i

Each blade element develops an elementary torque:

$$d\Gamma(j) = r(j) \cdot dF_t(j) \quad (3.20)$$

By integrating Equation 3.20 along the blade length and by using Equation 3.18, one obtains the total torque developed by the wind turbine rotor:

$$d\Gamma(j) = \int r(j) \cdot dF_t(j) \quad (3.21)$$

This result is obtained assuming that the rotor has an infinite number of blades. As in fact there is a finite number of blades, N_B, Equation 3.21 is amended by a correction factor, called Prandtl's coefficient:

$$\sigma_P = 1 - \frac{0.93}{N_B \cdot \sqrt{\left(\lambda_r(j)^2 + 0.445\right)}} \quad (3.22)$$

In conclusion, the wind turbine torque computation procedure can be synthesized, using the expressions presented above, as in Algorithm 3.1 (Diop *et al.* 1999; Nichita *et al.* 2006).

Algorithm 3.1. Computation of wind torque according to blade element theory

---
#0. *Input* (constant) *data*: number of blades, blade length, air density, number of finite elements, chord variation along the blade, pitch variation along the blade, aerodynamic characteristics C_z and C_x depending on the incidence angle.
#1. *Input variables*: wind velocity, rotational speed, pitch angle (at the hub).
#2. For each element *j*, compute:
– elementary tip speed ratio $\lambda_r(j)$;
– elementary distance to the hub *r(j)*, total pitch corresponding to the element, $\beta(j)$;
– elementary incidence angle, *i(j)*;
– elementary axial and tangential interference factors, *a(j)* and *b(j)*;
– relative wind speed, *w(j)*;
– lift and drag coefficients, $C_z(i)$ and $C_x(i)$, and also their report $1/\varepsilon(i)$;
– elementary torque (function of many above computed variables).
#3. Apply the Prandtl's coefficient.
#4. Numerically integrate the elementary torque equation (using a suitably chosen method).

Fixed-pitch Case

The aerodynamic subsystem model describes (by averaging) the interaction of the turbine rotor with the air masses (wind); this subsystem is modelled by the mechanical torque provided by the rotor motion (Wilkie *et al.* 1990; Miller *et al.* 2003). Following the assumptions listed above, for a fixed-pitch (β) wind turbine, this torque depends on the low-speed shaft rotational speed and on the wind speed:

$$\Gamma_{wt} = \Gamma_{wt}(\Omega_l, v)\big|_{\beta=\text{constant}} \quad (3.23)$$

According to Equation 2.31, one obtains

$$\Gamma_{wt} = \frac{P_{wt}}{\Omega_l} = \frac{1}{2} \cdot \pi \cdot \rho \cdot v^2 \cdot R^3 \cdot C_\Gamma(\lambda), \quad (3.24)$$

where $C_\Gamma = C_p/\lambda$ is the torque coefficient (introduced by Equation 2.32). More detailed aerodynamic models can be developed, emphasizing the rotational sampling, spatial filtering or induction lag (Rodriguez-Amenedo et al. 1998; Vihriälä 2002; Molenaar 2003), leading to a more complex expression of the developed aerodynamic torque, Γ_{wt}. Furthermore, the power coefficient characteristic should take into account the effects due to Reynolds number and air density variations. For low-power wind turbines, these effects, together with the structural dynamics, can be neglected, and simplified models are preferred (Wilkie et al. 1990).

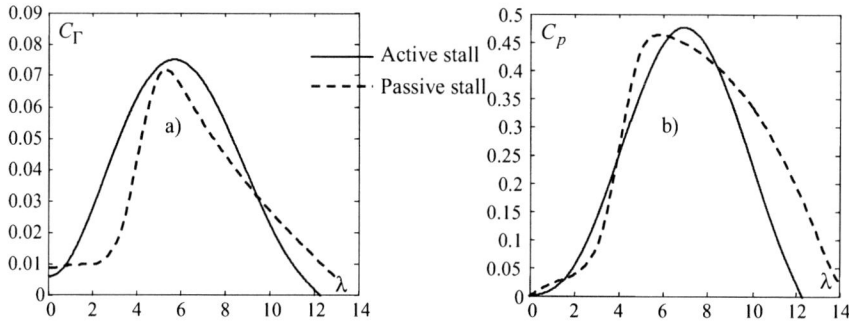

Figure 3.11. Typical HAWT $C_\Gamma - \lambda$ (**a**) and $C_p - \lambda$ (**b**) curves

The torque coefficient can be described by a polynomial function of the tip speed ratio, λ (Figure 3.11a – Nichita 1995; Miller et al. 2003):

$$C_\Gamma(\lambda) = a_6 \cdot \lambda^6 + a_5 \cdot \lambda^5 + a_4 \cdot \lambda^4 + a_3 \cdot \lambda^3 + a_2 \cdot \lambda^2 + a_1 \cdot \lambda + a_0 \quad (3.25)$$

Parameters a_i, $i = 0...6$, are usually determined by fitting the look-up table representing an experimental torque characteristic in a least squares sense. Typical static C_Γ and C_p variations with respect to tip speed ratio for two bladed HAWT are given in Figure 3.11. Different blade profiles imply different aerodynamic characteristic shapes; for example, in Figure 3.11 are depicted $C_\Gamma - \lambda$ and $C_p - \lambda$ curves belonging to a passive-stall regulated rotor (dotted line) and to an active-stall regulated rotor (continuous line).

The corresponding torque and power characteristics with respect to the LSS speed (obtained using Equations 3.24 and 2.31 respectively), parameterised by the wind speed, are shown in Figure 3.12.

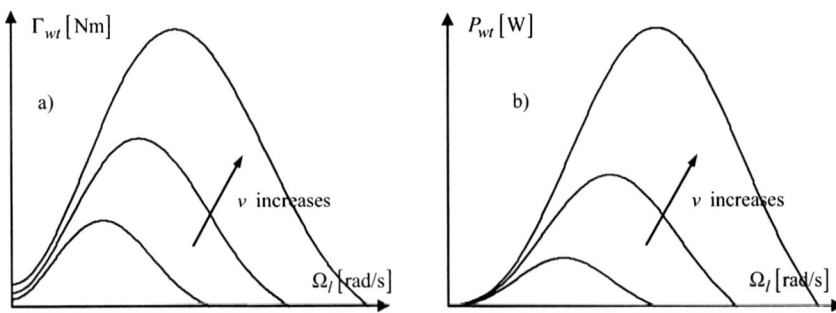

Figure 3.12. Typical HAWT torque (**a**) and power (**b**) characteristics (LSS)

3.2.3 Wind Torque Computation Based on the Wind Speed Experienced by the Rotor

The fixed-point wind speed model, as experienced by the turbine's hub, has been presented in Section 3.2.1. It represents only the initial information for determining the effective wind speed model, as experienced across the turbine's blades. The wind speed's effective properties are also influenced by a series of effects due to the rotor motion. Two of these effects induce deterministic variations of the wind torque, namely:
- the *tower shadow* effect, which takes place when one of the blades passes the tower while moving and results in a decreasing fluctuation of the wind torque;
- the *wind shear* effect, due to the wind speed periodic variation with the height to ground, producing periodic wind torque variations.

The two effects cumulatively yield periodic wind torque variations with a frequency that is an integer multiple of the blades' rotational speed.

The interaction between wind and turbine requires that the spectral properties of the fixed-point wind speed variations be strongly modified, as compared with those of the fixed-point wind turbulence model. These changes are significant, especially for large wind turbines. Rodriguez-Amenedo *et al.* (1998) present a model for obtaining the wind torque, Γ_{wt} (Figure 3.13), which allows a simplified description of the wind speed fluctuations due to wind-turbine interaction.

In the following, the main dynamic subsystems appearing in the block diagram in Figure 3.13 are described.

The *spatial filter* performs the averaging of the wind speed's variations across the area swept by the rotor. The fixed-point spectrum is modified in such a way that a spectral representation of the average wind speed across the rotor is obtained. The filter's transfer function is (Wilkie *et al.* 1990; Rodriguez-Amenedo *et al.* 1998)

$$H_{sf}(s) = \frac{\sqrt{2} + b_{sf} \cdot s}{\left(\sqrt{2} + b_{sf}\sqrt{a_{sf}} \cdot s\right) \cdot \left(1 + \frac{b_{sf}}{\sqrt{a_{sf}}} \cdot s\right)}, \qquad (3.26)$$

where a_{sf} is an empirical factor ($a_{sf}=0...55$) and b_{sf} is a parameter describing the intercorrelation between wind speed evolutions in different points across the rotor, which can be expressed as

$$b_{sf} = \gamma_{sf} \cdot \left(\frac{R}{v_s}\right), \qquad (3.27)$$

with $\gamma_{sf} = 1.3$, R being the blade length and v_s being the average wind speed experienced by the hub.

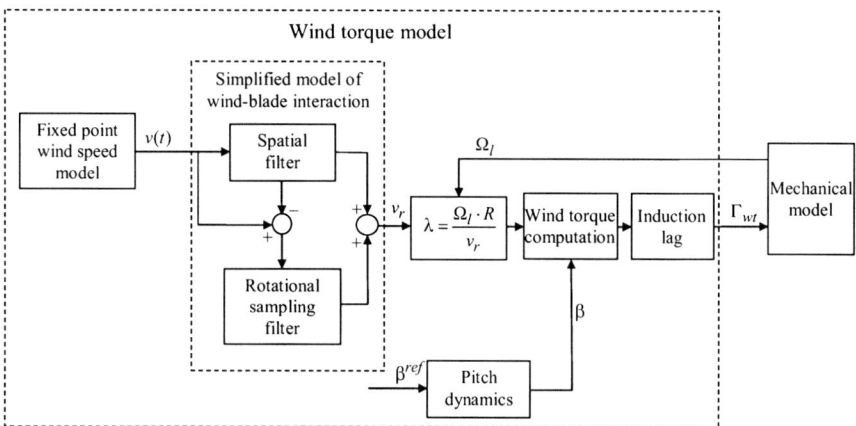

Figure 3.13. Simplified wind torque model

The *rotational sampling filter* includes a number of effects, among which are the ones mentioned above, that induce deterministic changes of the wind torque. Thus, the rotational sampling effect is due to blades' rotating motion inside the turbulence field of the fixed-point wind speed. Properties of wind fluctuations in a certain point on the blade are modified as compared with the properties of fixed-point wind turbulence (De La Salle *et al.* 1990; Leithead *et al.* 1991). Figure 3.14 *qualitatively* emphasizes the nature of these changes: a power transfer from medium to high frequencies takes place, *i.e.*, to integer multiples of the rotational speed, Ω_l. Therefore, when compared to the fixed-point wind model (whose power density is plotted with continuous line in Figure 3.14), the spectral model related to a certain point on the blade exhibits a power density diminishing at medium frequency for peaking at frequencies $i \cdot f_p$, $i=1,2...$, where $f_p = N_b \cdot \Omega_l/(2\pi)$, with N_b being the number of blades (represented with dashed line in Figure 3.14).

In Chapter 6 one can find details concerning the computation of the rotational sampling filter in a certain point on the blade, using a given model of the fixed-point wind turbulence.

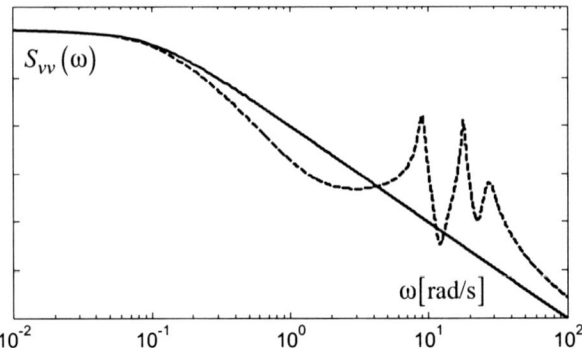

Figure 3.14. Wind power spectral density: fixed-point (*solid line*) vs. rotationally-sampled (*dashed line*)

As the procedure is quite complicated, the simplified model presented in Figure 3.13 is usually preferred for computing the wind speed and the wind torque. Thus, a single filter is used, which determines a single power density concentration, at f_p frequency. In literature the following transfer function of the rotational sampling filter is proposed (Wilkie *et al.* 1990; Rodriguez-Amenedo *et al.* 1998):

$$H_{rs}(s) = \frac{(s + N_b \cdot \Omega_l + \varepsilon) \cdot (s + N_b \cdot \Omega_l - \varepsilon)}{(s + \sigma)^2 + (N_b \cdot \Omega_l)^2}, \quad (3.28)$$

where parameters ε and σ determine the magnitude of the power density concentration at $N_b \cdot \Omega_l$, and also the characteristic's selectivity at this frequency.

The wind torque model from Figure 3.13 also contains the subsystem denoted by "Induction lag", which describes an aerodynamic effect, experimentally proven when the wind speed or the pitch angle is abruptly varying. This effect is derivative and can be modelled as

$$H(s) = \frac{cs+1}{ds+1}, \quad c > d, \quad (3.29)$$

where both parameters, c and d, depend on the average wind speed (Wilkie *et al.* 1990; Rodriguez-Amenedo *et al.* 1998).

Figure 3.15a,b illustrates how the wind speed is obtained using the model shown in Figure 3.13.

The Bode diagrams of the shaping filter used in fixed-point wind model and of the one used for capturing the rotational sampling effect are comparatively depicted in Figure 3.15a. The power spectral density of the wind fluctuation embedding the rotational sampling effect is given in Figure 3.15b.

Figure 3.16a,b suggests a comparison between the two wind profiles, *i.e.*, fixed-point *vs.* rotationally-sampled, on the same time window. Figure 3.16c zooms out the wind fluctuation with rotational sampling effect.

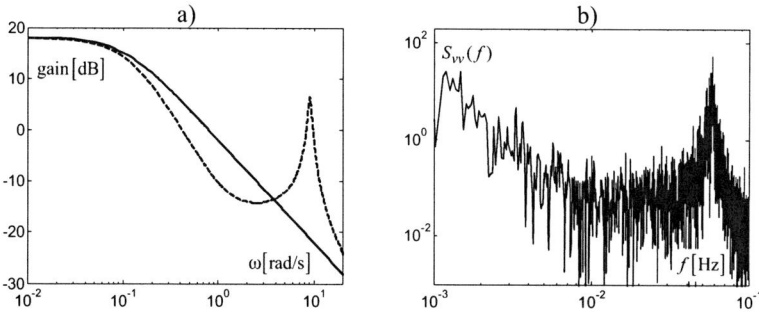

Figure 3.15. a Bode characteristics of shaping filters: fixed-point (*solid line*), rotational sampling (*dashed line*). **b** Rotationally-sampled power spectral density

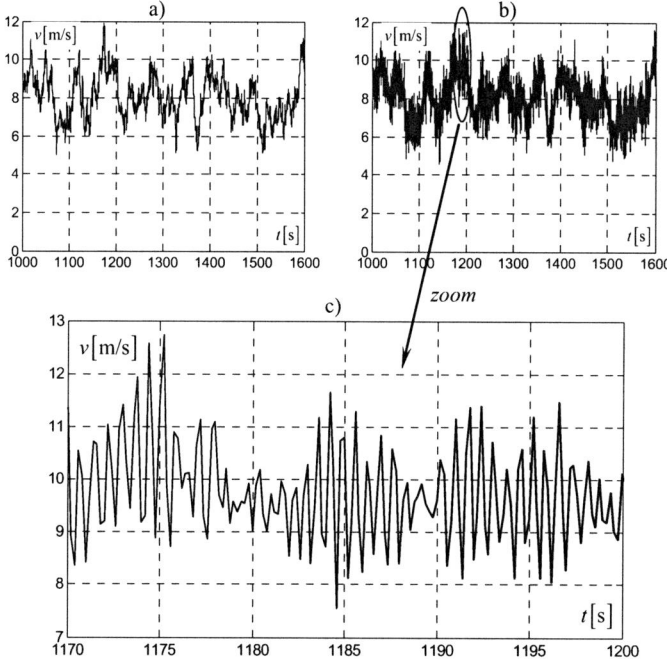

Figure 3.16. a Fixed-point wind speed. **b** Wind speed with the effect of rotational sampling. **c** Zoom on the rotationally-sampled wind speed

Let us now remark that the effects described above are in fact important only for large scale WECS. For the low/medium-power WECS, the rotor diameter is sufficiently small for one to neglect the tower effect and structural dynamics of the turbine, especially if it is equipped with a teeter hub or some other equipment ensuring the damping of these undesired dynamic effects. Also, the wind speed can be considered constant for the entire area swept by the rotor. This means that the wind shear effect is negligible too. The rotational sampling effect is reduced;

therefore a fixed-point wind speed model can be employed, usually neglecting the blade torsional dynamics and induction lag. These remarks suggest that the aerodynamic model of a high-power wind turbine is far more complicated compared to that employed in low-power WECS.

3.3 Electrical Generator Modelling

Electrical generators are systems whose power regime is generally controlled by means of power electronics converters. From this viewpoint, irrespective of their particular topologies, controlled electrical generators are systems whose inputs are stator and rotor voltages, having as state variables the stator and rotor currents or fluxes (Leonhard 2001). They are composed of an electromagnetic subsystem, which outputs the electromagnetic torque, further referred to as Γ_G, and the electromechanical subsystem, through which the generator experiences a mechanical interaction. Figure 3.17 illustrates the modelling principle for the SCIG case. The necessity of using (d,q) models comes from vector control implementation, which has the advantage of ensuring torque variation minimization and thus better motion control.

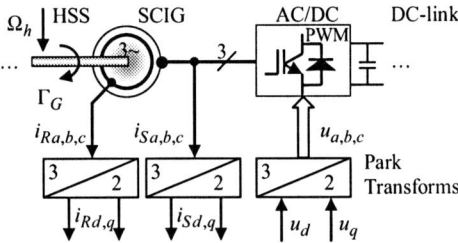

Figure 3.17. Generator modelling: identifying inputs, outputs and states – SCIG case

Below, the generator modelling is focused on capturing the evolution of the electromagnetic subsystem into a mathematical form. Thus, a set of equations involving the generator's electrical variables – voltages, fluxes and currents – results. In WECS the generator interacts with the drive train; hence, to this set of equations is usually added the high-speed shaft (HSS) motion equation in the form

$$J\frac{d\Omega_h}{dt} = \Gamma_{mec} - \Gamma_G, \qquad (3.30)$$

where the static and viscous frictions have been neglected, J is the equivalent inertia rendered to the HSS, Γ_{mec} is the mechanical torque, Ω_h is the HSS rotational speed and Γ_G is the electromagnetic torque resulting from the interaction between the stator and rotor fluxes and depending upon each particular configuration, as listed below. The modelling has assumed that the influence of the generator constructive features on its dynamics (*e.g.*, higher harmonics, asymmetries, *etc.*) is neglected and its parameters are constant.

3.3.1 Induction Generators

Doubly-fed Induction Generator (DFIG)
The doubly-fed induction generator's electromagnetic torque is expressed in (d,q) frame as (Leonhard 2001; Bose 2001):

$$\Gamma_G = 3/2\, pL_m \cdot \left(i_{Sq} \cdot i_{Rd} - i_{Rq} \cdot i_{Sd}\right), \qquad (3.31)$$

with p being the pole pairs number, L_m the stator-rotor mutual inductance, i_{Sd}, i_{Sq}, i_{Rd} and i_{Rq} are the stator, respectively rotor current (d,q) components, obtained by integrating the following differential equations:

$$\begin{cases} \dfrac{di_{Sd}}{dt} = \dfrac{V_{Sd}}{L_S} - \dfrac{R_S}{L_S}\cdot i_{Sd} - \dfrac{L_m}{L_S}\cdot\dfrac{di_{Rd}}{dt} + \omega_S\cdot\left(i_{Sq} + \dfrac{L_m}{L_S}\cdot i_{Rq}\right) \\[4pt] \dfrac{di_{Sq}}{dt} = \dfrac{V_{Sq}}{L_S} - \dfrac{R_S}{L_S}\cdot i_{Sq} - \dfrac{L_m}{L_S}\cdot\dfrac{di_{Rq}}{dt} - \omega_S\cdot\left(i_{Sd} + \dfrac{L_m}{L_S}\cdot i_{Rd}\right) \\[4pt] \dfrac{di_{Rd}}{dt} = \dfrac{V_{Rd}}{L_R} - \dfrac{R_R}{L_R}\cdot i_{Rd} - \dfrac{L_m}{L_R}\cdot\dfrac{di_{Sd}}{dt} + (\omega_S - \omega)\cdot\left(i_{Rq} + \dfrac{L_m}{L_R}\cdot i_{Sq}\right) \\[4pt] \dfrac{di_{Rq}}{dt} = \dfrac{V_{Rq}}{L_R} - \dfrac{R_R}{L_R}\cdot i_{Rq} - \dfrac{L_m}{L_R}\cdot\dfrac{di_{Sq}}{dt} - (\omega_S - \omega)\cdot\left(i_{Rd} + \dfrac{L_m}{L_R}\cdot i_{Sd}\right) \end{cases}, \qquad (3.32)$$

where
$\omega = p\cdot\Omega_h$ is the speed in electrical radians per second (where Ω_h is the generator rotational speed), $\omega_S = d\theta_S/dt$ (rad/s) is the stator field frequency, R_S, R_R are the stator and rotor resistances, L_S, L_R are the stator and rotor inductances; V_{Sd}, V_{Rd}, $V_{Rd} = V_{Rq} = 0$ are the stator, respectively rotor voltage (d,q) components; $\Phi_{Rd} = L_R \cdot i_{Rd}$, $\Phi_{Rq} = L_R \cdot i_{Rq}$ are the rotor flux (d,q) components.

By adopting the notation

$$\begin{cases} \mathbf{x} = \begin{bmatrix} x_1(t) & x_2(t) & x_3(t) & x_4(t) \end{bmatrix}^T = \begin{bmatrix} i_{Sd} & i_{Sq} & i_{Rd} & i_{Rq} \end{bmatrix}^T \\ \mathbf{u} = \begin{bmatrix} V_{Sd} & V_{Sq} & V_{Rd} & V_{Rq} \end{bmatrix}^T \end{cases} \qquad (3.33)$$

for the state and input vector respectively, the DFIG state model can be presented as a fourth-order model:

$$\begin{cases} \dot{\mathbf{x}} = \mathbf{A}(\Omega_h)\cdot\mathbf{x} + \mathbf{B}\cdot\mathbf{u} \\ y \equiv \Gamma_G = \dfrac{3pL_m}{2}(x_2 x_3 - x_1 x_4) \end{cases}, \qquad (3.34)$$

where $\sigma = 1 - L_m^2/(L_S L_R)$ and

$$A(\Omega_h) = \begin{cases} \begin{bmatrix} -\dfrac{R_S}{\sigma L_S} & \omega_S + \dfrac{p\Omega_h L_m^2}{\sigma L_S L_R} & \dfrac{L_m R_R}{\sigma L_S L_R} & \dfrac{p\Omega_h L_m}{\sigma L_S} \\ -\left(\omega_S + \dfrac{p\Omega_h L_m^2}{\sigma L_S L_R}\right) & -\dfrac{R_S}{\sigma L_S} & -\dfrac{p\Omega_h L_m}{\sigma L_S} & \dfrac{L_m R_R}{\sigma L_S L_R} \\ \dfrac{L_m R_S}{\sigma L_S L_R} & -\dfrac{p\Omega_h L_m}{\sigma L_R} & -\dfrac{R_R}{\sigma L_R} & \omega_S - \dfrac{p\Omega_h}{\sigma} \\ \dfrac{p\Omega_h L_m}{\sigma L_R} & \dfrac{R_S L_m}{\sigma L_S L_R} & \dfrac{p\Omega_h}{\sigma} - \omega_S & -\dfrac{R_R}{\sigma L_R} \end{bmatrix} \\ \\ B = \begin{bmatrix} \dfrac{1}{\sigma L_S} & 0 & \dfrac{-L_m}{\sigma L_S L_R} & 0 \\ 0 & \dfrac{1}{\sigma L_S} & 0 & \dfrac{-L_m}{\sigma L_S L_R} \\ \dfrac{-L_m}{\sigma L_S L_R} & 0 & \dfrac{1}{\sigma L_R} & 0 \\ 0 & \dfrac{-L_m}{\sigma L_S L_R} & 0 & \dfrac{1}{\sigma L_R} \end{bmatrix} \end{cases}$$

(3.35)

One can note that matrix **A** depends on the rotational speed, Ω_h, which is an interaction variable. The output variable is the electromagnetic torque, given by Equation 3.31. The torque characteristic of the induction machine is represented in Figure 3.18.

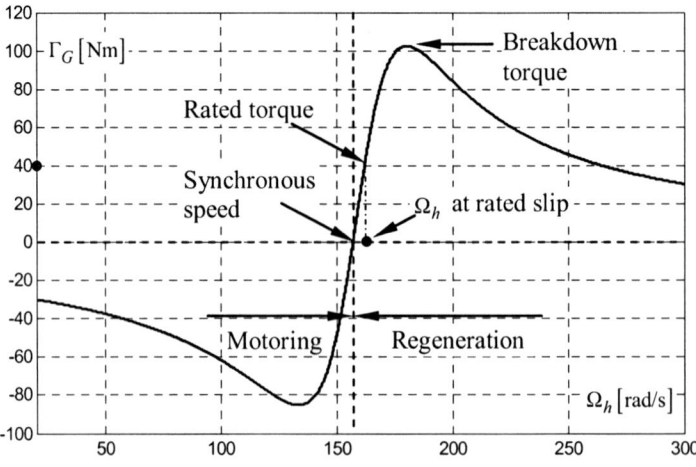

Figure 3.18. Mechanical characteristic of the induction machine

Squirrel-cage Induction Generator (SCIG)
The SCIG model can be obtained from the DFIG model by setting the d and q components of the rotor voltage to zero. Therefore, in Equation 3.33, by setting $V_{Rd} = V_{Rq} = 0$, the state and input vectors are respectively obtained:

$$\begin{cases} \mathbf{x} = \begin{bmatrix} x_1(t) & x_2(t) & x_3(t) & x_4(t) \end{bmatrix}^T = \begin{bmatrix} i_{Sd} & i_{Sq} & i_{Rd} & i_{Rq} \end{bmatrix}^T \\ \mathbf{u} = \begin{bmatrix} V_{Sd} & V_{Sq} \end{bmatrix}^T \end{cases} \quad (3.36)$$

The SCIG state model is then

$$\begin{cases} \dot{\mathbf{x}} = \mathbf{A}(\Omega_h) \cdot \mathbf{x} + \mathbf{B} \cdot \mathbf{u} \\ y \equiv \Gamma_G = \dfrac{3pL_m}{2}(x_2 x_3 - x_1 x_4) \end{cases}, \quad (3.37)$$

where $\sigma = 1 - L_m^2/(L_S L_R)$, matrix $\mathbf{A}(\Omega_h)$ is the same as in Equation 3.35 and

$$\mathbf{B} = \begin{bmatrix} \dfrac{1}{\sigma L_S} & 0 & \dfrac{-L_m}{\sigma L_S L_R} & 0 \\ 0 & \dfrac{1}{\sigma L_S} & 0 & \dfrac{-L_m}{\sigma L_S L_R} \end{bmatrix}^T \quad (3.38)$$

Reduced-order Model
Reduced-order modelling of induction machines is based upon noting that stator transients are much faster than the rotor ones (Krause *et al.* 2002). Therefore, a second-order model results, which describes only the rotor currents dynamic.

A first reduced-order modelling method (Iov 2003) starts from considering fluxes instead of currents as state variables. Thus, taking $\mathbf{\Psi} = \begin{bmatrix} \Psi_{Sd} & \Psi_{Sq} & \Psi_{Rd} & \Psi_{Rq} \end{bmatrix}^T \equiv \begin{bmatrix} \mathbf{\Psi}_S & \mathbf{\Psi}_R \end{bmatrix}^T$ as state vector and input vector $\mathbf{u}$ defined as in Equation 3.33, the rotor flux dynamic equations have the form

$$\dot{\mathbf{\Psi}}_R = \begin{bmatrix} \alpha\dfrac{R_R L_m^2}{\sigma L_S L_R^2} - \dfrac{R_R}{\sigma L_R} & \left(\alpha\dfrac{R_R L_m}{R_S L_R}+1\right)\omega_S - p\Omega_h \\ -\left(\alpha\dfrac{R_R L_m}{R_S L_R}+1\right)\omega_S + p\Omega_h & \alpha\dfrac{R_R L_m^2}{\sigma L_S L_R^2} - \dfrac{R_R}{\sigma L_R} \end{bmatrix} \cdot \mathbf{\Psi}_R$$
$$+ \begin{bmatrix} \alpha\dfrac{R_R}{R_S L_R} & \alpha\dfrac{\sigma R_R L_S}{L_R R_S^2} & 1 & 0 \\ -\alpha\dfrac{\sigma R_R L_S}{L_R R_S^2} & \alpha\dfrac{R_R}{R_S L_R} & 0 & 1 \end{bmatrix} \cdot \mathbf{u} \quad , (3.39)$$

where $\alpha = \dfrac{1}{1+(\sigma\omega_S L_S/R_S)^2}$. Stator fluxes are then obtained from purely algebraic equations:

$$\Psi_S = \begin{bmatrix} \alpha\dfrac{L_m}{L_R} & \alpha\dfrac{R_S L_m}{\sigma L_S}\cdot\omega_S \\ -\alpha\dfrac{R_S L_m}{\sigma L_S}\cdot\omega_S & \alpha\dfrac{L_m}{L_R} \end{bmatrix}\cdot\Psi_R + \begin{bmatrix} \alpha\dfrac{R_S}{\sigma L_S} & \alpha\dfrac{R_S^2}{\sigma^2 L_S^2}\cdot\omega_S \\ -\alpha\dfrac{R_S^2}{\sigma^2 L_S^2}\cdot\omega_S & \alpha\dfrac{R_S}{\sigma L_S} \end{bmatrix}\cdot\begin{bmatrix} V_{Sd} \\ V_{Sq} \end{bmatrix}$$

(3.40)

Finally, currents are also computed based on an algebraic dependence:

$$\begin{bmatrix} i_{Sd} \\ i_{Sq} \\ i_{Rd} \\ i_{Rq} \end{bmatrix} = \begin{bmatrix} L_S & 0 & L_m & 0 \\ 0 & L_S & 0 & L_m \\ L_m & 0 & L_R & 0 \\ 0 & L_m & 0 & L_R \end{bmatrix}^{-1}\cdot\Psi \qquad (3.41)$$

Starting from the fourth-order DFIG model given in Equation 3.34, one can obtain a reduced-order model by employing a second method. For this purpose, the singular perturbation method is applied, which takes into account that the stator currents influence the dynamic of the rotor currents by their steady-state values.

Consequently, the state vector **x** from Equation 3.33 is partitioned:

$$\mathbf{i}_S = \begin{bmatrix} i_{Sd} & i_{Sq} \end{bmatrix}^T \quad \mathbf{i}_R = \begin{bmatrix} i_{Rd} & i_{Rq} \end{bmatrix}^T, \qquad (3.42)$$

such that the state equation from Equation 3.34 to be put into the form

$$\begin{bmatrix} \dot{\mathbf{i}}_S \\ \dot{\mathbf{i}}_R \end{bmatrix} = \begin{bmatrix} A_{11} & | & A_{12} \\ -- & -- & -- \\ A_{21} & | & A_{22} \end{bmatrix}\cdot\begin{bmatrix} \mathbf{i}_S \\ \mathbf{i}_R \end{bmatrix} + \begin{bmatrix} B_1 \\ -- \\ B_2 \end{bmatrix}\cdot\mathbf{u}, \qquad (3.43)$$

where

$$A_{11} = \begin{bmatrix} -\dfrac{R_S}{\sigma L_S} & \omega_S + \dfrac{p\Omega_h L_m^2}{\sigma L_S L_R} \\ -\left(\omega_S + \dfrac{p\Omega_h L_m^2}{\sigma L_S L_R}\right) & -\dfrac{R_S}{\sigma L_S} \end{bmatrix} \quad A_{12} = \begin{bmatrix} \dfrac{L_m R_R}{\sigma L_S L_R} & \dfrac{p\Omega_h L_m}{\sigma L_S} \\ -\dfrac{p\Omega_h L_m}{\sigma L_S} & \dfrac{L_m R_R}{\sigma L_S L_R} \end{bmatrix}$$

$$A_{21} = \begin{bmatrix} \dfrac{L_m R_S}{\sigma L_S L_R} & -\dfrac{p\Omega_h L_m}{\sigma L_R} \\ \dfrac{p\Omega_h L_m}{\sigma L_R} & \dfrac{R_S L_m}{\sigma L_S L_R} \end{bmatrix} \quad A_{22} = \begin{bmatrix} -\dfrac{R_R}{\sigma L_R} & \omega_S - \dfrac{p\Omega_h}{\sigma} \\ \dfrac{p\Omega_h}{\sigma} - \omega_S & -\dfrac{R_R}{\sigma L_R} \end{bmatrix}$$

$$\mathbf{B}_1 = \begin{bmatrix} \dfrac{1}{\sigma L_S} & 0 & \dfrac{-L_m}{\sigma L_S L_R} & 0 \\ 0 & \dfrac{1}{\sigma L_S} & 0 & \dfrac{-L_m}{\sigma L_S L_R} \end{bmatrix} \quad \mathbf{B}_2 = \begin{bmatrix} \dfrac{-L_m}{\sigma L_S L_R} & 0 & \dfrac{1}{\sigma L_R} & 0 \\ 0 & \dfrac{-L_m}{\sigma L_S L_R} & 0 & \dfrac{1}{\sigma L_R} \end{bmatrix}$$

To obtain the steady-state values one must zero the derivatives. By zeroing $\dot{\mathbf{i}}_S$, Equation 3.43 becomes

$$\begin{cases} \mathbf{A}_{11}\mathbf{i}_S + \mathbf{A}_{12}\mathbf{i}_R + \mathbf{B}_1\mathbf{u} = 0 \\ \mathbf{A}_{21}\mathbf{i}_S + \mathbf{A}_{22}\mathbf{i}_R + \mathbf{B}_2\mathbf{u} = \dot{\mathbf{i}}_R \end{cases} \tag{3.44}$$

Steady-state stator currents result from the first relation of Equation 3.44 as

$$\mathbf{i}_S = -\mathbf{A}_{11}^{-1} \cdot (\mathbf{A}_{12}\mathbf{i}_R + \mathbf{B}_1\mathbf{u}), \tag{3.45}$$

where the inverse of matrix $\mathbf{A}_{11}$ is

$$\mathbf{A}_{11}^{-1} = \dfrac{1}{\left(\dfrac{R_S}{\sigma L_S}\right)^2 + \left(\omega_S + \dfrac{p\Omega_h L_m^2}{\sigma L_S L_R}\right)^2} \cdot \begin{bmatrix} -\dfrac{R_S}{\sigma L_S} & -\left(\omega_S + \dfrac{p\Omega_h L_m^2}{\sigma L_S L_R}\right) \\ \omega_S + \dfrac{p\Omega_h L_m^2}{\sigma L_S L_R} & -\dfrac{R_S}{\sigma L_S} \end{bmatrix}, \tag{3.46}$$

and are replaced in the second relation of Equation 3.44 to provide the DFIG reduced-order (second-order) dynamic, concerning only the d and q rotor currents:

$$\dot{\mathbf{i}}_R = \left(\mathbf{A}_{22} - \mathbf{A}_{21}\mathbf{A}_{11}^{-1}\mathbf{A}_{12}\right) \cdot \mathbf{i}_R + \left(\mathbf{B}_2 - \mathbf{A}_{11}^{-1}\mathbf{B}_1\right) \cdot \mathbf{u} \tag{3.47}$$

3.3.2 Synchronous Generators

Permanent-magnet Synchronous Generator (PMSG)
The PMSG is modelled under the following simplifying assumptions: sinusoidal distribution of stator winding, electric and magnetic symmetry, negligible iron losses and unsaturated magnetic circuit. Under these assumptions, the generator model in the so-called steady-state (or stator) coordinates is first obtained (Leonhard 2001). Another simpler model can be obtained in (d,q) rotor coordinates; conversion between (a,b,c) and (d,q) coordinates can be realized by means of the Park Transform (Leonhard 2001). Then, after neglecting the homopolar voltage, u_0, by virtue of symmetry, the (d,q) PMSG model becomes

$$\begin{cases} u_d = R i_d + L_d \dot{i}_d - \Phi_q \omega_S \\ u_q = R i_q + L_q \dot{i}_q + \Phi_d \omega_S \end{cases}, \tag{3.48}$$

where R is the stator resistance, u_d, u_q are d and q stator voltages, L_d, L_q are d and q inductances and ω_S is the stator (or else electric) pulsation,

$$\Phi_d = L_d i_d + \Phi_m \tag{3.49}$$

$$\Phi_q = L_q i_q \tag{3.50}$$

are d and q fluxes and Φ_m is the flux that is constant due to permanent magnets. Thus, the model at Equation 3.48 becomes

$$\begin{cases} u_d = R i_d + L_d \dot{i}_d - L_q i_q \omega_S \\ u_q = R i_q + L_q \dot{i}_q + (L_d i_d + \Phi_m) \omega_S \end{cases} \tag{3.51}$$

The electromagnetic torque is obtained as

$$\Gamma_G = p(\Phi_d i_q - \Phi_q i_d) = p\left[\Phi_m i_q + (L_d - L_q) i_d i_q\right], \tag{3.52}$$

where p is the number of pole pairs. If the permanent magnets are mounted on the rotor surface, then $L_d = L_q$ and the electromagnetic torque becomes

$$\Gamma_G = p \Phi_m i_q \tag{3.53}$$

When the machine operates as a *grid connected generator*, Equation 3.51 becomes

$$\begin{cases} u_d = -R i_d - L_d \dot{i}_d + L_q i_q \omega_S \\ u_q = -R i_q - L_q \dot{i}_q - (L_d i_d - \Phi_m) \omega_S \end{cases} \tag{3.54}$$

The stator frequency, ω_S, is proportional to the shaft rotational speed, $\omega_S = p \cdot \Omega_h$, which depends on how the electrical generator interacts mechanically. The state and input vector are identified respectively as

$$\mathbf{x} = \begin{bmatrix} x_1(t) & x_2(t) \end{bmatrix}^T \equiv \begin{bmatrix} i_d(t) & i_q(t) \end{bmatrix}^T \quad \mathbf{u} = \begin{bmatrix} u_d & u_q \end{bmatrix}^T \tag{3.55}$$

Hence, the grid connected PMSG state model is obtained in the form

$$\begin{cases} \dot{\mathbf{x}} = \begin{bmatrix} -\dfrac{R}{L_d} x_1 + p \dfrac{L_q}{L_d} x_2 \Omega_h \\ -\dfrac{R}{L_d} x_1 - p \dfrac{L_d x_1 - \Phi_m}{L_q} \Omega_h \end{bmatrix} + \begin{bmatrix} -\dfrac{1}{L_d} & 0 \\ 0 & -\dfrac{1}{L_q} \end{bmatrix} \cdot \mathbf{u} \\ y \equiv \Gamma_G = p \Phi_m x_2 \end{cases} \tag{3.56}$$

Suppose now that the PMSG supplies an *isolated symmetric tri-phased resistive load*, R_l. Then the following relations hold:

$$\begin{cases} (L_d + L_s)\dot{i}_d = -(R + R_l)i_d + p(L_q + L_s)i_q \Omega_h \\ (L_q + L_s)\dot{i}_q = -(R + R_l)i_q - p(L_d + L_s)i_d \Omega_h + p\Phi_m \Omega_h \end{cases}$$

In this case, the state and input vectors are respectively

$$\mathbf{x} = \begin{bmatrix} x_1(t) & x_2(t) \end{bmatrix}^T \equiv \begin{bmatrix} i_d(t) & i_q(t) \end{bmatrix}^T \quad \mathbf{u} \equiv R_l \quad (3.57)$$

The PMSG state model can be put into the form

$$\begin{cases} \dot{\mathbf{x}} = \begin{bmatrix} \dfrac{1}{L_d + L_s}\left(-Rx_1 + p(L_q - L_s)x_2 \Omega_h\right) \\ \dfrac{1}{L_q + L_s}\left(-Rx_2 - p(L_d + L_s)x_1 \Omega_h + p\Phi_m \Omega_h\right) \end{bmatrix} + \begin{bmatrix} -\dfrac{1}{L_d + L_a} & 0 \\ 0 & -\dfrac{1}{L_q + L_s} \end{bmatrix} \cdot \mathbf{x} \cdot \mathbf{u} \\ y \equiv \Gamma_G = p\Phi_m x_2 \end{cases}$$

(3.58)

Wound-rotor Synchronous Generator (WRSG)
Modelling of the wound-rotor synchronous generator is performed under the same simplifying assumptions as in the case of the permanent-magnet synchronous generator.

The model is based upon the classical representation of a three stator winding synchronous machine, one field winding and two damping windings (Leonhard 2001), being described using the following variables and parameters:

- i_d, i_q the (d,q) currents;
- v_d, v_q the (d,q) voltages;
- i_{fd}, v_{fd} the field current and voltage, respectively;
- i_{kd}, i_{kq} the damping currents on d and q axis;
- R_S the stator resistance;
- R_{fd} the field winding resistance;
- R_{kd}, R_{kq} the resistances of the damping winding;
- L_d, L_q the (d,q) inductivities;
- L_{fd} the rotor inductivity;
- L_{kd}, L_{kq} the inductivities of damping windings;
- L_{md}, L_{mq} magnetization inductivities on d and q axis;

The electromagnetic torque has the expression

$$\Gamma_G = (L_q - L_d)i_d i_q + L_{md}i_{fd}i_q + L_{md}i_{kd}i_q - L_{mq}i_d i_{kq} \qquad (3.59)$$

Noting the state and input vector respectively as

$$\begin{cases} \mathbf{x} = \begin{bmatrix} x_1(t) & x_2(t) & x_3(t) & x_4(t) & x_5(t) \end{bmatrix}^T \equiv \begin{bmatrix} i_q & i_d & i_{fd} & i_{kd} & i_{kq} \end{bmatrix}^T \\ \mathbf{u} = \begin{bmatrix} v_d & v_q & v_{fd} & 0 & 0 \end{bmatrix}^T \end{cases} \qquad (3.60)$$

and matrices

$$\begin{cases} \mathbf{M} = \begin{bmatrix} L_d & 0 & -L_{md} & -L_{md} & 0 \\ 0 & L_q & 0 & 0 & -L_{mq} \\ L_{md} & 0 & -L_{fd} & -L_{md} & 0 \\ L_{md} & 0 & -L_{md} & -L_{kd} & 0 \\ 0 & L_{mq} & 0 & 0 & -L_{kq} \end{bmatrix} \\ \mathbf{N}(\Omega_h) = \begin{bmatrix} -R_S & \Omega_h L_q & 0 & 0 & -\Omega_h L_{mq} \\ -\Omega_h L_d & -R_S & \Omega_h L_{md} & \Omega_h L_{kd} & 0 \\ 0 & 0 & R_{fd} & 0 & 0 \\ 0 & 0 & 0 & R_{kd} & 0 \\ 0 & 0 & 0 & 0 & R_{kq} \end{bmatrix} \end{cases}, \qquad (3.61)$$

the model of a wound-rotor synchronous generator can be presented as

$$\begin{cases} \dot{\mathbf{x}} = \mathbf{M}^{-1}\mathbf{N}(\Omega_h)\cdot\mathbf{x} - \mathbf{u} \\ y \equiv \Gamma_G = (L_q - L_d)x_1 x_2 + L_{md}x_1 x_3 + L_{md}x_1 x_4 - L_{mq}x_2 x_5 \end{cases} \qquad (3.62)$$

The motion equation is similar to Equation 3.30:

$$\frac{d\Omega_h}{dt} = \frac{1}{J}(\Gamma_{mec} - \Gamma_G - F\Omega_h), \qquad (3.63)$$

where Ω_h, Γ_{mec} and J denote the same as above and F is the viscous friction constant.

3.4 Drive Train Modelling

The drive train (S_3 in Figure 3.1) can be viewed as a system by means of which subsystems S_1 and S_2 interact mechanically. Thus, the wind torque, Γ_{wt}, and the electromagnetic torque, Γ_G, are the inputs, whereas the rotational speed is the output. The modelling is done under the assumption that the mechanical transmission has a constant efficiency for the whole speed range; the influence of the constructive features (*e.g.*, vibrations, gear type, gear backlash, *etc.*) on its behaviour is considered parasitic and will be neglected. Furthermore, the mechanical and electrical systems are considered to be perfectly balanced, any fault is taken as a special operating regime.

3.4.1 Rigid Drive Train

The main element of a rigid drive train is the single-stage rigidly-coupled speed multiplier, of (fixed) ratio i and efficiency η (Figure 3.19). In this case, the model consists of a first-order motion equation, rendered either at the low-speed or at the high-speed shaft.

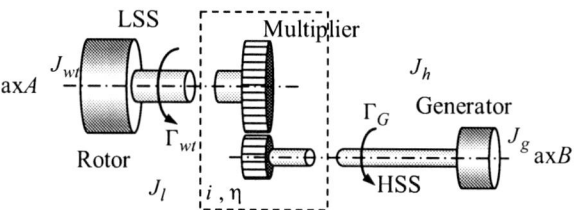

Figure 3.19. Rigid drive train

Due to the speed multiplier, the generator experiences an i times reduced torque and an i times increased speed, $\Omega_h = i \cdot \Omega_l$; the system equivalent inertia is reduced when rendered at the high-speed shaft, J_h (Munteanu 1997). Under the previously stated modelling assumptions, the WECS dynamic may be expressed rendered at either the HSS or at the LSS in two equivalent forms (Nichita 1995; Welfonder *et al.* 1997):

$$J_h \cdot \frac{d\Omega_h}{dt} = \frac{\eta}{i} \cdot \Gamma_{wt}(\Omega_l, v) - \Gamma_G(\Omega_h, c), \qquad (3.64)$$

$$J_l \cdot \frac{d\Omega_l}{dt} = \Gamma_{wt}(\Omega_l, v) - \frac{i}{\eta} \cdot \Gamma_G(\Omega_h, c), \qquad (3.65)$$

where
$\Gamma_{wt}(\Omega_l,v)$ is the aerodynamic torque, parameterized by the wind speed, v;
$\Gamma_G(\Omega_h,c)$ is the electromagnetic torque, parameterized by a generically called load variable, denoted by c;
J_h, J_l are inertias rendered at HSS and LSS respectively, being computed as

$$J_h = (J_1 + J_{wt}) \cdot \frac{\eta}{i^2} + J_2 + J_g, \qquad (3.66)$$

$$J_l = J_{wt} + J_1 + (J_g + J_2) \cdot \frac{i^2}{\eta}, \qquad (3.67)$$

where

J_1, J_2 are inertias of the multiplier gearings;

J_{wt}, J_g are inertias of turbine rotor and of electrical generator respectively.

As the drive train is rigid, the dynamical torque is given only by the first-order linear rotational speed variation. Figure 3.20a contains a Simulink® implementation of the motion equation rendered at the high-speed shaft (Equation 3.66). This equation is usually referred to as being the single-mass WECS model.

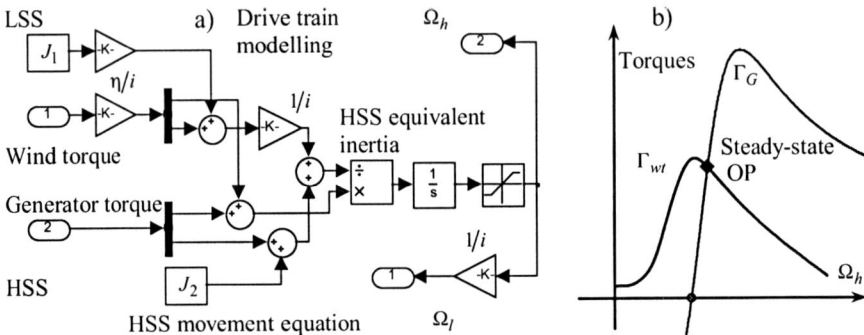

Figure 3.20. **a** Motion equation rendered at the high-speed shaft – Simulink® implementation; **b** steady-state regime of WECS

Figure 3.20b presents the steady-state regime of the interaction between the turbine rotor and the induction generator by means of the high-speed shaft turning at Ω_h. The steady-state operating point of WECS is thus determined as the cross-point of the rotor mechanical characteristic with that of the generator.

3.4.2 Flexible Drive Train

Elements of a flexible drive train are depicted in Figure 3.21.

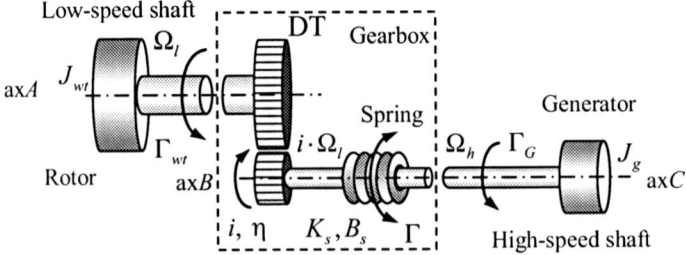

Figure 3.21. Schematics of a flexible drive train

Different from the rigid coupling, the two parts of the high-speed shaft, axB and axC in Figure 3.21, are now turning at different speeds, $i \cdot \Omega_l$ and Ω_h respectively, where i is the transmission ratio of the gearbox. The elastic energy variations yield a new state variable, the internal torque, Γ. Denoting by J_g the inertia of axC and by J_B the inertia of axB inertia, it holds that

$$J_B = \eta/i^2 \cdot J_{wt},$$

where η is the transmission efficiency and J_{wt} is the low-speed shaft inertia. The flexible drive train model is composed of axB and axC motion equations and the dynamic of the internal torque (De Battista and Mantz 1998; Akhmatov 2003):

$$\begin{cases} \dot{\Omega}_l = 1/J_{wt} \cdot \Gamma_{wt} - i/(J_{wt} \cdot \eta) \cdot \Gamma \\ \dot{\Omega}_h = 1/J_g \cdot \Gamma - 1/J_g \cdot \Gamma_G \\ \dot{\Gamma} = K_s \cdot (i \cdot \Omega_l - \Omega_h) + B_s \cdot \left(i \cdot \dot{\Omega}_l - \dot{\Omega}_h \right) \end{cases}$$

Finally, a third-order linear model results, having $\mathbf{x} = \begin{bmatrix} \Omega_l & \Omega_h & \Gamma \end{bmatrix}^T$ as states, $\mathbf{u} = \begin{bmatrix} \Gamma_{wt} & \Gamma_G \end{bmatrix}^T$ as inputs and $\mathbf{y} = \begin{bmatrix} \Omega_l & \Omega_h \end{bmatrix}^T$ as outputs:

$$\begin{cases} \dot{\mathbf{x}} = \begin{bmatrix} 0 & 0 & -\dfrac{1}{i \cdot J_B} \\ 0 & 0 & \dfrac{1}{J_g} \\ i \cdot K_s & -K_s & -B_s \cdot \left(\dfrac{1}{J_B} + \dfrac{1}{J_g} \right) \end{bmatrix} \cdot \mathbf{x} + \begin{bmatrix} \dfrac{1}{J_{wt}} & 0 \\ 0 & -\dfrac{1}{J_g} \\ \dfrac{i \cdot B_s}{J_{wt}} & \dfrac{B_s}{J_g} \end{bmatrix} \cdot \mathbf{u} \\ \mathbf{y} = \begin{bmatrix} 1 & 0 & 0 \\ 0 & 1 & 0 \end{bmatrix} \cdot \mathbf{x} \end{cases}, \quad (3.68)$$

where K_s and B_s are respectively the stiffness and the damping coefficients of the spring.

3.5 Power Electronics Converters and Grid Modelling

Power Electronics Interface
The power electronics converter realizing the interface between WECS and the electrical grid has some key roles in the wind turbine variable-speed operation. It realizes a certain decoupling between the two above elements and allows the effective power flow control.

The interest here will be focused on one of the most popular converter structures, namely the back-to-back AC–AC converter, employing two PWM-controlled voltage-source inverters (VSI), as shown in Figure 3.22. The voltages and frequencies differ on one side of the converter *vs.* the other. In fact, this configuration consists of a converter (rectifier) connected to the electrical generator, which will be further denoted as turbine-side converter, and of a converter connected to the mains, called in the following as grid-side inverter. The two power electronics converters are connected through the so-called DC-link, which represents a direct current circuit having a smoothing filter, *e.g.*, a capacitor.

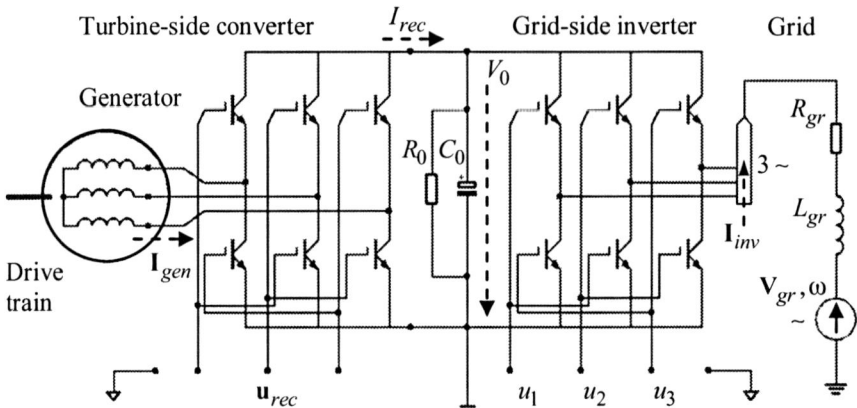

Figure 3.22. WECS converter for grid interfacing – SCIG case

Therefore, the general control objectives associated with variable-speed operation are split, namely in generator speed control and electrical power transfer control respectively.

The first objective is associated with the turbine-side converter connected to the turbine-generator coupling, and aims at directly controlling the captured power by variable-speed operation. The system subjected to control is therefore composed of the WT aerodynamics, mechanical transmission, electrical machine and the turbine-side converter.

The second control scope is associated with the interaction between the grid-side inverter and the grid, focused on the active and reactive power control. Here, the system subjected to control is composed of the electrical grid, the grid-side inverter and the DC-link.

In the generator vector control, the time constants introduced by the turbine-side converter can easily be neglected in relation to the others in the WECS; also, the power loss can be ignored. Furthermore, since the switching frequency of the power converters is large enough to be filtered by the inductances involved in their circuits, the influence of the higher than zero-order harmonics can be neglected if a global WECS modelling is envisaged. If an average modelling of the electrical tri-phased system (only the zero-order harmonic is taken) is considered, then a generator (d,q) model can be developed. In this case, a (d,q) lossless average model of a converter can be reduced to a gain for the direct and quadrature voltage components. This is the reason why turbine-side inverter modelling is not treated here.

3.5 Power Electronics Converters and Grid Modelling

Grid-side Inverter and Grid Modelling

Electrical grid modelling aims at describing the interaction with the WECS. For general WECS control purposes, a simplified average global model having finite power is used.

If a "strong" and balanced grid is assumed, the triphased electrical grid is generally represented by the sinusoidal mains voltages of amplitude V_m, $\mathbf{V}_{gr} = \begin{bmatrix} V_{gr1} & V_{gr2} & V_{gr3} \end{bmatrix}^T$ and the short-circuit impedances, $R_{gr} + j\omega L_{gr}$ (see Figure 3.22). The DC-link feeding the inverter is described by voltage V_0, capacity C_0 and resistance R_0.

Let $\mathbf{I}_{inv} = \begin{bmatrix} I_{inv1} & I_{inv2} & I_{inv3} \end{bmatrix}^T$ denote the inverter's output currents and $\mathbf{u} = \begin{bmatrix} u_1 & u_2 & u_3 \end{bmatrix}^T$ the commutation functions related to the control inputs. These latter take two values: 1 if the corresponding switch is on and −1 if it is off. The corresponding notations in the case of the turbine-side inverter are $\mathbf{I}_{gen}$ for its current inputs and $\mathbf{u}_{rec}$ for the control inputs.

The following equations, obtained by writing Kirchhoff's theorem for each grid phase, describe the system behaviour (Bose 2001):

$$L_{gr} \cdot \dot{\mathbf{I}}_{inv} = \mathbf{V}_{gr} - \frac{V_0}{2} \cdot \mathbf{M} \cdot \mathbf{u} - R_{gr} \cdot \mathbf{I}_{inv}, \qquad (3.69)$$

where $\mathbf{M} = \frac{1}{3} \cdot \begin{bmatrix} 2 & -1 & -1 \\ -1 & 2 & -1 \\ -1 & -1 & 2 \end{bmatrix}$. By introducing some new commutation functions $\tilde{\mathbf{u}} \triangleq \mathbf{M} \cdot \mathbf{u}$, one can deduce that the $\mathbf{M}$ transform makes it possible that the influence of the commutation functions being decoupled on the three phases.

At the same time the following equation holds within the DC-link (Bose 2001):

$$C_0 \cdot \frac{dV_0}{dt} = \frac{1}{2} \cdot \mathbf{I}_{inv}^T \cdot \tilde{\mathbf{u}} - \frac{V_0}{R_0} - I_{rec} \qquad (3.70)$$

where $I_{rec} = \frac{1}{2} \cdot \mathbf{I}_{gen}^T \cdot \mathbf{u}_{rec}$ is the turbine-side inverter output feeding the DC-link.

In order to obtain the average (d,q) model, only the zero-order harmonic is taken. Let $\boldsymbol{\beta} = \begin{bmatrix} \beta_1 & \beta_2 & \beta_3 \end{bmatrix}^T$ be the zero-order harmonics of $\tilde{\mathbf{u}}$. By applying the Park Transform in synchronous reference frame ($V_{grd} = V_m$ and $V_{grq} = 0$) to Equations 3.69 and 3.70 one obtains (Rabelo and Hofmann 2003)

$$\begin{cases} L_{gr} \cdot \dfrac{dI_{invd}}{dt} = V_m - R_{gr} \cdot I_{invd} + L_{gr}\omega \cdot I_{invq} - \dfrac{V_0}{2} \cdot \beta_d \\ L_{gr} \cdot \dfrac{dI_{invq}}{dt} = -R_{gr} \cdot I_{invd} - L_{gr}\omega \cdot I_{invq} - \dfrac{V_0}{2} \cdot \beta_q \\ C_0 \cdot \dfrac{dV_0}{dt} = \dfrac{3}{2} \cdot \left[I_{invd} \cdot \beta_d + I_{invq} \cdot \beta_q \right] - \dfrac{V_0}{R_0} - I_{rec} \end{cases} \quad (3.71)$$

Model at Equation 3.71, describing the interaction between the power inverter and the grid, is useful especially for active/reactive power control. Other models can also be used, depending on a well-defined control goal (Lubosny 2003). For the medium/high-power case, the controlled generator and the power converters can be considered without any dynamic when used in a global modelling approach.

3.6 Linearization and Eigenvalue Analysis

The aim of this subsection is to present linearized models of different WECS configurations resulted from combining the above presented types of WECS parts (aerodynamics, drive train, electrical generator). The wind speed is an input variable. For grid-connected WECS, the other inputs are the stator (or also the rotor) voltages, whereas for autonomous WECS the load is an input.

3.6.1 Induction-generator-based WECS

In the following, the linearization procedure and eigenvalue analysis are illustrated in the case of a *variable-speed fixed-pitch rigid-drive-train* induction-generator-based WECS. The global nonlinear model results from adding to the DFIG model (Equation 3.34) the motion equation in the form of the rigid drive train equation rendered at the high-speed shaft (Equation 3.64), where the wind torque is given by Equation 3.24. Consequently, the state and input vectors are

$$\begin{cases} \mathbf{x} = \begin{bmatrix} x_1 & x_2 & x_3 & x_4 & x_5 \end{bmatrix}^T = \begin{bmatrix} i_{Sd} & i_{Sq} & i_{Rd} & i_{Rq} & \Omega_h \end{bmatrix}^T \\ \mathbf{u} = \begin{bmatrix} V_{Sd} & V_{Sq} & V_{Rd} & V_{Rq} & v \end{bmatrix}^T \end{cases} \quad (3.72)$$

Linearized Model

An operating point is characterized by a sextuple $\left(\overline{i_{Sd}}, \overline{i_{Sq}}, \overline{i_{Rd}}, \overline{i_{Rq}}, \overline{\Omega_h}, \overline{v} \right) \equiv \left(\overline{\mathbf{x}}, \overline{v} \right)$. The following notations are adopted to represent variations of variables around such an operating point:

$$\begin{cases} \Delta\mathbf{x} = \begin{bmatrix} \Delta i_{Sd} & \Delta i_{Sq} & \Delta i_{Rd} & \Delta i_{Rq} & \Delta\Omega_h \end{bmatrix}^T \\ \Delta\mathbf{u} = \begin{bmatrix} \Delta V_{Sd} & \Delta V_{Sq} & \Delta V_{Rd} & \Delta V_{Rq} & \Delta v \end{bmatrix}^T \end{cases} \quad (3.73)$$

3.6 Linearization and Eigenvalue Analysis

Linearization begins with obtaining a linearized model of the wind torque. Points of WECS usual operation are placed on the falling part of the wind torque characteristic, $\Gamma_{wt}(\Omega_l, v)$, where this one crosses the load characteristic, Γ_l, rendered at the high-speed shaft; Figure 3.23 depicts the case of a linear (or else called static) load characteristic. Generally, the load characteristic is controllable; let c be the control variable, therefore $\Gamma_l \equiv \Gamma_l(\Omega_h, c)$. In the case of a linear load characteristic, c is either its slope, K_l, or its initial abscissa, Ω_{l0}.

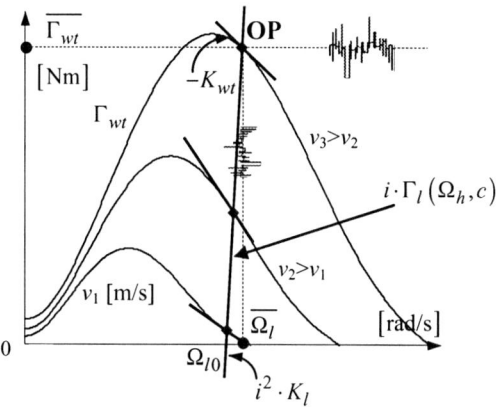

Figure 3.23. WECS operating point as cross point of the aerodynamic characteristic and (linear) load characteristic

In the following, for a generic functional variable x associated to the static operating point the following notation will be adopted:

$$\overline{x} = x\big|_{\text{static operating point}} \; ; \; \Delta x = x - \overline{x} \; ; \; \widetilde{\Delta x} = \frac{\Delta x}{\overline{x}} \tag{3.74}$$

Being defined in a *static* operating point, $\overline{x}$ is called steady-state value. Curves $\Gamma_{wt}(\Omega_l, v)$ and $\Gamma_l(\Omega_h, c)$ can be linearized around such an operating point, $(\overline{\Omega_l}, \overline{\Gamma_{wt}})$, by considering the first two terms of their Taylor's series (Ekelund 1997; Munteanu et al. 2005; Munteanu 2006). Using well-known linearization procedures leads to results showing that the linearized model exhibits a first-order dynamic like suggested in Figure 3.24. In this figure the gains are

$$K_{\Gamma v} = \overline{\frac{\partial \Gamma_{wt}}{\partial v}} \bigg/ \left(i^2 \cdot \overline{\frac{\partial \Gamma_l}{\partial \Omega_l}} - \overline{\frac{\partial \Gamma_{wt}}{\partial \Omega_l}} \right) \quad K_{lc} = \overline{\frac{\partial \Gamma_l}{\partial c}} \bigg/ \left(i^2 \cdot \overline{\frac{\partial \Gamma_l}{\partial \Omega_l}} - \overline{\frac{\partial \Gamma_{wt}}{\partial \Omega_l}} \right)$$

The time constant of the linearized model is

$$T = i \cdot J_h \bigg/ \left(i^2 \cdot \overline{\frac{\partial \Gamma_l}{\partial \Omega_l}} - \overline{\frac{\partial \Gamma_{wt}}{\partial \Omega_l}} \right),$$

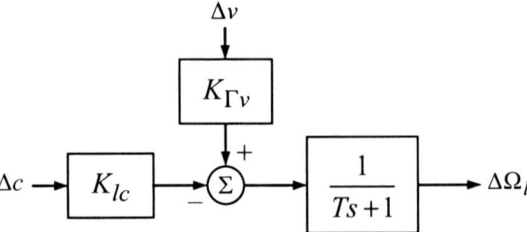

Figure 3.24. Linearized first-order dynamic of a rigid-drive-train WECS with static load

where J_h is given by Equation 3.66 and i is the drive train's ratio.

If the load characteristic is a linear one (see Figure 3.23), then, by noting $K_{wt} = -\overline{\partial \Gamma_{wt}/\partial \Omega_l}$ and $K_l = \overline{\partial \Gamma_l/\partial \Omega_l}$, one obtains the time constant in the form $T = i \cdot J_h / (K_{wt} + i^2 \cdot K_l)$. Therefore, the linearized dynamic depends on the system inertia and on the slopes of the two torque curves, K_{wt} and K_l.

When the load characteristic is specified as being the torque controlled characteristic of the generator, supposing that the electromagnetic torque can be instantaneously obtained, then $\Gamma_l \equiv \Gamma_G$ and $K_l = 0$. Thus, the simplest linearized model of WECS results in the form of the transfer function from the electromagnetic torque, Γ_G, to the low-speed shaft rotational speed, Ω_l:

$$H_{lin}(s) = -\frac{K_1}{s + K_2}, \qquad (3.75)$$

where $K_1 = 1/J_h$ and $K_2 = K_{wt}/(i \cdot J_h)$.

Therefore, supposing that Γ_G is instantaneously obtained (its dynamic is hundreds of times smaller than the time constant of the linearized system, $1/K_2$), then the rigid-drive-train WECS slow dynamic can be approximated as being of first-order (Figure 3.25).

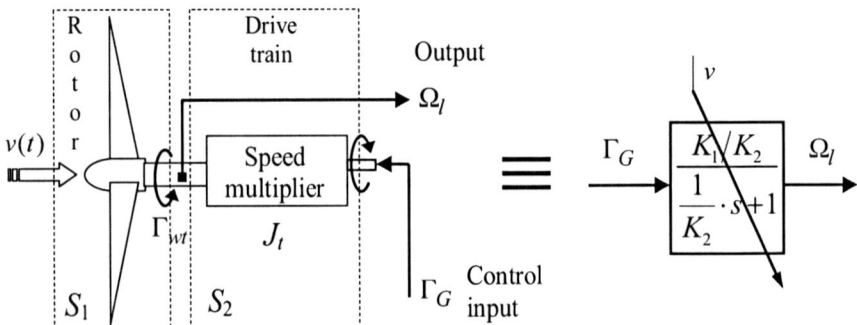

Figure 3.25. Simplest linearized model of rigid-drive-train torque-controlled WECS (aerodynamic subsystem and drive train): a first-order dynamic

Parameters of the linearized system described by $H_{lin}(s)$ depend on the mechanical constructive parameters. The time constant, equalling $i \cdot J_h/K_{wt}$, depends on the rotor inertia and does not significantly vary with the wind speed. Both the gain and the time constant also depend (through K_{wt}) on the wind speed (the system inertia decreases when the wind speed increases) and on the turbine load (torque $i \cdot \Gamma_G$). The WECS dynamics are faster as the generator characteristic is more abrupt; a zero-slope load characteristic (constant torque control) determines slow global dynamics.

In conclusion, the simplest linearized model of rigid-drive-train torque-controlled WECS is of first-order; its parameters depend on the operating point's position on the aerodynamic characteristic.

Another modelling method drives to models with linearized variations being obtained. Starting from the nonlinear torque characteristic (Equation 3.24), following calculations like in Ekelund (1997), one can obtain the linearized expression of the wind torque variations:

$$\Delta \Gamma_{wt} = \gamma \cdot \frac{\overline{\Gamma_{wt}}}{\overline{\Omega_l}} \cdot \Delta\Omega_l + (2-\gamma) \cdot \frac{\overline{\Gamma_{wt}}}{\overline{v}} \cdot \Delta v, \qquad (3.76)$$

and of its normalized variations:

$$\overline{\Delta \Gamma_{wt}} = \gamma \cdot \overline{\Delta\Omega_l} + (2-\gamma) \cdot \overline{\Delta v} \qquad (3.77)$$

where

$$\gamma = \gamma(\overline{\lambda}) = C'_p(\overline{\lambda}) \cdot \overline{\lambda}/C_p(\overline{\lambda}) - 1 \qquad (3.78)$$

is the *torque parameter*, which strongly depends on the operating point, $C'_p(\lambda) = dC_p(\lambda)/d\lambda$ and $\overline{\lambda} = R \cdot \overline{\Omega_l}/\overline{v}$. Using Equation 3.24 for expressing $\overline{\Gamma_{wt}}$, one obtains another form of Equation 3.76:

$$\Delta\Gamma_{wt} = \frac{\gamma\pi\rho R^3}{2} C_\Gamma(\overline{\lambda}) \cdot \frac{\overline{v}^2}{\overline{\Omega_l}} \cdot \Delta\Omega_l + \frac{(2-\gamma)\pi\rho R^3}{2} C_\Gamma(\overline{\lambda}) \cdot \overline{v} \cdot \Delta v \qquad (3.79)$$

The linearized WECS state model results from replacing the linear relation of the wind torque variations (Equation 3.79) into the drive train motion equation (3.64), then adding the result to the linearized form of the DFIG model at Equation 3.34. Considering that Γ_G is a negative torque in generator regime, then Equation 3.64 must be written with the sign "+" in order to describe a stable system. In this case, algebraic calculations lead to the form at Equation 3.80, where matrices are given by Equation 3.81. One can note that the linear model matrices depend on the steady-state operating point, so ultimately on the wind speed:

$$\dot{\mathbf{x}} = \mathbf{A}(\overline{\mathbf{x}},\overline{v}) \cdot \mathbf{x} + \mathbf{B}(\overline{\Omega_h},\overline{v}) \cdot \mathbf{u}, \qquad (3.80)$$

$$\mathbf{A}(\bar{x},\bar{v}) = \begin{bmatrix} -\dfrac{R_S}{\sigma L_S} & \omega_S + \dfrac{p\overline{\Omega_h}L_m^2}{\sigma L_S L_R} & \dfrac{L_m R_R}{\sigma L_S L_R} & \dfrac{p\overline{\Omega_h}L_m}{\sigma L_S} & \dfrac{pL_m}{\sigma L_S}\left(\dfrac{L_m\overline{i_{Sq}}}{L_R}+\overline{i_{Rq}}\right) \\[6pt] -\left(\omega_S+\dfrac{p\overline{\Omega_h}L_m^2}{\sigma L_S L_R}\right) & -\dfrac{R_S}{\sigma L_S} & -\dfrac{p\overline{\Omega_h}L_m}{\sigma L_S} & \dfrac{L_m R_R}{\sigma L_S L_R} & -\dfrac{pL_m}{\sigma L_S}\left(\dfrac{L_m\overline{i_{Sd}}}{L_R}+\overline{i_{Rd}}\right) \\[6pt] \dfrac{L_m R_S}{\sigma L_S L_R} & -\dfrac{p\overline{\Omega_h}L_m}{\sigma L_R} & -\dfrac{R_R}{\sigma L_R} & \omega_S-\dfrac{p\overline{\Omega_h}}{\sigma} & -\dfrac{p}{\sigma}\left(\dfrac{L_m\overline{i_{Sq}}}{L_R}+\overline{i_{Rq}}\right) \\[6pt] \dfrac{p\overline{\Omega_h}L_m}{\sigma L_R} & \dfrac{R_S L_m}{\sigma L_S L_R} & \dfrac{p\overline{\Omega_h}}{\sigma}-\omega_S & -\dfrac{R_R}{\sigma L_R} & -\dfrac{p}{\sigma}\left(\dfrac{L_m\overline{i_{Sd}}}{L_R}+\overline{i_{Rd}}\right) \\[6pt] -\dfrac{3}{2J_h}pL_m\overline{i_{Rq}} & \dfrac{3}{2J_h}pL_m\overline{i_{Rd}} & \dfrac{3}{2J_h}pL_m\overline{i_{Sq}} & -\dfrac{3}{2J_h}pL_m\overline{i_{Sd}} & \dfrac{\gamma}{iJ_h}\cdot\dfrac{\pi\rho R^3}{2}C_\Gamma(\bar{\lambda})\cdot\dfrac{\bar{v}^{-2}}{\overline{\Omega_h}} \end{bmatrix}$$

$$\mathbf{B}(\overline{\Omega_h},\bar{v}) = \begin{bmatrix} \dfrac{1}{\sigma L_S} & 0 & \dfrac{-L_m}{\sigma L_S L_R} & 0 & 0 \\[6pt] 0 & \dfrac{1}{\sigma L_S} & 0 & \dfrac{-L_m}{\sigma L_S L_R} & 0 \\[6pt] \dfrac{-L_m}{\sigma L_S L_R} & 0 & \dfrac{1}{\sigma L_R} & 0 & 0 \\[6pt] 0 & \dfrac{-L_m}{\sigma L_S L_R} & 0 & \dfrac{1}{\sigma L_R} & 0 \\[6pt] 0 & 0 & 0 & 0 & \dfrac{(2-\gamma)}{iJ_h}\cdot\dfrac{\pi\rho R^3}{2}C_\Gamma(\bar{\lambda})\cdot\bar{v} \end{bmatrix}$$

(3.81)

Eigenvalue Analysis
Figure 3.26 qualitatively shows how the poles of a fifth-order uncontrolled low-power (less than 10 kW) induction-generator-based WECS model typically migrate when the steady-state wind speed, $\bar{v}$, which mainly characterizes the operating point, changes in a large domain from the cut-in to cut-out value. In Figure 3.26b,c details zooming the rotor poles and respectively motion pole evolution are represented. One can deduce that, whereas the stator poles are not affected at all by the operating point variations, the rotor poles correspond to a slightly faster dynamic as the wind speed increases. In the same time, the motion transients are slower, as suggested by the zoom in Figure 3.26c.

Figure 3.27 presents typical pole distributions when the induction generator is controlled using two methods.

Thus, under the U/f=constant strategy at constant wind speed, a typical pole distribution appears like in Figure 3.27a. One can note that the electrical variables have a faster dynamic (as the complex stator and rotor poles' distribution denotes), whereas the motion pole denotes a slightly slower dynamic as the U/f ratio increases. Figure 3.27b displays the poles' migration for the constant torque control when the optimal operating point is tracked (on the optimal regimes characteristic, ORC). In this case, the dynamics can be reduced at that of a single real pole and it is faster as the wind speed increases.

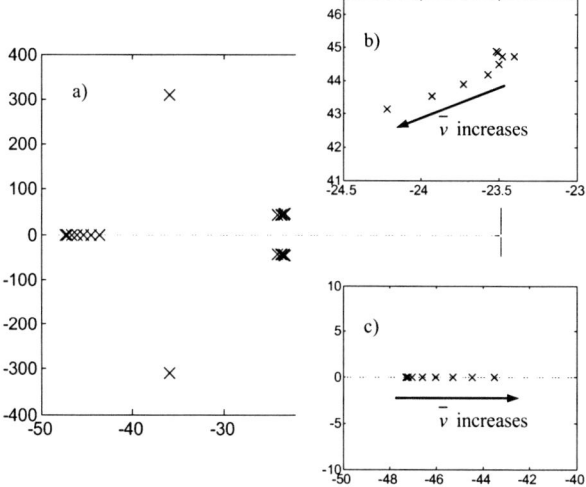

Figure 3.26. Migration of poles for a fifth-order uncontrolled induction-generator-based WECS model as the operating point (wind speed) varies: **a** stator, rotor and motion poles. **b** Zoom on rotor complex poles' distribution; **c** zoom on the motion real pole migration

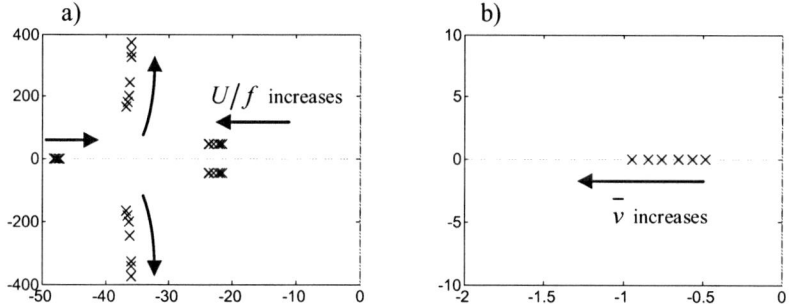

Figure 3.27. a Migration of poles for a fifth-order controlled induction-generator-based WECS through a U/f=constant strategy at constant wind speed as the U/f ratio varies. **b** Migration of the dominant pole through constant torque control on the ORC

The above remarks allow us to draw a conclusion to be further exploited for control purposes. Namely, the global dynamics of a controlled WECS depend on the shape of the generator characteristic, which further depends on the chosen control structure. More precisely, the relative position of the two mechanical characteristics – that of the turbine and that of the generator – determines the WECS response time. Thus, for example, dynamics of a constant torque control (when the load characteristic has zero slope in the (Ω_h, Γ_G) plane) can be up to 10 times slower than those of the case where a U/f = constant control (when the same characteristic has a large slope) is adopted. Therefore, the possibility of controlling the slope of the load characteristic is a key issue in WECS control.

3.6.2 Synchronous-generator-based WECS

The case of a *variable-speed fixed-pitch rigid-drive-train* synchronous-generator-based WECS is discussed here, where its autonomous operation on a symmetric tri-phased resistive load, R_l, is considered.

Linearized Model
In order to obtain the state model of an autonomous synchronous-generator-based WECS, one uses the model of a permanent-magnet synchronous generator supplying an isolated resistive load, R_l (Equation 3.58), together with the motion equation in the form of Equation 3.30, where the wind torque, Γ_{wt}, plays the role of the mechanical torque, Γ_{mec}. Γ_{wt} is a polynomial function of the tip speed (Equation 3.25 replaced in Equation 3.24).

For optimal control purposes, it is not necessary to use such a complicated expression of the torque coefficient, C_Γ, as that of Equation 3.25, which captures all operating regimes, including the starting one. Instead, one can rather employ a second-order polynomial approximation of C_Γ vs. the tip speed, λ:

$$C_\Gamma(\lambda) = a_0 + a_1\lambda + a_2\lambda^2, \qquad (3.82)$$

which satisfactorily describes the appearance of C_Γ near optimal regimes. Taking into account the tip speed given by Equation 2.30 and wind torque variation upon the torque coefficient (Equation 3.24), Equation 3.82 corresponds to the following expression of the wind torque depending on the wind speed, v, and on the generator's rotational speed, Ω_h:

$$\Gamma_{wt} = d_1 v^2 + d_2 v \Omega_h + d_3 \Omega_h^2, \qquad (3.83)$$

where

$$d_1 = 1/2\pi\rho R^3 \cdot a_0, \quad d_2 = 1/2\pi\rho R^4 \cdot a_1, \quad d_3 = 1/2\pi\rho R^5 \cdot a_2, \qquad (3.84)$$

with R being the blade length and ρ being the air density.

In relation to the model at Equation 3.58, a new input is added, namely the wind speed, and also a new state variable, the rotational speed, Ω_h, which is also chosen as output variable. Thus, with the new notation:

$$\begin{cases} \mathbf{x} = \begin{bmatrix} x_1(t) & x_2(t) & x_3(t) \end{bmatrix}^T \equiv \begin{bmatrix} i_d(t) & i_q(t) & \Omega_h \end{bmatrix}^T \\ \mathbf{u} = \begin{bmatrix} u_1(t) & u_2(t) \end{bmatrix}^T \equiv \begin{bmatrix} R_l & v \end{bmatrix}^T \\ y \equiv \Omega_h \end{cases}, \qquad (3.85)$$

the WECS model has the form

3.6 Linearization and Eigenvalue Analysis

$$\begin{cases} \dot{\mathbf{x}} = \begin{bmatrix} \dfrac{1}{L_d + L_s}\left(-Rx_1 + p\left(L_q - L_s\right)x_2 x_3\right) \\ \dfrac{1}{L_q + L_s}\left(-Rx_2 - p\left(L_d + L_s\right)x_1 x_3 + p\Phi_m x_3\right) \\ \dfrac{1}{J}\left(d_3 x_3^2 - p\Phi_m x_2\right) \end{bmatrix} + \begin{bmatrix} -\dfrac{1}{L_d + L_a} x_1 u_1 \\ -\dfrac{1}{L_q + L_s} x_2 u_1 \\ \dfrac{1}{J}\left(d_1 u_1^2 + d_2 u_2 x_3\right) \end{bmatrix} \\ \mathbf{y} = \begin{bmatrix} 0 & 0 & 1 \end{bmatrix} \cdot \mathbf{x} \end{cases}$$

(3.86)

The third-order nonlinear model (Equation 3.86) is linearized around an arbitrarily chosen operating point. Letting $\Delta \mathbf{x} = \begin{bmatrix} \Delta x_1 & \Delta x_2 & \Delta x_3 \end{bmatrix}^T$ and $\Delta \mathbf{u} = \begin{bmatrix} \Delta R_l & \Delta v \end{bmatrix}^T$ be the variations of the state variables and of inputs around this point leads to

$$\begin{cases} \Delta \dot{\mathbf{x}} = \begin{bmatrix} a_1 + a_2 \overline{R_l} & a_3 \overline{x_3} & a_3 \overline{x_2} \\ b_1 \overline{x_3} & b_2 + b_3 \overline{R_l} & b_1 \overline{x_1} + b_4 \\ 0 & c_4 & c_2 \overline{v} + c_3 \overline{x_3} \end{bmatrix} \cdot \Delta \mathbf{x} + \begin{bmatrix} a_2 \overline{x_1} & 0 \\ b_3 \overline{x_2} & 0 \\ 0 & c_1 \overline{v} + c_3 \overline{x_3} \end{bmatrix} \cdot \Delta \mathbf{u} \\ \Delta \mathbf{y} = \begin{bmatrix} 0 & 0 & 1 \end{bmatrix} \cdot \Delta \mathbf{x} \end{cases}$$

(3.87)

where

$$\begin{cases} a_1 = \dfrac{-R}{L_d + L_s};\ a_2 = \dfrac{-1}{L_d + L_s};\ a_3 = p\dfrac{L_q - L_s}{L_d + L_s}; \\ b_1 = -p\dfrac{L_d + L_s}{L_q + L_s};\ b_2 = \dfrac{-R}{L_q + L_s};\ b_3 = \dfrac{-1}{L_q + L_s};\ b_4 = \dfrac{p\Phi_m}{L_q + L_s}; \\ c_1 = \dfrac{2d_1}{J};\ c_2 = \dfrac{d_2}{J};\ c_3 = \dfrac{2d_3}{J};\ c_4 = -p\dfrac{\Phi_m}{J} \end{cases}$$

(3.88)

Eigenvalue Analysis

The linearized model given by Equations 3.87 and 3.88 has been used for studying the zero-pole distribution of the PMSG-based WECS in various operating points, defined by the resistance load, as well as by the wind speed. Three cases have been analyzed, described by the way the operating point variation takes place, namely:
1. variable wind speed and constant load;
2. variable load and constant wind speed;
3. variable wind speed and variable load, provided that the $v - R_l$ pair ensures the optimal operation, i.e., $\lambda = \Omega \cdot R/v = \lambda_{opt}$.

Case 1. Wind speed takes the values 4 m/s, 7 m/s and 10 m/s and the load resistance is constant (80 Ω). By analyzing the zero-pole distribution presented in Figure 3.28, the following remarks can be made:

- the dynamic system has real poles when wind speed is low;
- for v=7 m/s, there are two dominant poles and an insignificant third one;
- the dominant poles become complex when the wind speed is high.

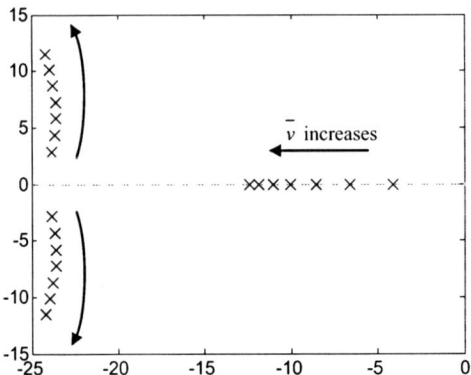

Figure 3.28. Zero-pole distribution for various wind speed values

Case 2. When the load resistance varies, for a constant wind speed (*e.g.*, in the middle of its variation range), the zero-pole distribution reveals the same nature of poles under different load regimes (see Figure 3.29). Dynamics of both electrical and mechanical variables become slower as the load resistance increases.

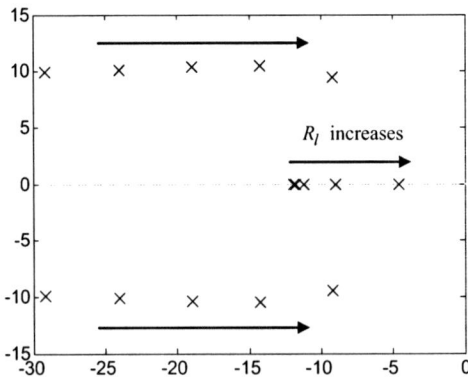

Figure 3.29. Zero-pole distribution for various load values

Case 3. In this case both the wind speed and the load resistance vary, namely such that the operating point move along the optimal regimes characteristic (ORC). The zero-pole distribution is obtained for each of these optimal operating points, as is depicted in Figure 3.30. One can note that the system exhibits second-order dynamics whose cut-off frequency and damping factor strongly depend on the position of the operating point on the ORC.

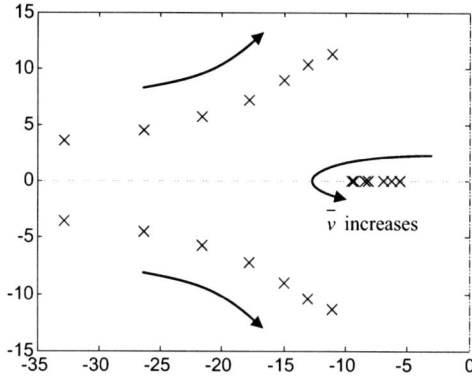

Figure 3.30. Zero-pole distribution for optimal regimes

3.7 Case Study (1): Reduced-order Linear Modelling of a SCIG-based WECS

A 2-MW fixed-pitch rigid-drive-train SCIG-based WECS will illustrate the reduced-order modelling and linearization. Its parameters are given in Appendix A and Table A.6. A WECS model is called reduced here if only the SCIG rotor currents dynamic (Equation 3.47) instead of both stator and rotor currents dynamic (Equation 3.37) is coupled to the motion equation in form of Equation 3.30.

Let us consider a typical steady operating point in the partial load region, e.g., for the wind speed $\overline{v} = 10$ m/s. Transient behaviours of the state variables for the third-order vs. the fifth-order model in response to wind speed step changes do not differ significantly. In Figure 3.31 one can compare these transients in response to V_{Sq} step changes of 100 V.

As previously shown, WECS dynamic behaviour around a steady operating point may be considered linear. Equation 3.81 allows for an eigenvalue analysis being carried out. The same thing can be obtained in MATLAB®/Simulink® without explicitly computing matrices from Equation 3.81. First, one must obtain the steady-state values by numerical simulation on the nonlinear model, and then call the linmod function with these values as arguments.

Simulation has given for the steady operating regime at $\overline{v} = 10$ m/s the following values of the fifth-order model state variables: $\overline{i_{Sd}} = -1.442$ kA, $\overline{i_{Sq}} = 0.9254$ kA, $\overline{i_{Rd}} = 0.0317$ kA, $\overline{i_{Rq}} = -0.9487$ kA and $\overline{\Omega_h} = 157.2$ rad/s. The eigenvalue set computed by using matrix **A** from Equation 3.81 (left-hand side) and the one numerically provided by linmod (right-hand side) are given below ($j = \sqrt{-1}$). Differences are negligible (of order 10^{-4}):

$-7.7106 + j \cdot 313.9884$
$-7.7106 - j \cdot 313.9884$
$-4.5486 + j \cdot 27.0832$
$-4.5486 - j \cdot 27.0832$
-9.0671

$-7.7101 + j \cdot 313.9882$
$-7.7101 - j \cdot 313.9882$
$-4.5489 + j \cdot 27.08313$
$-4.5489 - j \cdot 27.08313$
-9.0676

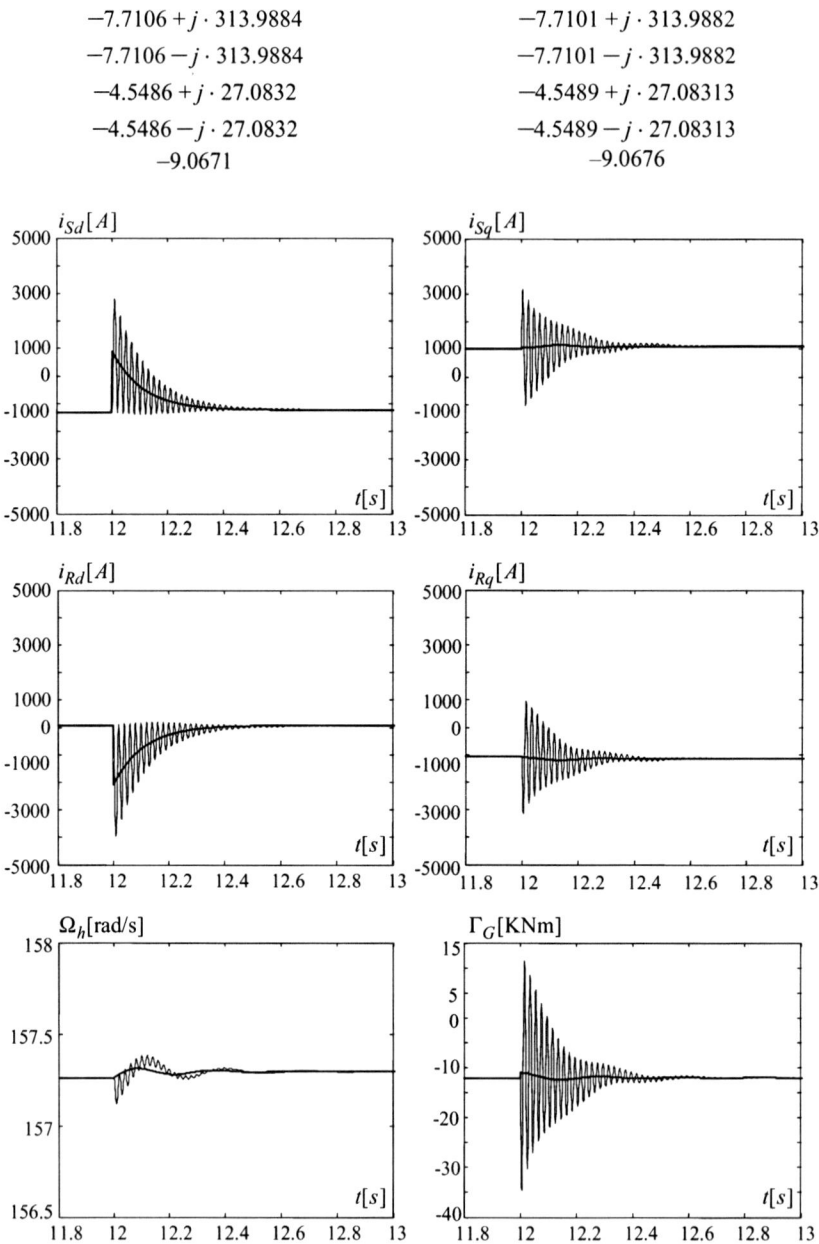

Figure 3.31. Transients of state variables of a 2-MW SCIG-based WECS in response to voltage supply (V_{Sq}) step changes: fifth-order (*thin line*) vs. third-order (*thick line*) model

Details about the MATLAB®/Simulink® implementation of this case study can be found in the folder case_study_1 from the software material.

4

Basics of the Wind Turbine Control Systems

4.1 Control Objectives

Taking into account the ideas presented in the previous chapters, one can highlight the objectives of the WECS control (see Section 2.7). The list bellow selects the most important:
- controlling the wind captured power for speeds larger than the rated;
- maximising the wind harvested power in partial load zone as long as constraints on speed and captured power are met;
- alleviating the variable loads, in order to guarantee a certain level of resilience of the mechanical parts;
- meeting strict power quality standards (power factor, harmonics, flicker, *etc.*);
- transferring the electrical power to the grid at an imposed level, for wide range of wind velocities;

There can be three main control subsystems (see Figure 4.1).

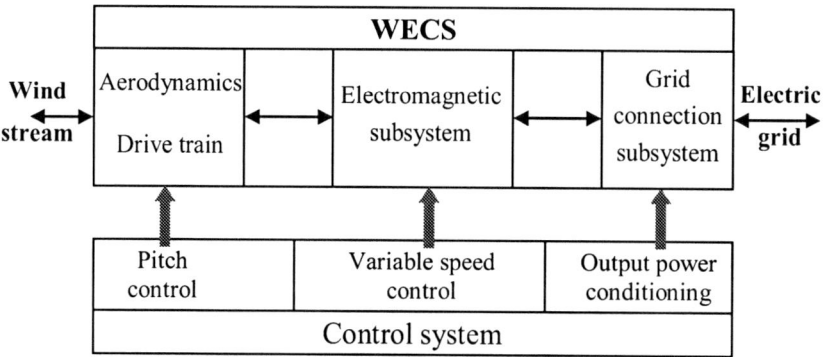

Figure 4.1. Main control subsystems of a WECS

The first control subsystem affects the pitch angle following aerodynamic power limiting targets. The second implements the generator control, in order to obtain the variable-speed regime and the third controls the transfer of the full (or a fraction) of electric power to the electric grid, with effects on WECS output power quality.

The control structures result from defining one or more of the above goals stated in relation to a certain mathematical model of WECS. The controller determines the desired global dynamic behaviour of the system, such that ensuring power regulation, energy maximization in partial load, mechanical loads alleviation and reduction of active power fluctuations.

4.2 Physical Fundamentals of Primary Control Objectives

Consider that the turbine operates in partial load at fixed pitch – often named "fine pitch" – that gives good aerodynamic performance and which can be considered pitch reference. When the wind velocity exceeds the rated, the turbine is operating in what is called full-load regime and the captured power – which potentially can vary with the wind speed cubed – must be aerodynamically limited (controlled). This is the formulation of the primary objective of the WECS control. There are several techniques usually used in order to fulfil this objective, which are reviewed next (Burton et al. 2001).

The wind turbine aerodynamic behaviour fundamentals can be analysed in Figures 3.9 and 3.10. Some elements also result from the associated analysis using blade element theory (see Algorithm 3.1).

One can remark that the key variable in aerodynamics behaviour, the incidence angle, increases with wind velocity and decreases with increase of rotational speed and pitch angle. Consequently, the aerodynamic efficiency $C_z(i)/C_x(i)$, will be affected by the incidence angle evolution as presented in Figure 4.2.

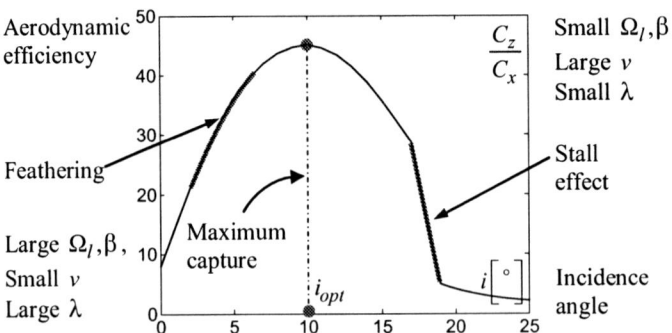

Figure 4.2. Feathering and stall effects on the aerodynamic efficiency curve

When the turbine experiences high winds, the aerodynamic power can be reduced by controlling the incidence angle through the rotational speed and/or pitch angle. The appearance of the aerodynamic efficiency curve in Figure 4.2 suggests two courses of action. Decrease of the incidence angle, which corresponds to increased values of rotational speed and/or pitch angle, leads to an aerodynamic process of losing power called *blade feathering*. The second method involves increasing the incidence angle (low values of rotational speed and/or pitch angle) and aims at diminishing the aerodynamic power by a process called *blade stall*. One can note a sensibly larger slope of the aerodynamic efficiency in stall than in feathering.

In conclusion, the power limitation to the rated is possible either by generator control at variable speed (varying the WECS rotational speed, Ω_l), or by pitch angle (β) control, both in the framework of one of the above-mentioned strategies. The generator can be speed/torque controlled using power electronics converters. The pitch control can be achieved using collective or individual (on each blade) actuator systems for rotating the blades around their axes. The usual technique is the full-span pitch control (the entire blade is pitched), but the power control can be effective even only when the partial-span pitch control is employed (the outer 15% of the blade is pitched).

4.2.1 Active-pitch Control

Power limitation in high winds is typically achieved by using pitch angle control. This action, also called active-pitch control (or *pitch-to-feather*), corresponds to changing the pitch value such that the leading edge of the blade is moved into the wind (increase of β), thus inducing blade feathering effect. The range of blade pitch angles required for power control in this case is large, about 35° from the pitch reference. Therefore, for limiting power excursions, the pitching system has to act rapidly, with fast pitch change rates, *i.e.*, 5°/s. Therefore, one could expect large gains within the power control loop. The power control structure employed is the same as for the active-stall control machines (see Figure 4.3).

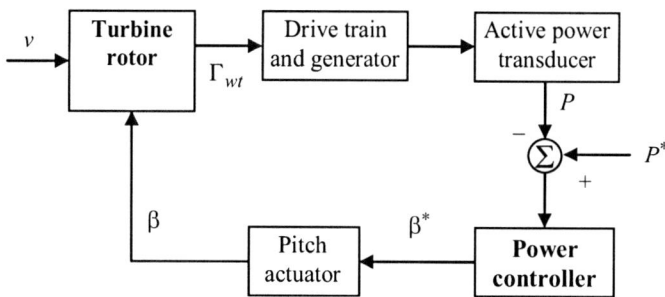

Figure 4.3. Power control structure

4.2.2 Active-stall Control

Active-stall control (also called negative-pitch control) reduces the aerodynamic power by diminishing the blade pitch angle, β, in order to increase the incidence angle. The blades are pitched towards stall, in the contrary direction to the pitch-control case, by turning the leading edge downwind.

Only small changes of pitch angle are required to maintain the power output at its rated value, as the range of incidence angles required for power control is much smaller in this case than in the case of pitch control.

Compared to the pitch-to-feather technique, the travel of the pitch mechanism is very much reduced; significantly greater thrust loads are encountered, but the thrust is much more constant, inducing smaller mechanical loads. The employed power control structure is briefed in Figure 4.3 (Burton *et al.* 2001).

4.2.3 Passive-pitch Control

This alternative control of blade pitch uses self-operated (direct action) controllers to obtain the imposed pitch changes at higher wind speeds. In this case, the hub mounting of blades twist under the action of certain loads on the blades (*e.g.*, centrifugal loads). The regulator combines in a single mechanical device the sensor driven by the blade loads, the controller itself and the pitch actuator. The energy required for control action is entirely supplied by the transducer, which also actuates on the final control element by a system of mechanical transmission, without amplifying the control signal.

Different types of loads can be used for both sensing the full-load regime and as control signals. Among them, the centrifugal loads used to control the blades passively are an efficient solution. In this case, the centrifugal forces drive some rotating masses which act on a screw cylinder to push a preloaded spring. When the rotational speed exceeds the critical value, the screw cylinder becomes free to act on the blades' mounting hub, thus pitching the blades. Such centrifugal regulators are used especially for small wind turbines, in order to limit the output power by stall effect.

4.2.4 Passive-stall Control

This technique is the simplest form of power control, providing aerodynamic efficiency reduction by stall effect in high winds without changes in blade geometry. As the wind velocity increases, even at constant speed, Ω, and constant pitch, β, the stall regime can still be obtained (because of incidence angle increasing). The key factor in this method is a special design of the blade profile, providing an accentuated stall effect around rated power without the undesired collateral aerodynamic behaviour.

As the wind speed increases above its rated value, the output power reaches a certain ceiling, dropping to some lower value as the turbine enters a deeper stall regime. However, in even stronger winds, the captured power continues to increase uncontrollably with the wind velocity and emergency brakes are necessary to ensure turbine safety.

Figure 4.4 suggests a comparison between passive-stall control and active-pitch control.

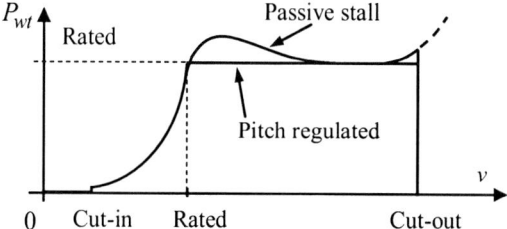

Figure 4.4. Comparison between passive-stall and active-pitch control features

4.3 Principles of WECS Optimal Control

This section is dedicated to the basics of WECS energy conversion optimization in the partial load regime.

4.3.1 Case of Variable-speed Fixed-pitch WECS

Control of variable-speed fixed-pitch WECS in the partial load regime generally aims at regulating the power harvested from wind by modifying the electrical generator speed; in particular, the control goal can be to capture the maximum power available from the wind. For each wind speed, there is a certain rotational speed at which the power curve of a given wind turbine has a maximum (C_p reaches its maximum value).

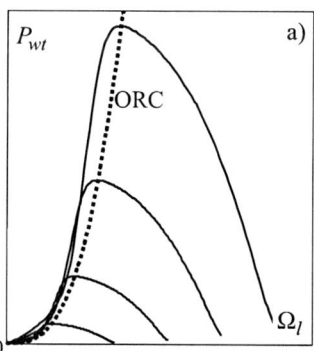 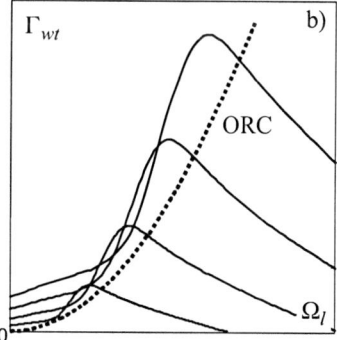

Figure 4.5. Optimal regimes characteristic, ORC: **a** in the $\Omega_l - P_{wt}$ plane; **b** in the $\Omega_l - \Gamma_{wt}$ plane

All these maxima compose what is known in the literature as the *optimal regimes characteristic, ORC* (see Figure 4.5a – Nichita 1995). In the $\Omega_l - \Gamma_{wt}$ plane, the ORC is placed at the right of the torque maxima locus (Figure 4.5b).

By keeping the static operating point of the turbine around the ORC one ensures an optimal steady-state regime, that is, the captured power is the maximal one available from the wind. This is equivalent to maintaining the tip speed ratio at its optimal value, λ_{opt} (Figure 4.6) and can be achieved by operating the turbine at variable speed, corresponding to the wind speed (Connor and Leithead 1993).

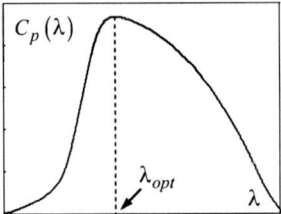

Figure 4.6. The unimodal power coefficient curve, expressing the aerodynamic efficiency

Basically, the control approaches encountered in the WECS control field vary in accordance with some assumptions concerning the known models/parameters, the measurable variables, the control method employed and the version of WECS model used. Depending on how rich the information is about the WECS model, especially about its torque characteristic, the optimal control of variable-speed fixed-pitch WECS is based upon the following approaches.

Maximum Power Point Tracking (MPPT)

This approach is adequate when parameters λ_{opt} and $C_{p\,max} = C_p(\lambda_{opt})$ are not known. The reference of the rotational speed control loop is adjusted such that the turbine operates around maximum power for the current wind speed value. In order to establish whether this reference must be either increased or decreased, it is necessary to estimate the current position of the operating point in relation to the maximum of $P_{wt}(\Omega_l)$ curve. This can be done in two ways:

- the speed reference is modified by a variation $\Delta\Omega_l$, the corresponding change in the active power, ΔP, being determined in order to estimate the value $\partial P_{wt}/\partial \Omega_l$. The sign of this value indicates the position of the operating point in relation to the maximum of characteristic $P_{wt}(\Omega_l)$. If the speed reference is adjusted in ramp with a slope proportional to this derivative, then the system evolves to optimum, where $\partial P_{wt}/\partial \Omega_l = 0$;

- a probing signal is added to the current speed reference; this signal is a slowly variable sinusoid; its amplitude does not significantly affect the system operation, but still produces a detectable response in the active power evolution. In order to obtain the position of the operating point in relation to the maximum, one compares the phase lag of the probing sinusoid and that of the sinusoidal component of active power. If the phase lag is zero/π, then the current operating point is placed on the ascending/descending part of $P_{wt}(\Omega_l)$, therefore, the slope of the speed reference must increase/decrease. Around the maximum, the probing signal does not produce any detectable response and the speed reference does not have to change.

4.3 Principles of WECS Optimal Control

In this simplified presentation of MPPT techniques, factors like the influence of wind turbulence and system dynamics that distort the information concerning the operating point position have been neglected. A more detailed description and analysis of performances can be found in Sections 5.1.1 and 5.2.

Shaft Rotational Speed Optimal Control Using a Setpoint from the Wind Speed Information

This solution can be applied if the optimal value of the tip speed ratio, λ_{opt}, is known. The turbine operates on the ORC if

$$\lambda(t) = \lambda_{opt}, \tag{4.1}$$

which supposes that the shaft rotational speed is closed-loop controlled such that to reach its optimal value:

$$\Omega_{l_{opt}}(t) = \frac{\lambda_{opt}}{R} \cdot v(t) \tag{4.2}$$

This approach has some drawbacks related to the wind speed being measured by an anemometer mounted on the nacelle, which offers information on the fixed-point wind speed. But this information differs from the wind speed experienced by the blade (see Section 3.2.3), mainly because of the time lag between the two signals.

Active Power Optimal Control Using a Setpoint from the Shaft Rotational Speed Information

This method is used when both λ_{opt} and $C_{p_{max}} = C_p(\lambda_{opt})$ are known. From the expression of the power extracted by a turbine (Equation 2.31), it follows that

$$P_{wt} = \frac{1}{2}C_p(\lambda)\rho\pi R^2 v^3 = \frac{1}{2} \cdot \frac{C_p(\lambda)}{\lambda^3}\rho\pi R\Omega_l^3 \tag{4.3}$$

By replacing $\lambda(t) = \lambda_{opt}$ and $C_p = C_p(\lambda_{opt})$, one obtains the power reference for the second region of the power–wind speed curve:

$$P_{wt_{opt}} = P_{ref} = K \cdot \Omega_{l_{opt}}^3, \tag{4.4}$$

where

$$K = \frac{1}{2} \cdot \frac{C_p(\lambda_{opt})}{\lambda_{opt}^3}\rho\pi R^5 \tag{4.5}$$

This approach supposes an active power control loop being used, whose reference is determined based upon Equation 4.5. This method is widely employed, especially for medium and high-power WECS.

In both methods an ORC tracking control loop is used. The wind turbulence can significantly influence the dynamic performances. In this context, various control laws can be employed, suitable to any particular WECS configuration. In most of cases, classical PI or PID control are preferred. Advanced control techniques can be used in order to ensure better performance, especially for guaranteeing robustness to modelling uncertainties. For example, the sliding mode control is a well-performing solution when dealing with uncertainties of dynamical features. Chapter 5 of this book presents in detail some of these WECS optimal control design methods.

The wind turbulence induces deviations of the operating point around ORC. As can be seen in Figure 4.5a, these excursions contribute in a non-symmetrical manner to the decrease of the conversion efficiency, because the wind power decreases with different slopes at the left and at the right side of ORC (Bianchi *et al.* 2006). Figure 4.7 shows, along with ORC (with thick line), the loci of 99% and respectively 95% from the maximum power. Let us consider an ORC tracking loop. If one assumes equal dynamical errors in relation to the optimal reference, then positive errors would affect the conversion efficiency more than would negative errors of the same absolute value. Therefore, it is opportune to define the optimal regime as being not exactly the ORC, but the ORC slightly slipped to the right.

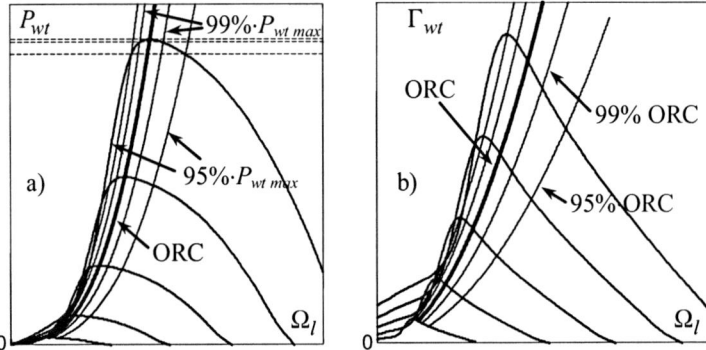

Figure 4.7. ORC, along with loci of 99% and 95% from the maximum wind power: **a** in the $\Omega_l - P_{wt}$ plane; **b** in the $\Omega_l - \Gamma_{wt}$ plane

4.3.2 Case of Fixed-speed Variable-pitch WECS

In this case optimization involves using characteristic $C_p(\lambda,\beta)$, where β is the pitch angle. In general, one finds the optimum of $C_p(\lambda,\beta)$ by means of look-up table interpolation. Under constant-speed operation, *e.g.*, for a given value of the rotational speed, this curve depends on v and β. In Figure 4.8 a family of curves $C_p(\lambda,\beta)$, parameterised by the wind speed, v, is shown (Hansen *et al.* 2003). Optimization is achieved by changing the angle β, such that the operating point to be placed at the maximum of the $C_p(\beta)$ corresponds to the current wind speed value.

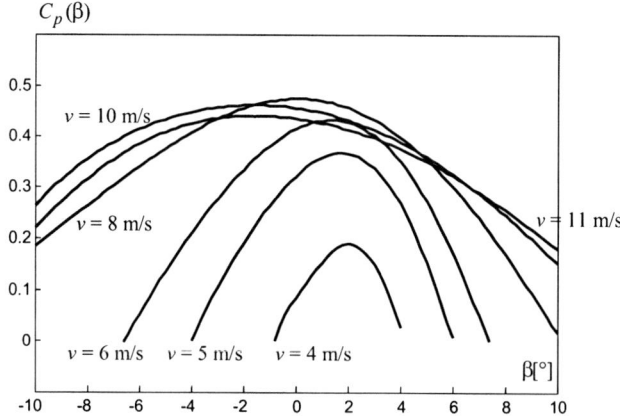

Figure 4.8. Family of power coefficient vs. pitch angle curves parameterised by the wind speed

In Figure 4.9 one can see the optimal control characteristic, defined by

$$\beta_{opt} = F(v), \quad \underset{\beta}{\text{Max}}\, C(\beta, v) = C(\beta_{opt}, v),$$

corresponding to the family of curves $C_p(\beta)$ given in Figure 4.8. Value β_{opt} represents the reference of the pitch control system. For the characteristic $\beta_{opt}(v)$ from Figure 4.9, the region of large wind speeds is drawn with a dashed line, because here the pitch angle corresponds to power limiting.

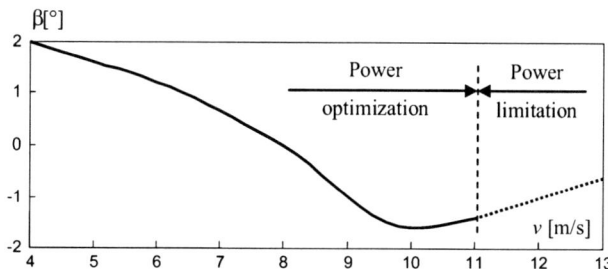

Figure 4.9. Optimal pitch angles vs. the wind speed

In this region the pitch angle is modified by the rated power regulation loop, as detailed further in Section 4.4.2.

The minimal performances that the pitch actuator system must ensure depend on the wind speed value. As Figure 4.8 suggests, the sensibility of variable C_p in relation to β is large for low wind speed. For large wind speeds, $C_p(\beta)$ is a relatively flat curve around the maximum; therefore the sensibility of C_p is reduced.

The main data used in optimal control are the active power and the wind speed. They are employed in the form of averaged instead of instantaneous values (the method of moving average is recommended by Hansen *et al.* 2003).

4.4 Main Operation Strategies of WECS

This section presents principles of WECS control in the general case, when the operating point covers both the partial-load regime and the full-load one. In this case, the control system must ensure, in addition to the power coefficient maximization, the following essential constraint:

$$P_{wt}(t) \leq P_{max} \quad (4.6)$$

When the limitation at Equation 4.6 is enabled, the system works in the full-load regime. In order to avoid acoustic disturbances, especially for large wind turbines, the following constraint is frequently imposed:

$$\Omega_l(t) \leq \Omega_{max} \quad (4.7)$$

If the speed limitation takes place in partial load, then optimization is possible only for variable-pitch turbines. A third constraint that can be imposed involves the wind torque:

$$\Gamma_{wt}(t) \leq \Gamma_{max} \quad (4.8)$$

The latter constraint can be considered when the system operates in partial load, at underrated rotational speed and mechanical power. It is easily implemented through the generator control system. If the speed limitation is imposed first, then fulfilling the condition at Equation 4.6 implicitly ensures that the condition at Equation 4.8 is fulfilled.

Of the three constraints, the most important are those concerning the wind power and the low-speed shaft rotational speed.

4.4.1 Control of Variable-speed Fixed-pitch WECS

In order to illustrate the optimization functions in the partial load regime, the power limiting in the full-load regime and, if needed, the rotational speed limiting, a low-power variable-speed permanent-magnet-synchronous-generator (PMSG)-based WECS is considered, having the structure shown in Figure 4.10. The system is controlled by means of control input x applied to the DC–DC converter.

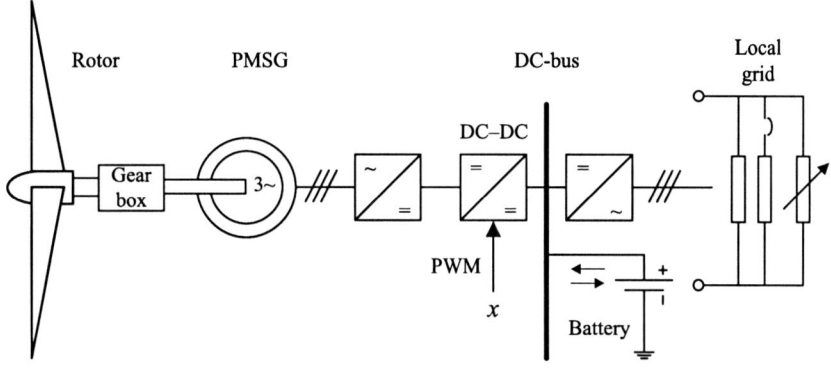

Figure 4.10. Variable-speed PMSG-based WECS

The family of wind torque characteristics $\Gamma_{wt}(\Omega_l, v_i)$, $v_i = 5, 7, 9, 11$ (m/s) (with solid line) is presented in Figure 4.11. The rated wind speed is $v_n = 11$ m/s. In the same figure one can note the ORC (drawn with dotted line) and the family of electromagnetic torque curves of the PMSG, $\Gamma_G(\Omega_l, x)$ (with dashed line), parameterised by the control input of the optimization loop, x. The WECS operating points are identified as the cross-points between $\Gamma_{wt}(\Omega_l, v)$ and $\Gamma_G(\Omega_l, x)$, which are represented by circles in Figure 4.11.

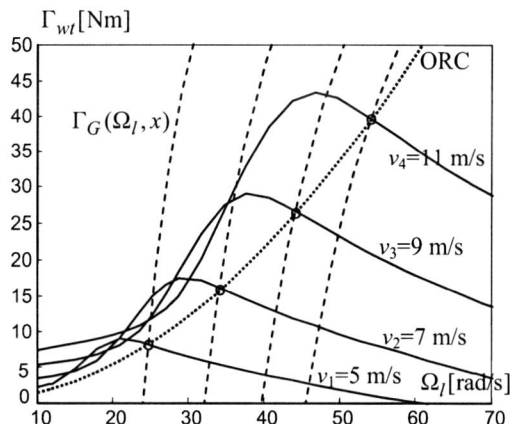

Figure 4.11. Wind and electromagnetic torque characteristics in relation to the ORC

Let us initially suppose that the rotational speed constraint, $\Omega_l(t) \leq \Omega_{max}$, is met and, in addition to optimization in partial load regime, only the power constraint, $P_{wt}(t) \leq P_{max}$, is imposed. A linear increase of wind speed, from the cut-in value, v_S, to the rated value, v_n, is also supposed. Under these conditions, the output of the optimization controller will continuously move the PMSG torque characteristic to the right, until it reaches the limit of the partial load regime. If the wind speed continues to grow beyond this limit, then it is necessary to change the control goal from optimization to limiting power at its rated value, P_n. This new goal is achieved by bringing the turbine in a regime where the power coefficient can be controlled to diminish. As shown in Section 4.2, there are two such regimes, i.e., stalling and blade feathering, which are analyzed in the following.

Power Limitation by Stall Control
In this case the control law is switched such that the operating point moves at the right side of ORC, which corresponds to stall. When the wind velocity increases linearly from the cut-in value, v_S, to a value that sensibly exceeds the rated value, v_n, then the operating point describes a trajectory in the $\Omega_l - \Gamma_{wt}$ plane as in Figure 4.12 (thick line). Zone I from this trajectory corresponds to the dynamic starting regime; in zone II, between A and B, the controller within the power optimization loop has Equation 4.4 as reference, thus keeping the operating point on the ORC. In zone III, between B, C and D the power is kept constant, at its rated value, by means of the power limiting controller.

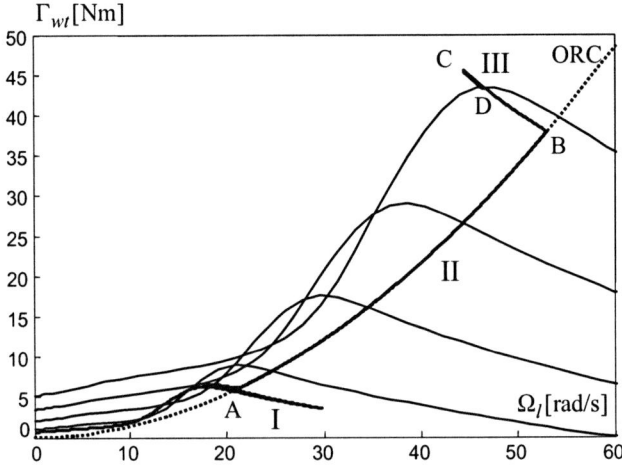

Figure 4.12. WECS operating zones when the wind velocity covers the entire variation range, putting in evidence the stall-control region

Figure 4.13 shows the evolution of the control input $x(t)$ received by the chopper. When moving on the trajectory A–B from Figure 4.13a, applying $x(t)$ results in moving to the right the curve $\Gamma_G(\Omega_h, x)$ (see Figure 4.11), whereas in the $\lambda - C_p$ plane the operating point is on the maximum. In point B the control goal is switched from optimization to power limitation at the rated value. For this second goal, the gradient of the control input $x(t)$ changes its sign and the operating point moves on trajectory B–C–D from Figure 4.12. In this way, the curve $\Gamma_G(\Omega_h, x)$ is moved to the left (Figure 4.11), while the operating point moves between AB and CD in the $\lambda - C_p$ plane (Figure 4.13b).

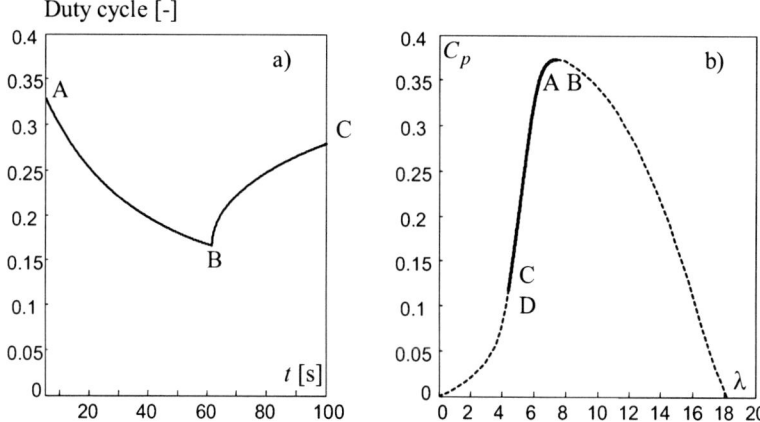

Figure 4.13. Switching control goal from optimization to rated power regulation by stall control: **a** control input received by chopper; **b** evolution in the $\lambda - C_p$ plane

The wind torque and the low-speed shaft rotational speed vary as in Figure 4.14; these variations explain how the operating point moves on trajectory B–C–D in Figure 4.12.

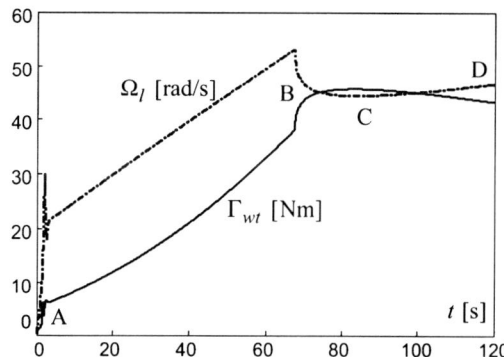

Figure 4.14. Evolutions of wind torque and of low-speed shaft rotational speed as the wind velocity covers the entire operating range (power limitation by stall control)

An important property of power limitation by stall control is that, if the constraint $\Omega_l(t) \leq \Omega_{max}$ is not required to be met in zone II, then practically it will not be required in zone III either. However, as the wind speed continues to increase the low-speed shaft rotational speed could increase following the B–C–D trajectory beyond the limit reached in point B and further, such that the admissible limit is reached.

Let us take the case where constraint $\Omega_l(t) \leq \Omega_{max}$ is active in zone II. Figure 4.15a,b presents how the operating point evolves in the $\Omega_l - \Gamma_{wt}$ plane and $v - \Omega_l$ plane respectively. The segment from origin to point A denotes the starting regime, on segment A–B the turbine operates on the ORC, between B and C the system is in partial load under rotational speed limitation, $\Omega_l(t) = \Omega_{max}$, and, finally, the segment C–D–E corresponds to power limitation – on this segment the evolutions of the rotational speed and of the wind torque match the segment B–C–D given in Figure 4.14.

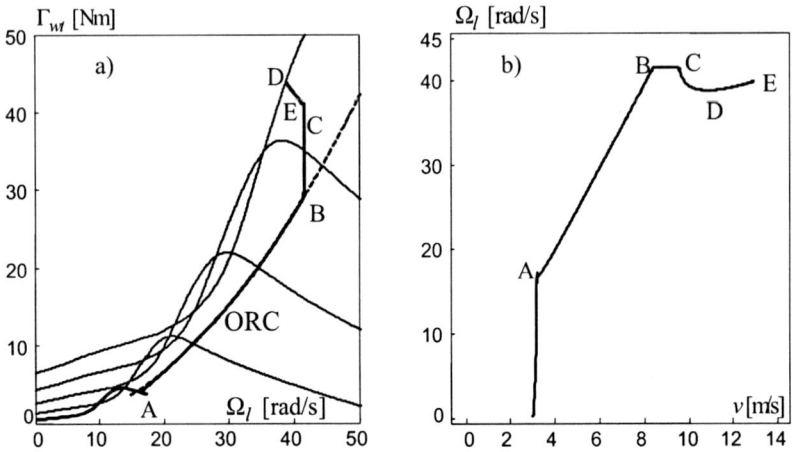

Figure 4.15. Switching from energy maximization to power limitation by stall control by an intermediary region of rotational speed limitation: **a** in the $\Omega_l - \Gamma_{wt}$ plane; **b** in the $v - \Omega_l$ plane

The evolutions of the power coefficient and of the tip speed ratio on trajectory A–B–C–E are presented in Figure 4.16a,b respectively.

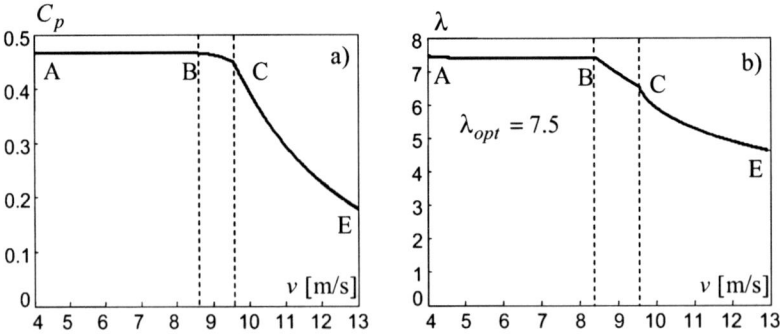

Figure 4.16. Power coefficient (**a**) and tip speed ratio (**b**) illustrating three operating regimes (power maximization, rotational speed limitation and power stall control)

Power Limitation by Blade-Feathering Control
In this case a power regulation loop is built, whose reference is given by Equation 4.4 when $P_{wt} < P_n$ and is set to the rated power, P_n, when the active

power tends to increase over the rated value. Under power limitation regime, operation of the system is illustrated by Figures 4.17–4.19.

By comparing these graphs with those presented for the stall control (Figures 4.12, 4.13, and 4.14 respectively), some important differences can be noted, as follows. The control input, $x(t)$, keeps decreasing when switching from optimization to power limiting (Figure 4.18a). The generator torque characteristic, $\Gamma_G(\Omega_h, x)$, moves in the same direction and the operating point moves on the trajectory B–C from Figure 4.17.

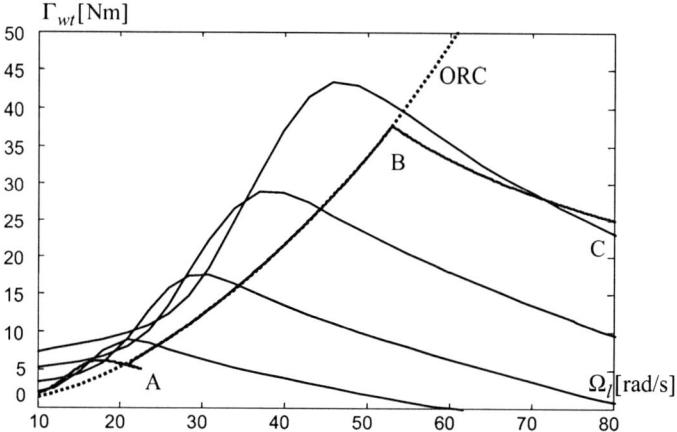

Figure 4.17. WECS operating zones for the entire operation range, putting in evidence the feathering control region

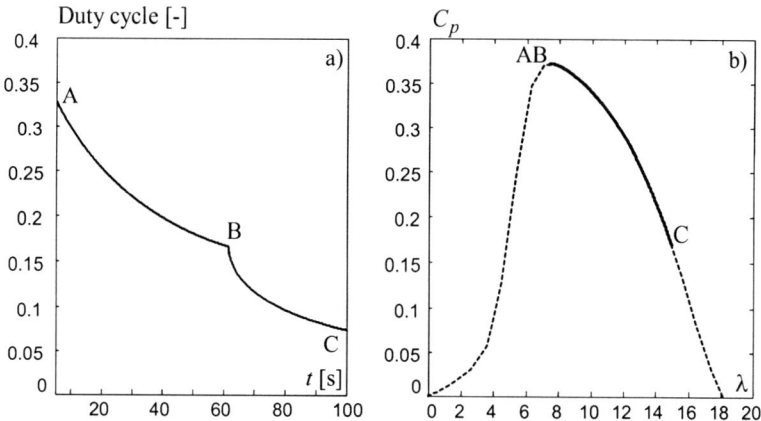

Figure 4.18. Switching between optimization and rated power limitation by feathering control: **a** control input received by chopper; **b** evolution in the $\lambda - C_p$ plane

The power coefficient is less drastically reduced by feathering control than by stall control; consequently, the horizontal movement of the operating point in the plane $\lambda - C_p$ is more important (trajectory A–B–C in Figure 4.19). The shaft rotational speed is thus significantly increasing (Figure 4.19).

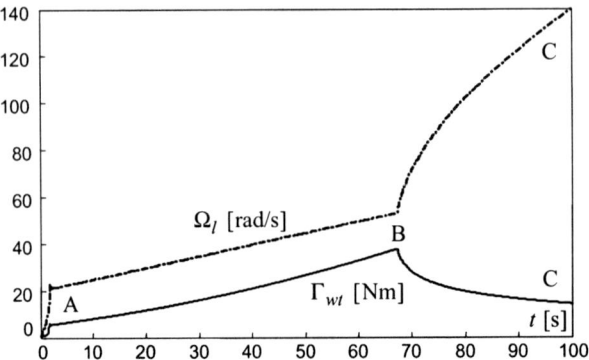

Figure 4.19. Evolutions of wind torque and of low-speed shaft rotational speed as the wind velocity covers the entire operating range (power limitation by feathering control)

The blade-feathering control cannot achieve the rated rotational speed regulation in partial load. Even if this constraint would not be imposed in this regime, the limitation $P_{wt}(t) \leq P_{max}$ is realized by increasing the rotational speed beyond admissible limits. Therefore, the stall control is the only feasible solution for complying with both rotational speed and power limitation constraints in the case of variable-speed fixed-pitch WECS.

4.4.2 Control of Variable-pitch WECS

Case of Fixed-speed Variable-pitch WECS
As previously shown, basic control functions are power optimization in zone II (see Section 4.3.2), and power limitation, the latter being achieved by either active-stall control, or by pitch-to-feather control. Figure 4.20 qualitatively presents the pitch angle, β, vs. the wind speed, v, when the system operates at the rated power, $P = P_n$ (Burton *et al.* 2001).

Point P corresponds to passing from partial to full-load regime, respectively

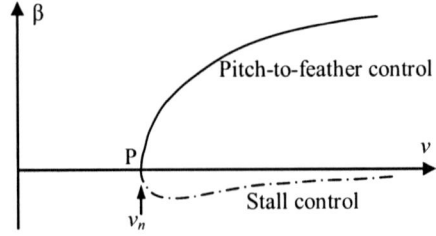

Figure 4.20. Delimitation between pitch-to-feather and stall control in the $v - \beta$ plane

from optimization to power limitation. Values of the pitch angle are usually negative under stall control and power regulation takes place by small changes of the pitch. In contrast, angle β can vary considerably under pitch-to-feather control, e.g., reaching values of 25° or 35°.

Let us consider the curve family of a fixed-speed WECS, $C_p(\beta)$, parameterised by the wind speed (Figure 4.8), for which the optimal control input, $\beta(v)$, is given in Figure 4.9. The pitch control law for the entire operating range, $\beta(v)$, as well as the corresponding evolution of the power coefficient, are respectively presented in Figure 4.21 (Hansen *et al.* 2003).

At present, the pitch-to-stall control of high-power WECS has limited use because the blade oscillations are less damped at large values of the incidence angle. Therefore, the pitch-to-feather control is currently preferred for power limitation in full-load.

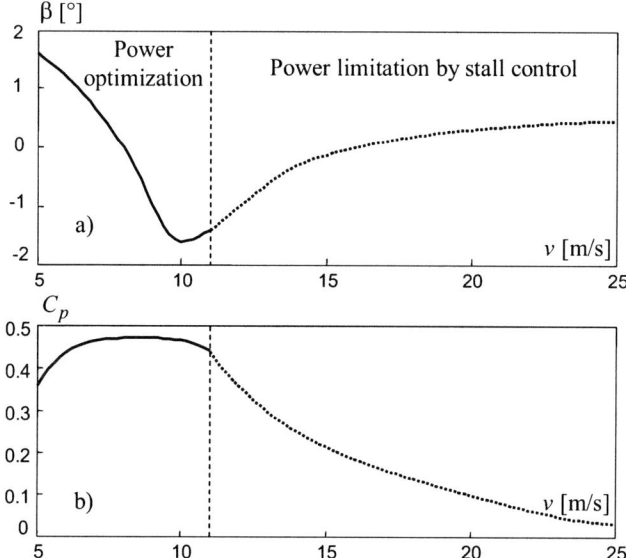

Figure 4.21. Pitch control law for the entire operating range (**a**) and associated evolution of the power coefficient (**b**)

Power regulation is habitually achieved by classical PI controllers, which provide reference values for the pitch actuators. Limitation of both pitch angle value and of its gradient is ensured by the pitch servomechanisms.

Taking into account the limitations imposed to actuator, as well as the nonlinear aerodynamic features of blades, the PI power controller is equipped with two kinds of control procedures:
– an anti-windup mechanism, which deactivates the integral component of the control law, whenever the pitch angle and/or its gradient go into limitation (Sørensen *et al.* 2005);
– a gain scheduling procedure, by means of which the controller parameters are

adjusted such that changes of the aerodynamic features of blades induced by changing the pitch angle do not affect the regulation quality.

In the second case, the proportional component of the control law, K_P, is computed based upon the analysis of the power sensitivity in relation to β. Thus, the pitch sensitivity of the system computes as (Hansen *et al.* 2005):

$$S_\beta(\lambda,\beta) = \frac{dP_{wt}}{d\beta} = \frac{1}{2}\rho\pi R^2 v^3 \frac{dC_p(\lambda,\beta)}{d\beta}$$

In order to preserve the properties of the power control system, the open-loop gain, expressed by

$$K_{olc} = K_P \frac{dP_{wt}}{d\beta}$$

must be kept constant. To this end, the proportional gain, K_P, must be replaced by

$$K_{olc} \cdot \left(\frac{dP_{wt}}{d\beta}\right)^{-1}$$

In the case of constant power reference, sensitivity S_β increases as the wind speed increases; when the wind speed is constant, the same sensitivity decreases as the power reference increases (Sørensen *et al.* 2005). Obviously, when S_β increases, gain K_P must be reduced.

Case of Variable-speed Variable-pitch WECS
The widely spread case of power optimization by variable-speed operation is considered in the following. The WECS modes of operation, taking into account rotational speed and power limitations, are depicted in Figure 4.22 (Hansen *et al.* 2004). These modes are detailed next.

1. The operation at constant speed, namely at its minimal operating value, Ω_{min}, corresponds to segment A–B in Figure 4.22. The tip speed ratio is $\lambda(v) = \Omega_{min} \cdot R / v$ and optimization is realised by pitch control.
2. The variable-speed fixed-pitch operation, where the operating point describes the ORC, corresponds to trajectory B–C.
3. The rotational speed limitation, namely at its rated value, Ω_n, is represented by trajectory C–D in Figure 4.22. In this region power optimization is achieved following the procedure specific to constant-speed operation.
4. Finally, segment D–E from Figure 4.22a represents the power limitation at its rated value. In the plane $\Omega_l - P_{wt}$ (Figure 4.22b), the operating point remains in the same position. Power limitation takes place by pitch control. In dynamic regime, the DFIG rotational speed can go beyond its rated value, until it reaches a maximally admissible limit, Ω_{dM}. Existence of a nonzero dynamic speed error within reasonable limits has a positive effect: it reduces the dynamic torque variations at wind gusts. Therefore, the operational range of DFIG speed is $\Omega_{min} \div \Omega_{dM}$.

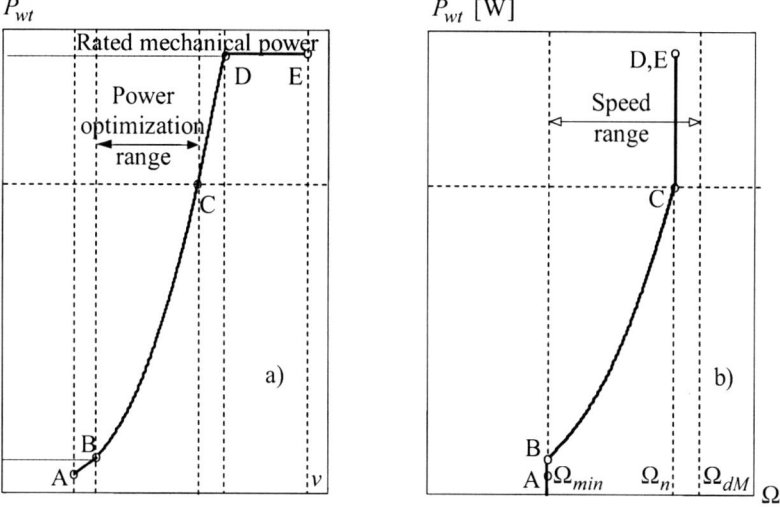

Figure 4.22. Operation modes of variable-speed variable-pitch WECS: **a** static wind power curve *vs.* the wind speed; **b** wind power *vs.* shaft rotational speed

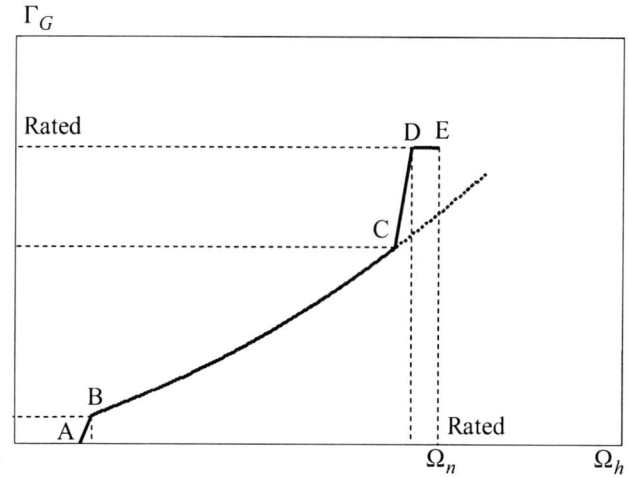

Figure 4.23. Variable-speed variable-pitch WECS operation: generator torque versus rotational speed

Burton *et al.* (2001) presents the variable-speed variable-pitch WECS operation by generator torque control for underrated power and by pitch control for power limiting. Figure 4.23 shows the trajectory A–B–C–D–E, illustrating the operation modes in the plane $\Omega_h - \Gamma_G$. On segment B–C the torque control ensures the ORC operation, like in the previously presented version. But on segments A–B and C–D the rotational speed is not constant any more; its variation determines important variations of the torque reference. On segment D–E, the control system has constant reference, equal to the rated torque. When the wind speed increases, the torque is regulated by increasing the speed to the rated one, Ω_n. If the wind speed continues to increase, the rotational speed attempts to go above rated and the system enters the power limitation regime, achieved by pitch control; in this way, the rotational speed is kept at the rated value. In this case, the operating point in the plane $\Omega_h - \Gamma_G$ is maintained in E. On the rest of the trajectory, A–B–C–D–E, the system is operated by controlling the generator torque, Γ_G. In region B–C, the torque loop must ensure operation on ORC. Taking account of Equations 4.4 and 4.5 and of the relations $\Omega_h = i \cdot \Omega_l$ and $P_{wt} = \Gamma_{wt} \cdot \Omega_l$, the optimal torque results:

$$\Gamma_{G\,opt} = \frac{1}{2} \cdot \frac{C_p(\lambda_{opt})}{\lambda_{opt}^3 \cdot i^3} \rho \pi R^5 \cdot \Omega_h^2 \qquad (4.9)$$

If considering losses due to the friction torque, Γ_f, one can consider that the torque reference is

$$\Gamma_{G_{ref}} = K_G \cdot \Omega_h^2 - \Gamma_f, \quad K_G = \frac{1}{2} \cdot \frac{C_p(\lambda_{opt})}{\lambda_{opt}^3 \cdot i^3} \rho \pi R^5 \qquad (4.10)$$

Because coordinates of points A, B, C and D in the plane $\Omega_h - \Gamma_G$ are known, the torque reference can be deduced as depending on the generator speed all along the trajectory A–B–C–D–E.

4.5 Optimal Control with a Mixed Criterion: Energy Efficiency – Fatigue Loading

As previously detailed, the optimal control problem associated with wind energy conversion systems essentially consists of optimizing the energy conversion, namely of maximizing the energy captured from the wind. This is equivalent to operating the system at its optimal tip speed ratio, λ_{opt}, as depicted in Figures 4.5a and 4.6. If λ_{opt} is specified by the wind turbine producer, the optimal control may be implemented by tracking the desired value of the low-speed shaft speed, $\Omega_l^{ref} = v \cdot \lambda_{opt}/R$. But in most cases the value of λ_{opt} is not known and the MPPT methods – detailed in Chapter 5 of this book – are preferred, which are based on an on-off controller using minimal information from the system. On the other hand, the operating point oscillates largely around the energy maximum, which is harmful to the power quality and to the mechanical reliability.

The above methods have as an exclusive goal the maximization of the energy efficiency, while ignoring the possible drawbacks related to the system's reliability, due to some large control input efforts. In Novak and Ekelund, (1994) and Ekelund (1997) it is stated that keeping λ_{opt} in turbulent winds is possible only with large generator torque variations (significantly high mechanical stress).

4.5 Optimal Control with a Mixed Criterion: Energy Efficiency – Fatigue Loading

Therefore, the supplementary mechanical fatigue of the drive train can be reduced by imposing the minimization of the generator torque variations, $\Delta\Gamma_G(t)$, used as control input, *around the optimal operating point*, as Ekelund (1997) has expressed by a combined optimization criterion:

$$I = \underbrace{E\left\{\alpha\cdot\left[\lambda(t)-\lambda_{opt}\right]^2\right\}}_{I_1} + \underbrace{E\left\{\Delta\Gamma_G^2(t)\right\}}_{I_2} \rightarrow \min, \qquad (4.11)$$

where $E\{\cdot\}$ is the statistical average symbol. The first term illustrates the energy efficiency maximisation, whereas the second term expresses the minimisation of the torque control variations. The trade-off between the two terms is adjusted by means of the weighting coefficient, α. The solution proposed to deal with the dependency of the linearized dynamical system's parameters on the average wind speed was a gain-scheduling adaptive structure, together with an observer for state reconstruction, which uses the high-speed shaft rotational speed, Ω_h, as measurable output. Using this idea, one can built an adaptive control structure as presented in Figure 4.24, which allows different control goals to be formulated, depending on a given operating point.

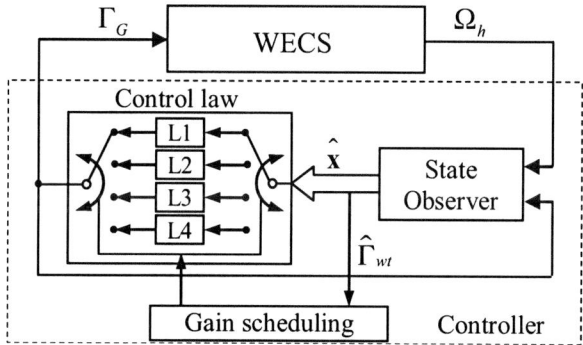

Figure 4.24. Adaptive control structure, using the idea from Ekelund (1997)

Apart from its complexity, this adaptive structure has as drawback the necessity of quite precise knowledge of the rotor characteristic. For example, computing the torque parameter, $\gamma(t)$, is necessary for each operating point, requiring the power coefficient curve, $C_p(\lambda)$, to be known, which is seldom an available information.

To decide which controller to apply to the plant is a poorly formalized step of any classic gain scheduling design. Options vary from simply switching between the controllers associated to various operating points to quite sophisticated interpolation strategies (Shamma 1996).

In Munteanu *et al.* (2005) the energy-reliability criterion to be minimized is stochastically defined for the normal operating (partial load) region, the aim being to eliminate, if possible, the adaptive structures and to reduce the required feedback information. Thus, an optimal control structure is presented, which optimizes the criterion at Equation 4.11 without using adaptive structures. This approach – named *the frequency separation principle* – relies upon separating the turbulence (high-frequency) and the low-frequency wind speed components derived from the Van der Hoven wind model (Nichita *et al.* 2002) by using a low-pass filter. The two components – the low-frequency and the turbulent ones, as to Equation 3.1 – excite the plant's dynamics in two distinct spectral ranges. Correspondingly, the proposed structure is formed by two loops, separately driven by the low-frequency and the turbulence component respectively. This structure desensitizes the closed loop system subject to the steady-state operating point, as detailed in Chapter 6.

The mechanical loads alleviation is an issue in all the WECS operating modes (cases of variable/fixed-speed fixed/variable-pitch WECS). At high winds, when the system works in power limitation, for example, a reference solution is given by LQG optimal control (Boukhezzar *et al.* 2007). The flexible-drive-train WECS is linearized around the nominal operating point; the state vector, $\mathbf{x}$, is composed of variations of the essential variables in relation to their nominal values. The optimal controller determines the pitch variation, $\Delta\beta$, such that to minimize a classic performance index, of the form

$$I = E\left\{\mathbf{x}^T\mathbf{Q}\mathbf{x} + \alpha \cdot (\Delta\beta)^2\right\}, \qquad (4.12)$$

where parameter α weights the control effort. A larger control approach of WECS operating in full load, which takes account of the tower dynamics, is presented in (Lescher *et al.* 2006). Here, the demand of simultaneously achieving the power limitation and mechanical loads alleviation is formulated as a multi-purpose optimization problem, rendered at minimizing $\mathcal{H}_2/\mathcal{H}_\infty$ norms.

4.6 Gain-scheduling Control for Overall Operation

Bianchi *et al.* (2005, 2006) propose to deal with multi-purpose optimal control of WECS controllers by using gain-scheduling techniques for linear parameter varying systems (LPV – Apkarian and Adams 1998). The gain scheduling control can thus be systematically designed in order to ensure stability and performance features over the entire operating range. Indeed, this approach allows what the authors call the "control strategy locus" being unitary defined as a 3D trajectory in the wind speed – rotational speed – pitch angle space, over the whole domain of wind speed variation (Figure 4.25a). In the case of fixed-pitch WECS this locus comes down to a planar representation (Figure 4.25b). In both figures one can note the three regions: energy optimization in low winds (I) and rated power regulation in high winds (III), the two relied through a region of rated rotational speed regulation (II).

The control objectives are not defined for each operating regime any more, but there is a single goal expressing the trade-off between tracking the control strategy locus and minimizing the control efforts. This task can be cast into a convex optimisation problem with linear matrix inequalities constraints. The design procedure is similar to $\mathcal{H}_\infty$ synthesis and gives rise to a single controller instead of a family of controllers. A third goal – *i.e.*, mitigation of cyclic loads inside the drive train – as well as the power conditioning is separately fulfilled.

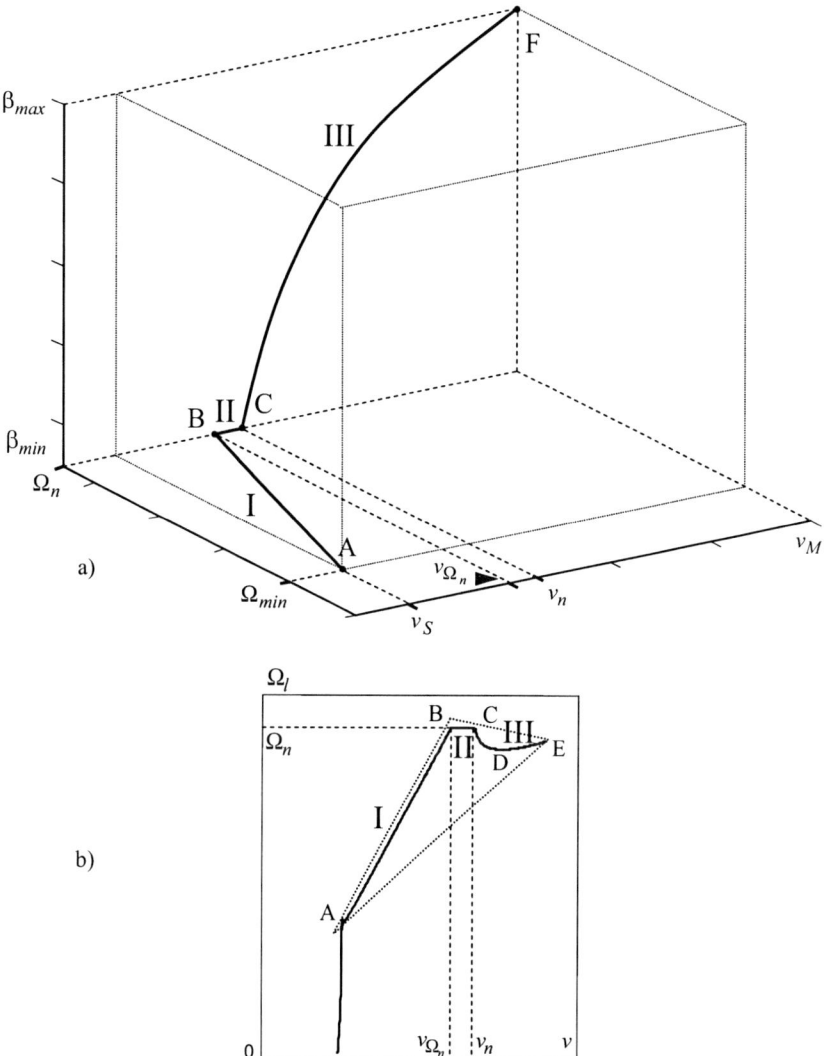

Figure 4.25. Control objective defined as tracking the control strategy locus: **a** variable-speed variable-pitch WECS; **b** variable-speed fixed-pitch WECS

As an example, the main guidelines for LPV modelling and gain scheduling design dedicated to fixed-pitch flexible-drive-train WECS as developed by Bianchi *et al.* (2006) are briefly presented hereafter.

An LPV model denotes a set of linear time varying (LTV) models parameterised by a set of time varying parameters, $\theta(t)$, having values in a bounded set Θ. In this case, an LPV model results from putting together the linearized flexible drive train model (Equation 3.68) and the model of the controlled generator. Because the generator is usually controlled through constant flux strategy, its (habitually linear) torque characteristic moves by remaining parallel with itself under the control action. Therefore, a good approximation of this characteristic is (see also Figure 3.23)

$$\Gamma_G = K_l \cdot (\Omega_h - \Omega_{h0}),$$

where Ω_{h0} is the zero-torque speed and plays the role of the control input. In conclusion, an LPV model for fixed-pitch flexible-drive-train WECS can be written in the form

$$\begin{cases} \dot{x} = A(\theta) \cdot x + B_v(\theta) \cdot \Delta v + B(\theta) \cdot u \\ y = C \cdot x + D \cdot u \end{cases} \quad (4.13)$$

where the states, outputs, set of parameters and control input are respectively:

$$\begin{cases} x = \begin{bmatrix} \Delta\Omega_l & \Delta\Omega_h & \Delta\Gamma \end{bmatrix}^T \quad y = \begin{bmatrix} \Delta\Omega_h & \Delta\Gamma_G \end{bmatrix}^T \\ \theta = \begin{bmatrix} \bar{v} & \bar{\Omega}_l \end{bmatrix}^T \qquad u \equiv \Delta\Omega_{h0} \end{cases} \quad (4.14)$$

The system has two kinds of inputs: the control input, **u**, and the disturbance Δv, i.e., the wind turbulence. Following the $\mathcal{H}_\infty$ synthesis, the control objectives must be put into the equivalent form of minimizing the norm of an input-output operator. Therefore, input **w**, output **z**, called performance output or error, and weighting functions must be selected. The open-loop system together with the chosen weighting functions composes the augmented plant, whose mathematical description is suggested in Figure 4.26 (matrices derive from Equation 4.13 and from choosing the weighting functions).

At the next step, an LPV controller $(A_c(\theta), B_c(\theta), C_c(\theta), D_c(\theta))$ that minimizes the induced $\mathcal{L}_2$-norm of operator $T_{zw}: w \to z$ must be found out. In this way it is ensured that the closed-loop system, having **w** as input and **z** as output (Figure 4.26), is stable and the $\mathcal{L}_2$-norm of the performance output, **z**, goes small for all inputs **w** having less than unity norm.

The controller is effectively computed by solving a convex optimisation problem with constraints of LMI type. Since the LPV model at Equation 4.13 is affine in parameters θ and these latter have values in a set Θ covered by a convex polytope (with three vertices; see Figure 4.25b with dotted line), it is sufficient that the LMI constraints be checked at the vertices of the polytope (Apkarian *et al.* 1995). The design procedure, consisting essentially of finding Lyapunov functions, results in a set of three linear time invariant controllers; the LPV controller is finally computed as a weighted linear combination of the three.

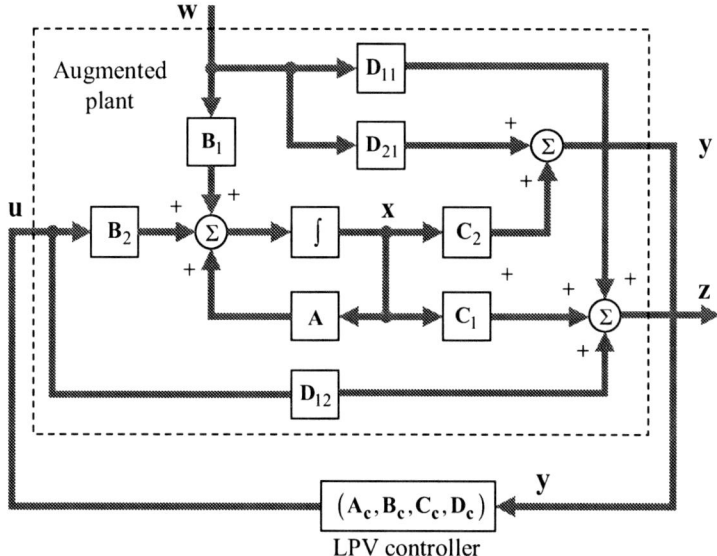

Figure 4.26. Block diagram of a variable-speed fixed-pitch WECS with LPV controller

4.7 Control of Generators in WECS

The generator motion control is obtained by introducing in the conversion chain some power electronics devices. Therefore, unlike fixed-speed wind turbines, the rotational speed can be varied within a large range, allowing the positioning of the operating point on the ORC, irrespectively from the wind speed value. In spite of better aerodynamic efficiency one can note that the insertion of the power electronics converters in the WECS involves a supplementary power loss. But this technology has even more advantages, like output electrical power conditioning.

As the electrical machine exhibits a dual behaviour, its motion control as generator is very similar to the one in motoring regime, that is, one can use the same hardware and the same control logic. The turbine providing the rotary power is more or less rigidly coupled to the electrical generator, thus allowing the steady-state speed to be imposed by generator torque control. Simply speaking, by acting on the generated electrical voltages, the generator is loaded more or less, so its electromagnetic torque varies. The steady-state operating point of the rotor-generator coupling then moves to another position, and the turbine rotor speed changes. Depending on the available measurements, different control loops can be built. Most of them envisage torque (current) control or speed control.

4.7.1 Vector Control of Induction Generators

A very popular structure used for the induction machines motion control is the one containing an AC–AC voltage source back-to-back inverter. This converter also provides insulation between the machine and the grid voltages by means of a DC circuit. The control input applied to the AC–AC converter results from a vector control structure (Bose 2001) which employs the induction machine modelling in the (d,q) frame. Based on this technique, very good performances regarding the torque settling time and oscillations can be obtained, making this kind of control suitable for wind conversion applications.

The principle of the vector control will be illustrated here only for the SCIG-based case of WECS, as its model is very similar to the DFIG one. However, because the generator's control inputs are different in the two mentioned cases (stator vs. rotor voltages) there are some slight differences in conceiving the controller (Leonhard 2001) mostly regarding the flux orientation.

SCIG Motion Control

The blueprint of the SCIG-based WECS connection to the electrical grid is given in Figure 4.27. The machine side power converter (**A** in Figure 4.27) allows the high-speed shaft torque to be varied by controlling the generator torque, thus enabling the variable-speed regime of the turbine. The grid-side converter (**B** in Figure 4.27) is used for transferring the produced electrical power (by DC-link voltage regulation) to the mains, while controlling the voltage and the frequency of the output voltages. Some other parameters regarding the electrical power injected to the grid – *i.e.*, the power factor or harmonics) can also be improved. In accordance to the control objective – *i.e.*, the SCIG motion control – the two power converters are considered ideal (*i.e.*, the imposed triphased voltages are physically obtained without dynamic and power loss) and are separately controlled. As it concerns mostly the grid interfacing, the control of inverter **B** is not covered in this section – its role will be detailed in Section 4.8.

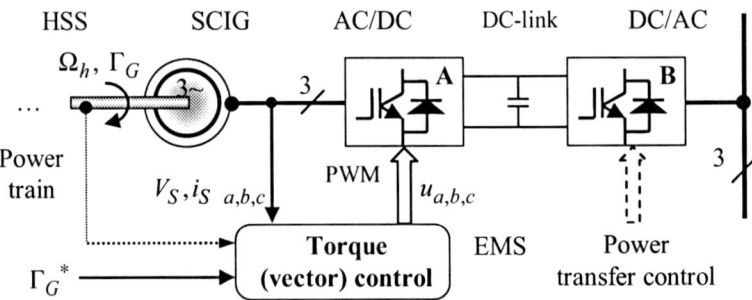

Figure 4.27. Structure of the electromagnetic subsystem, EMS – SCIG case

The torque control of the generator is realised by suitably controlling inverter **A** (Peña *et al.* 2000; Bose 2001). Depending on the desired electromagnetic torque, the controller must deliver the V_d, V_q voltages and stator flux position θ_S, which can be used in Park transform for computing the stator voltages' amplitude and frequency ($V_{S\,a,b,c}$ and ω_S).

It is envisaged to control separately the electromagnetic torque and the flux of the induction machine. This method basically relies on the alignment of the rotor flux, Φ_R, on the *d*-axis of the (d,q) rotating frame. The electromagnetic torque

4.7 Control of Generators in WECS

expression, containing the two rotor fluxes Φ_{Rd}, Φ_{Rq}, and the stator currents, derives from Equation 3.31 and is written as

$$\Gamma_G = \frac{3}{2} \cdot p \cdot \frac{L_m}{L_R} \cdot \left(i_{Sq} \cdot \Phi_{Rd} - \Phi_{Rq} \cdot i_{Sd} \right)$$

Consider now that the rotor field is oriented on the *d*-axis; this implies the cancellation of the Φ_{Rq} component ($\Phi_{Rq} \approx 0$) and ensures a new expression for the torque, as in the first expression at Equation 4.15. The Equations 3.34 and 3.35, giving the rotor currents dynamics, can also be rewritten to yield the last two expressions from Equation 4.15:

$$\begin{cases} \Gamma_G = \frac{3}{2} \cdot p \cdot \frac{L_m}{L_R} \cdot i_{Sq} \cdot \Phi_{Rd} \\ \frac{L_r}{R_r} \cdot \frac{d\Phi_{Rd}}{dt} + \Phi_{Rd} = L_m \cdot i_{Sd} \\ \omega_s = p \cdot \Omega + \frac{L_m}{\Phi_{Rd}} \cdot \frac{R_r}{L_r} \cdot i_{Sq} \end{cases} \quad (4.15)$$

Using Equations 4.15 in the SCIG model (Equation 3.37) one can develop the MIMO mathematical model of the field-oriented machine, having as input vector $[V_{Sd} \; V_{Sq}]^T$ and as state vector $[\Phi_{Rd} \; i_{Sq}]^T$ (Leonhard 2001):

$$\begin{cases} \Phi_{Rd}(s) = M(s) \cdot \left(V_{Sd}(s) + N(s) \cdot I_{Sq}(s) \right) \\ I_{Sq}(s) = P(s) \cdot \left(V_{Sq}(s) - R(s) \cdot \Phi_{Rd}(s) \right) \end{cases}, \quad (4.16)$$

where

$$\begin{cases} M(s) = \dfrac{1}{\left[(R_S + \sigma L_S \cdot s) \cdot \dfrac{1 + L_R/R_R \cdot s}{L_m} + \dfrac{L_m}{L_R} \cdot s \right]}, \quad N(s) = \omega_s \cdot \sigma L_S \\ P(s) = \dfrac{1}{R_S + \sigma L_S \cdot s}, \quad R(s) = \omega_s \cdot \left(\sigma L_S \cdot \dfrac{1 + L_R/R_R \cdot s}{L_m} + \dfrac{L_m}{L_R} \right) \end{cases} \quad (4.17)$$

A rough analysis of this model shows a coupling between the *d*- and *q*-axis – see Figure 4.28a. Based on this model one can separately control the two states and finally the generator torque value using simple PI controllers – dimensioned to meet some dynamic specifications – and a decoupling correction (Bose 2001; Williams and Antsaklis 1996) in order to attain orthogonality between $V_{Sd} \to \Phi_{Rd}$ and $V_{Sq} \to i_{Sq}$ channels (see Figure 4.28b).

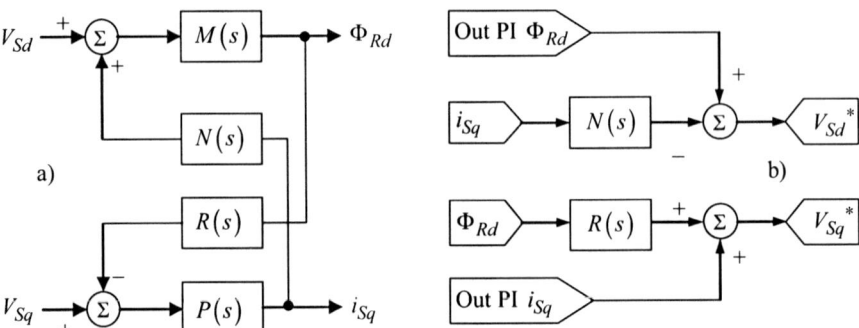

Figure 4.28. a Rotor field-oriented SCIG model. **b** *d-q* decoupling by state feedback plus pre-compensation

Figure 4.29 contains a widely employed vector control structure for torque controlling a squirrel cage induction machine. It basically consists of two decoupled loops: the first is a rotor flux loop ensuring the field orientation of the induction machine in order to control Φ_{Rd} and the second is a torque loop which imposes the electromagnetic torque in order to control i_{Sq}.

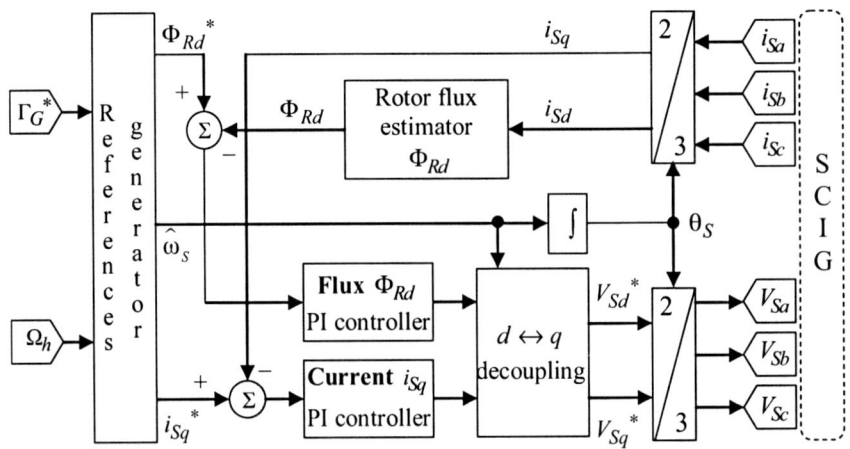

Figure 4.29. SCIG (indirect) vector control

As this control structure allows very fast and accurate torque response, an I/O model of the torque controlled SCIG can be assimilated to a first-order element, with very fast dynamic. A wise choice of the PI parameters in the vector control scheme allows a millisecond rated EMS time constant, T_G, to be achieved, so the phenomena which govern the operation of subsystem S_2 (from Figure 3.1) happen much more rapidly *vs.* the global WECS dynamics. This allows one to neglect this subsystem's dynamic in the total dynamics, when this is especially required.

DFIG Motion Control

The DFIG grid connection relies upon the same hardware as the well-known static Kramer drive (see Figure 4.30). Unlike the SCIG, the DFIG is more flexible, being able to operate as a generator (at negative torque) both in sub-synchronous (positive slip) and over-synchronous (negative slip) regimes. Considering an ideal power electronics converter, the generator output power, P_{grid} is the sum of the stator and rotor powers:

$$\begin{cases} P_{stator} = \dfrac{P_{grid}}{1-s} \\ P_{rotor} = -s \cdot P_{stator} \end{cases}, \qquad (4.18)$$

with $s = 1 - \Omega_h/\omega_s$ being the slip. Therefore, in the sub-synchronous operation the rotor power is absorbed from the grid (positive), whereas in the over-synchronous operation it is fed to grid (negative). In both cases the stator feeds energy to grid. Equation 4.18 shows that the power transferred through the power electronics inverter is significantly smaller than the rated power of the generator. Compared to the previous described case, the inverter rated power is reduced, the speed varies in a narrower range and the amount of the reactive power injected to the grid is smaller.

The DFIG vector control is based on the same principles as in the SCIG case. One of the differences is that the alignment of the rotor flux is usually taken on the *d*-axis of the *(d,q)* frame turning synchronously with the stator field (Peña *et al.* 1996; Hofmann *et al.* 1997). The control inputs are the rotor voltages in the *(d,q)* frame and the general control structure replicates the one in Figure 4.29.

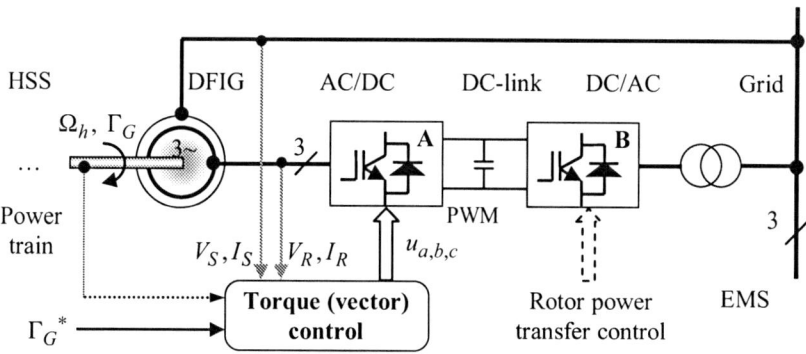

Figure 4.30. DFIG grid connection structure

To conclude, this control structure feeding an induction machine by means of a voltage source inverter allows a very fast control of the wind turbine high-speed shaft. It can also easily be embedded into a speed or power control loop, allowing thus the imposing of the WECS rotational speed (Bose 2001), or active power (Pöller 2003).

4.7.2 Control of Permanent-magnet Synchronous Generators

Generally speaking, permanent-magnet synchronous generators are used more frequently in low-to-medium power WECS applications. Their motion control has many similarities with that used for motoring applications and also with that of induction machine. Depending on the power electronics converter used in the specific application, the operation of the synchronous machine can be controlled in nested speed-torque loops, using different torque control algorithms (Bose 2001).

For medium-power WECS fully controlled rectifier-inverter pair is usually used, together with a vector torque control structure for variable-speed operation (Chen and Spooner 1998; Schiemenz and Stiebler 2000; Akhmatov 2003). For smaller systems a diode rectifier – chopper – inverter configuration is preferred as it provides handling simplicity with just a small decrease in cost effectiveness (Higuchi et al. 2000; Amei et al. 2002; Song et al. 2003; Knight and Peters 2005). Like in the induction machine case, the inverter is used for grid interfacing while the pair rectifier-chopper is used for PMSG output current/torque control (see Figure 4.31).

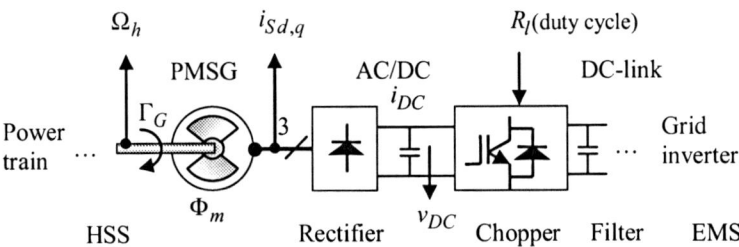

Figure 4.31. Structure of the electromagnetic system – low-power PMSG case

The controlled (step-up/step-down) chopper acts as an equivalent load resistor, R_l, in the stator circuit, having a value depending on its duty cycle. The DC current and therefore the generator load will change in response to variation of R_l. This means that the generator voltages in the (d,q) frame will vary with the same ratio and the currents i_{Sd} and i_{Sq} will be simultaneously affected.

Variable-speed operation of PMSG-based WECS is usually achieved through output power or rotational speed control loops. In Figure 4.32 the system response in rotational speed and electric power at changes of load resistor is given. This gives some information regarding the controller structure to be employed for controlling the electromechanical variables.

The control structures encountered in literature are very different. As the plane $\Omega_l - \Gamma_{wt}$ can be easily transposed in the $i_{DC} - v_{DC}$ plane, the control loops can envisage the electrical variables in the DC circuit. For imposing a certain value of the DC current, i_{DC}, a local control loop containing a PI current regulator is often employed.

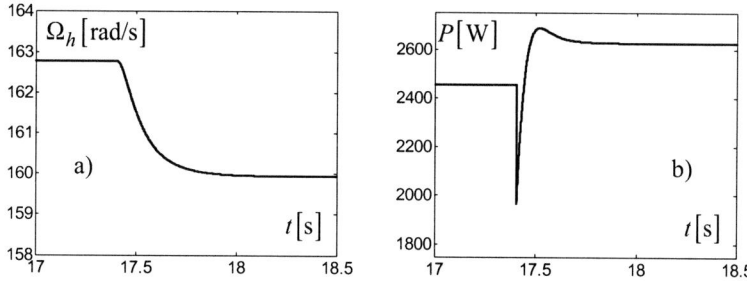

Figure 4.32. Step response in R_l changing: **a** rotational speed; **b** active power

4.8 Control Systems for Grid-connected Operation and Energy Quality Assessment

A major issue in the control and stability of electric power systems is to maintain the balance between generated and consumed power. Because of the fluctuating nature of wind speeds, the increasing use of wind turbines for power generation has caused more focus on the impact of the power production of the wind turbines on the power systems.

The power produced by wind turbines influences the grid in terms of both dynamic stability and power quality.

4.8.1 Power System Stability

The electrical power system, due to the increasing importance of power produced by WECS, becomes more vulnerable and dependent on wind power generation. Therefore, the capability of wind turbines (especially when they are geographically located in wind farms) to act as conventional power plants is becoming more and more important.

The power control requirements regarding different power system control and stability aspects can be summarized as follows (Sørensen *et al.* 2005):

1. *Power/frequency control ability* with focus on:
 - *primary control*—fast, automatic adjustment of power to frequency;
 - *secondary control*—slower, automatic or manual regulation of the power to the power reference imposed by the system operator at any time.
2. *Voltage control ability* with focus on voltage regulation and reactive power capability.
3. *Dynamic stability* with focus on the ability of wind turbines to remain connected to the grid during some specific grid faults.

Modern wind turbines/farms are required to provide advanced grid support, such as different control functions for both active power/frequency control and reactive power/voltage control.

For example, regarding active power/frequency control, the Danish TSO requirements involve different types of power control: absolute power limitation, delta limitation, balance control, stop control, ramp limitation, and fast down regulation to support system protection. On top of that, automatic frequency control is required.

The *balance control* is the method where the wind turbine/farm production is reduced by control to constant levels; see Figure 4.33. In this case, the reserve, *i.e.*, the difference between the (controlled) produced and available power, is not constant.

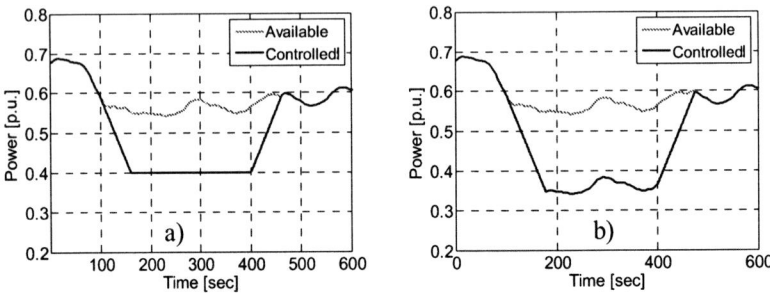

Figure 4.33. Balance (**a**) and delta (**b**) control

The *Delta control* is the control where the wind turbine/farm production is reduced with certain constant (delta) offset; see Figure 4.33. The advantage of delta control is that a certain amount (the delta) of reserve power is always available and can be utilized in automatic primary frequency control.

The *automatic frequency control* includes droop and dead band (Figure 4.34). In normal operation (without activated balance or delta control), the power setpoint P_0 will be equal to the maximum available power P_{max}, and the power-frequency characteristics in Figure 4.34 will only have droop for over-frequencies, because there is no reserve power available for under-frequencies.

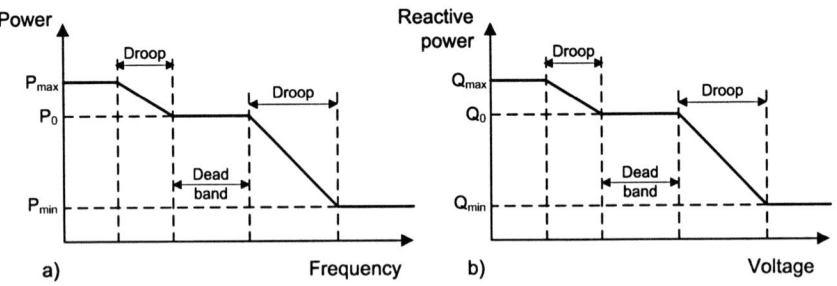

Figure 4.34. Primary frequency control (**a**) and automatic voltage control (**b**)

The reactive power/voltage control requirements for wind turbines/farms, according to the Danish TSO, demand that the reactive power from a large wind farm can be controlled to a specific interval, which is close to unity power factor. However, most wind turbines are also able to provide more advanced reactive

4.8 Control Systems for Grid-connected Operation and Energy Quality Assessment

power control, which can be useful as grid support. Depending on the technology and the electrical design, such wind turbines will normally have some additional capacity for reactive power, although the available reactive power normally depends on the active power as it does for any other generating units in the power system. This dependency is expressed in the PQ diagrams (Lund *et al.* 2007). The additional reactive power capacity can either be used to control constant reactive power or constant power factor, or it can be used in automatic voltage control.

A standard reactive power/voltage control strategy is a combined droop and dead-band control, as illustrated in Figure 4.34 (Sørensen *et al.* 2005).

As can be seen from the above, even if the control strategies for power system stability are usually implemented at wind farm level, they depend on the control capabilities, especially regarding the reactive power, of the individual wind turbines. The state-of-the art regarding the reactive power control of wind turbines is presented in the following.

Fixed-speed wind turbines have limited reactive power control. The fixed-speed wind turbine concept, also known as the "Danish concept", is presented in Figure 4.35. They use squirrel cage induction generators and they need to draw reactive power from the grid. Those wind turbines are equipped with a capacitor bank for reactive power compensation. Reactive power control is not possible.

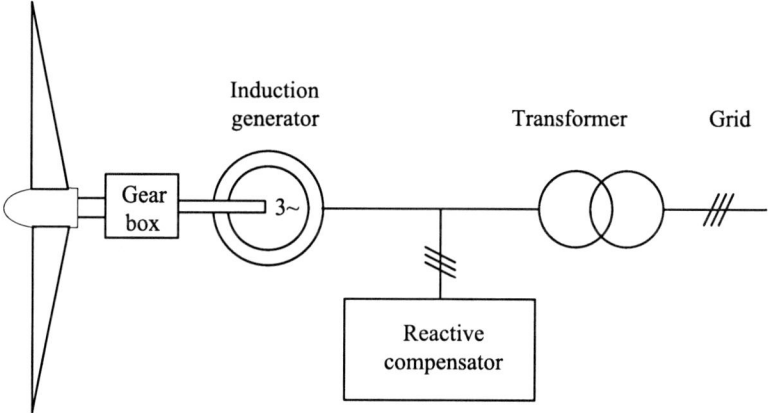

Figure 4.35. Fixed-speed wind turbine concept

Semi-variable-speed wind turbines are equipped with doubly-fed induction generators (DFIG), with the stator connected directly to the grid and the rotor connected to the grid through a back-to-back power converter. The speed can vary by ±30% of the synchronous speed. The main advantage of the DFIG wind turbine concept is that the power converter is rated at a fraction of the generator rated power. There is complete control of active P_{ref} and reactive Q_{ref} power through the back-to-back power converter, as illustrated in Figure 4.36.

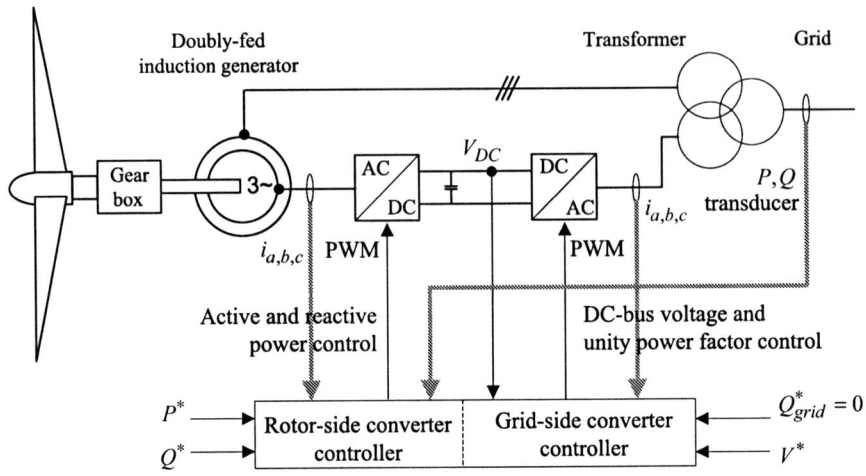

Figure 4.36. DFIG wind turbine

In this configuration there are two main control modules: the rotor-side converter control and the grid-side converter control.

The rotor-side converter operates in a stator-flux dq-reference frame, leading to a decoupled control of the active power (q-axis) and reactive power (d-axis) through the rotor current. The control consists of a cascade structure, with a very fast inner current-control loop and a slower outer power-control loop, as shown in Figure 4.37 (Pöller 2003; Iov 2003; Akhmatov 2003). The output of the current controllers is the width-pulse modulation factor, P_m (with its d and q-components, P_{md} and P_{mq}), which is the control variable of the PWM converters.

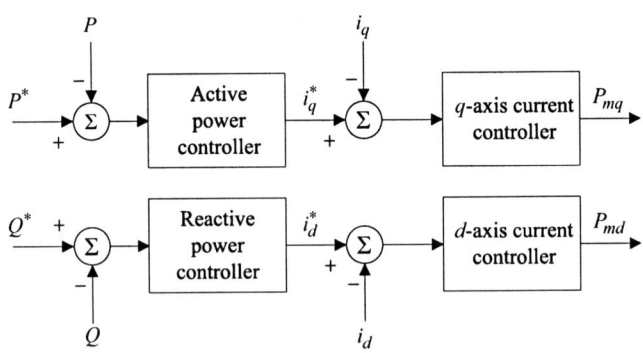

Figure 4.37. Rotor-side converter control

The grid-side converter keeps the DC-link voltage constant regardless of the magnitude and direction of the rotor power and assures the quality of the output voltage and current in accordance with the standards. Usually this converter operates at unity power factor (Iov 2003). The control concept of the grid-side converter is presented in Figure 4.38 (Pöller 2003).

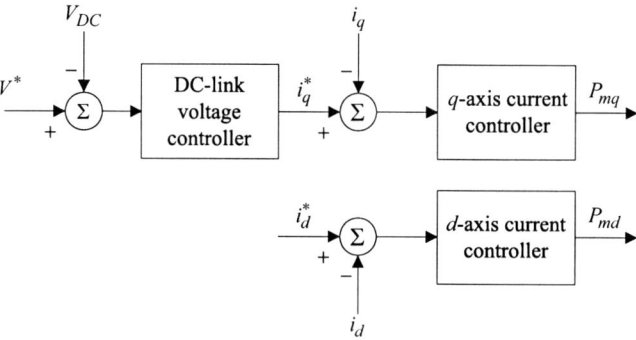

Figure 4.38. Grid-side converter control

Similar to the rotor-side converter control, the active and reactive components of the grid-side converter currents are controlled by a very fast inner control current loop. The DC-link voltage is controlled by a slower outer control loop that defines the q-axis current component setpoint. The aim is to maintain a constant DC-link voltage.

Variable-speed wind turbines can use both induction and synchronous generators and they have the stator connected to the grid through a full-scale back-to-back power converter, as presented in Figure 4.39.

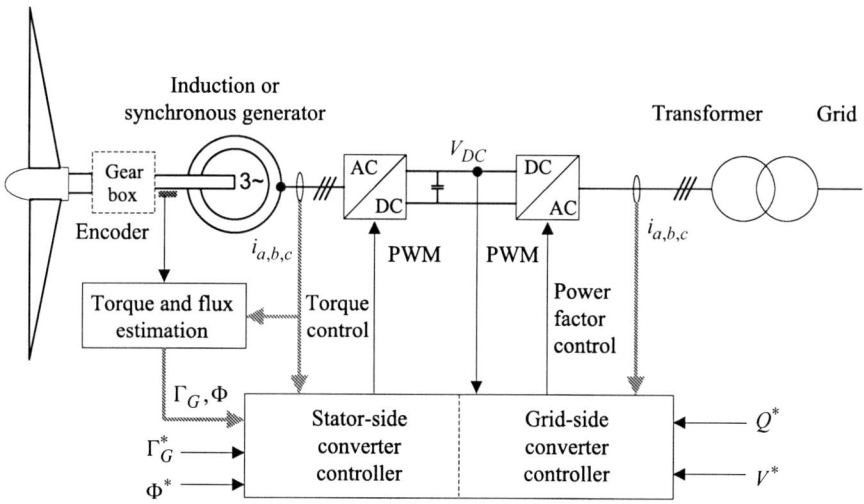

Figure 4.39. Variable-speed wind turbine with full-scale power converter

The active and reactive powers are fully controlled through the power converter. The control of the grid-side converter is similar to the one in the DFIG wind turbine (Figure 4.36). The stator-side converter control scheme is presented in Figure 4.40 (Iov 2003).

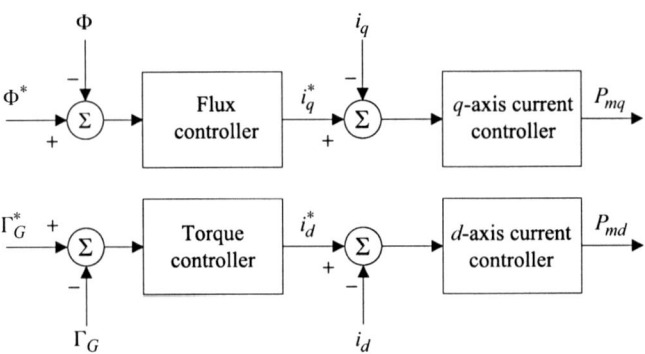

Figure 4.40. Stator-side converter control

Usually an indirect rotor-flux-oriented is used for the squirrel-cage induction generator, due to its simplicity. The flux (q-axis) and the torque (d-axis) are controlled trough the stator current. The torque reference is driven by a maximizing power extraction control, while the rotor magnetizing current reference is a function of the rotor speed or, for simplicity, it can be kept constant (Iov 2003).

4.8.2 Power Quality

The term "power quality" in relation to a wind turbine describes the electrical performance of the wind turbine electricity generating system (Ackermann 2005). The main influences of a wind turbine on the power and voltage quality are voltage changes and fluctuations – leading to flicker – and harmonics for wind turbines with power electronics.

Flicker
Wind turbines can have considerable fluctuations in the output power, due to the stochastic nature of the wind. As a result of that, the grid suffers voltage fluctuations and flicker.

Flicker is the effect of the voltage fluctuations on the brightness of incandescent lights and the subsequent annoyance to customers (Mirra 1988). The flicker produced by the wind turbines is mainly caused by the wind speed variations: the wind gradient (wind shear), the tower shadow effect (Larsson 2002) and the stochastic fluctuations due to the rotational sampling effect. From the combination of the above-mentioned phenomena, a power drop appears three times per revolution for a three-bladed wind turbine. This frequency is well known and is referred to, in the literature, as *3p*. Furthermore, frequency domain analyses of the output power of grid-connected wind turbines revealed that also the *6p, 9p, 12p* and *18p* components are visible too (Thiringer 1996).

The flicker level increases at higher wind speeds due to higher turbulence. For fixed-speed wind turbines, the flicker level increases around three times from lower to higher wind speeds, while for variable-speed wind turbines the flicker level increases with the wind speed up to the rated wind speed value, where the variable-speed system will smooth out the power fluctuations and, thereby, the flicker (Sun 2004). The flicker level produced by the wind turbines is strongly dependent on the technology used. Variable-speed wind turbines produce significantly lower level flicker than fixed-speed wind turbines. According to Papadopoulos *et al.* (1998), a four times reduction of the flicker level can be realized through variable-speed operation. Other factors that affect the flicker level are the short circuit capacity at the PCC of wind turbines, inversely proportional with the flicker level, and the grid impedance angle. If a proper value is chosen, the voltage changes from the varying active power flow will be cancelled by that from the varying reactive power flow and, therefore, the voltage fluctuations and the flicker level are reduced. The determining factor is the difference between the grid impedance angle and the wind turbine power factor angle (Papadopoulos *et al.* 1998).

The flicker mitigation of grid-connected wind turbines is realized by using auxiliary devices, such as reactive power compensation and energy storage equipments. Examples of using Static Var Compensator (SVC) for flicker mitigation can be found in the literature (Kubota *et al.* 2002). The use of STATCOM is shown to be superior of that of SVC, with respect to flicker mitigation (Larsson and Poumarede 1999; Saad-Saoud *et al.* 1998).

Flicker is usually evaluated, according to the International Electrotechnical Commission (IEC), over a 10-min period to give a "short-term severity value", P_{st}. For that, a flickermeter model, that incorporates weighting curves that represent the response of human eye to light variations produced in a 60W, 230V, 50Hz, double-coiled filament incandescent lamp, is presented. The output of the flickermeter is given as per unit flicker voltage, with one being the level that should cause noticeable and annoying light flicker, with the perception threshold for 50% of the human population (IEC 1997).

Harmonics

Variable-speed wind turbines used today are equipped with self-commutated inverter systems, mainly PWM inverters, using an insulated gate bipolar transistor (IGBT). The main advantage of this type of inverter is that both active and reactive power can be fully controlled, but it also has the disadvantage that it produces harmonic current, generally in the range of some kilohertz.

The measurement of the harmonic currents poses one of the biggest challenges to the measurement of power quality. They require great accuracy, even for higher frequencies, because the measurements refer to interharmonics that are in the range of 0.1 % of the rated current.

Harmonic voltages, u_h, where h demotes the harmonic order (*i.e.*, an integer multiple of 50 Hz), can be evaluated individually by their relative amplitude (Ackermann 2005):

$$u_h = \frac{U_h}{U_n}$$

According to EN (1995), the 10-min mean RMS values of each u_h have to be less than the limits given in Figure 4.41 during 95% of a week.

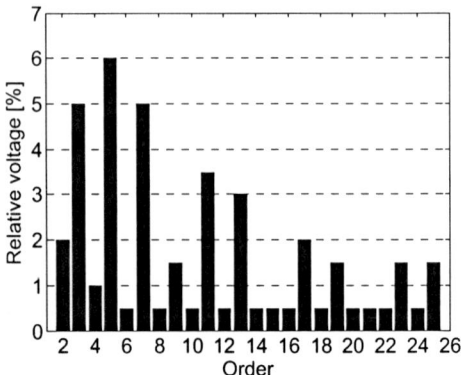

Figure 4.41. Individual harmonic voltages; percentage of the nominal phase-to-phase voltage, according to EN 50160 standard

Further, the total harmonic distortion (THD) of the voltage, calculated according to Equation 4.19, has to be $\leq 8\%$:

$$\text{THD} = \left[\sum_{h=2}^{40} (u_h)^2 \right]^{1/2} \tag{4.19}$$

With regard to higher-order harmonics, EN (1995) does not specify any limits but states that higher-order harmonics are usually small, though fairly unpredictable.

As a conclusion to the standards regarding harmonics from wind turbines, wind turbines with induction generators directly connected to the grid (without power converter) are not expected to distort the voltage waveform. However, if we consider variable-speed wind turbines with power converters, their emission of harmonic currents can be a problem regarding the power quality. In this case, filters are necessary to reduce the harmonics.

5

Design Methods for WECS Optimal Control with Energy Efficiency Criterion

5.1 General Statement of the Problem and State of the Art

As previously detailed, WECS optimal operation means roughly to extract the maximum power available in the wind stream, irrespective of the wind regimes, and takes place in the partial-load region. As the variable-pitch wind turbines operate at fixed pitch in this region, the energy optimization relies upon the same thing for both variable- and fixed-pitch cases, namely ensuring the optimal tip speed ratio, λ_{opt}. This means to track the optimal value of the shaft rotational speed, $\Omega_{l_{opt}} = v \cdot \lambda_{opt}/R$, (Equation 4.2), which is possible by variable-speed operation. Consequently, without loss of generality, only the fixed-pitch WECS case will be addressed.

Because of turbine inertia, the wind speed variations cannot be tracked precisely enough but with inadmissibly intense mechanical loads. Therefore, optimizing the dynamical regimes becomes necessary; the most important optimal control methods, split into two classes upon the type of model used, will be reviewed in the following.

In the literature, one can find various methods employed in variable-speed WECS controllers development, for various WECS types (including permanent-magnet or multipole synchronous generators, doubly-fed induction generators, *etc.*). This multitude of control approaches have emerged because of the erratic nature of the wind and of the life service reduction due to mechanical stress (Carlin *et al.* 2001), nonlinear variant behaviour of the WECS, poor reliability of some (essential) measurement information, unknown parameters/features, *etc.* Basically, the control approaches encountered in the WECS control field vary in accordance to some assumptions concerning the known models/parameters, the measurable variables, the control method employed and the version of WECS model used.

5.1.1 Optimal Control Methods Using the Nonlinear Model

Maximum Power Point Tracking (MPPT)
This technique is encountered in the literature under its acronym, MPPT. Its goal is to operate the WECS around the maximum power (within safety limits), using information from the static power characteristic and a minimum of information from the system. The power characteristic of the turbine rotor is completely unknown, but general features of WECS (such as rated power, rated rotational speed, total shaft inertia, *etc.*) are considered known. The high-speed shaft rotational speed, Ω_h, and the active power of the generator, P, are the only available measurements from the system.

Essentially, the approach is based on the computation of the power and rotational speed gradients, employed in a hill-climbing-like method. This information is used to determine $\partial P/\partial \Omega_h$ value, its sign corresponding to the position of the static operating point (OP) on the power curve in relation to the maximum of this curve (zero corresponds to the power maximum). The variable-speed control system (suggested in Figure 4.1) used in this case looks like in Figure 5.1a.

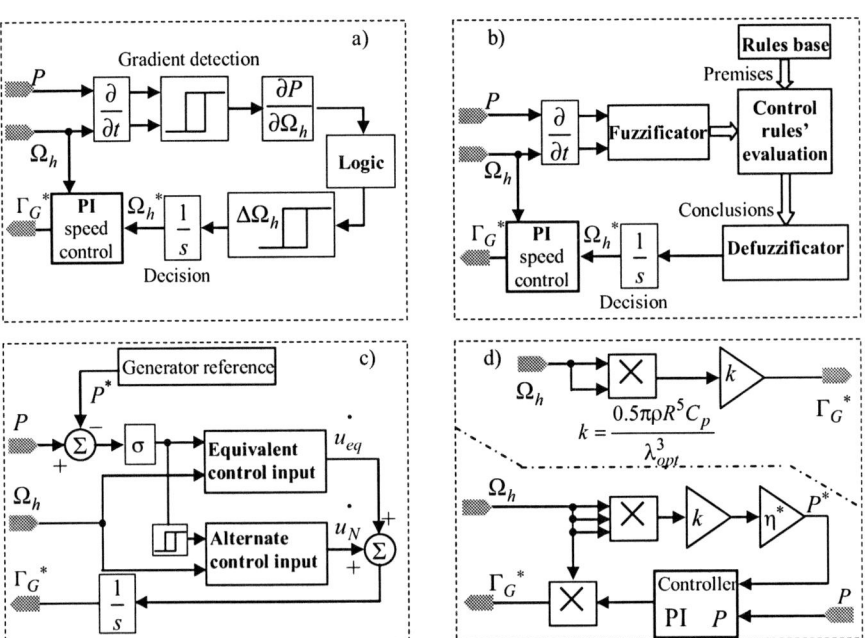

Figure 5.1. Nonlinear control approaches for WECS: **a** MPPT; **b** fuzzy; **c** SMC; **d** OOP controllers

An on-off control logic (XOR based), aiming at keeping $\partial P/\partial \Omega_h$ at small values, updates the speed reference of the generator vector control block with constant variations $\pm \Delta \Omega_h$ (Schiemenz and Stiebler 2000; Bhowmik and Spée

1998; Datta and Ranganathan 2003). Since the OP excursions around the optimal operating point (OOP) are quite large under turbulent winds, versions of this method employing wind speed estimation or adaptive control input can be used, depending on the closeness of OP to OOP and on its trend.

Even if it is robust subject to WECS parametric uncertainties and does not need much information, this method has the main inconvenient of using gradient estimations of some variables in dynamic conditions.

Fuzzy-logic Control
Maximizing the harvested power from the wind (in the same modelling assumptions) is aimed at in this case too, and the control law (evaluation of premises according to the control rules) is an extension of the MPPT method in the sense that the rules base is an extension of the MPPT-control logic. The rules base is therefore built for keeping the operating point around the optimal one (at a small value of $\partial P/\partial \Omega_h$). Value $\Delta\Omega_h$ is variable for a certain wind speed and depends on the distance between the optimal and the current operating point and on the speed variation of the latter (Simoes *et al.* 1997; Abo-Khalil *et al.* 2004; Zhang *et al.* 2004). In this way small variations of OP around OOP are obtained in steady-state operation. The dynamic response is also improved; the fuzzy controller is more flexible than an MPPT one. The most important drawback of this method is the context dependence (the wind site features, the turbine type, *etc.*), requiring quite consistent *a priori* knowledge. The corresponding variable-speed control system (Figure 4.1) used in this case is shown in Figure 5.1b.

Sliding-mode Control (SMC)
The sliding-mode or variable structure control (SMC/VSC) is a robust control method, suitable for nonlinear systems; the controller is a variable structure system which switches with high-frequency between several control laws. In particular, its output can be an on/off signal. As to well-known results from the variable structure systems' theory (Utkin 1971), the controller implements a nonlinear control (switching) law, in order to drive and maintain the system state trajectory on a desired (switching) hypersurface. Its robustness to disturbances and parametric uncertainties makes unnecessary a precise knowledge of the system; also, the VSC can be implemented using the power electronics already existing in the system. But for the applications using the electromagnetic torque as control input the mechanical stress increases due to chattering.

There is a certain difficulty about the VSC design, concerning the definition of a sliding surface with guaranteed properties of attractiveness and stability (DeCarlo *et al.* 1996; Young *et al.* 1999). VSC has been used on various WECS configurations for regulating the generated power as required by a general purpose user (Long *et al.* 1999; De Battista *et al.* 2000b; De Battista and Mantz 2004) or for power efficiency and torsional dynamics optimisation (De Battista *et al.* 2000a). Flexibility of the sliding-mode approach can be raised by adopting combined switching surfaces, in order to allow multi-criteria optimization (*i.e.*, captured power maximization and minimization of the electromagnetic torque variations).

Briefly, in the most general case the sliding surface depends explicitly on the system state, control input and time: $\sigma \equiv \sigma(\mathbf{x}(t),u(t),t)$. By adopting some simplifying assumptions (*i.e.*, by neglecting the blades' dynamics and the generator response time), the motion equation of the high-speed shaft in rigid-drive-train WECS (3.64) can be rewritten

$$\underbrace{\dot{\Omega}_h(t)}_{\dot{\mathbf{x}}(t)} = \underbrace{\frac{\Gamma_{wt}(\Omega_l(t),v(t))}{J_h \cdot i}}_{f(\mathbf{x},t)} + \underbrace{\left(-\frac{1}{J_h}\right)}_{b(\mathbf{x},t)} \cdot \underbrace{\Gamma_G(\Omega_h,t)}_{u(\mathbf{x},t)}, \quad (5.1)$$

where i is the drive train ratio. Equation 5.1 is the nonlinear state model in a form suitable to use for the sliding-mode design:

$$\dot{\mathbf{x}}(t) = f(\mathbf{x},t) + b(\mathbf{x},t) \cdot u(\mathbf{x},t) \quad (5.2)$$

The control input is obtained as a switching surface feedback, having the general form (see Figure 5.1c – DeCarlo *et al.* 1996):

$$\Gamma_G^* = u(\mathbf{x},t) = u_{eq}(\mathbf{x},t) + u_N(\mathbf{x},t), \quad (5.3)$$

where the first component is called *equivalent control input*, $u_{eq}(\mathbf{x},t)$, and satisfies the tracking condition $\dfrac{d\sigma(\mathbf{x}(t),u(t),t)}{dt} = 0$ and the second component, $u_N(\mathbf{x},t)$, is a dynamical one and results from the stability condition imposed to the closed-loop system. Stability is proved by finding an (energy) Lyapunov function, *i.e.*, positive definite with negative first derivative; typically, this is a quadratic form of the switching surface.

This method is effective and intrinsically robust, requiring relatively few information about the system and being insensitive to parametric variations. The chattering (DeCarlo *et al.* 1996), specific to VSC, is the main drawback in this case; it negatively influences the mechanical subsystem by inducing supplementary stress, and might excite the unmodelled dynamics (*e.g.*, the oscillating modes of the blades or of the flexible drive train), thus producing destructive oscillations. Ensuring a sufficiently high chattering frequency is absolutely necessary.

Direct Imposing of Optimal Operating Point (OOP) *Position*
From a static point of view, improving/maximizing energy capture below the rated power (in partial-load region) can be ensured by forcing the turbine rotor to operate at maximum power, P_{opt}, corresponding to the instantaneous wind speed. This means that the operating point is in its optimal position (OOP). Equivalently, this means imposing the electromagnetic torque which equals the wind torque corresponding to the maximum power available (denoted by $\Gamma_{wt_{opt}}$). The turbine works at maximal efficiency when turning at optimal tip speed ratio, λ_{opt}, so the maximum power is proportional to the rotational speed cubed, according to Relations 4.4 and 4.5.

Figure 5.1d (upper part) shows that the generator torque reference is obtained using a rotational speed measure and Equation 4.9. Also, if a power reference is preferred, a PI control is employed for modifying the generator mechanical characteristic and obtaining the torque reference, as depicted in the lower part of Figure 5.1d.

As the WECS control structure allows the wind speed to be tracked within admissible limits of mechanical loads, this method can be used as long as it depends only on slow wind speed variations, thus achieving a static optimisation. For turbulent winds, filtering of the variables is necessary, along with using compliant PI parameters, to ensure sufficiently slow closed-loop dynamics. This method is strongly sensitive to parametric variations.

Feedback Linearization Control
Because WECS are highly nonlinear systems, but with smooth nonlinearities, a possible optimal control design solution can be the feedback linearization control (Isidori 1989). This approach is suitable to electrical drive control (Lee *et al.* 2000); in the literature various applications to grid synchronous-generator-based energy conversion systems are also reported (Chapman *et al.* 1993; Savaresi 1999; Akhrif *et al.* 1999; Wang *et al.* 1993).

A frequent difficulty encountered by this approach is the synthesis computational complexity. In particular, in the case of WECS, the wind torque coefficient variation on the tip speed must be modelled by a high-order polynomial function, in order to capture all operating regimes (including the starting one). One must assume a simplified expression when deciding to use the feedback linearization – for example, to capture only the steady-state regime – otherwise, the design procedure is rendered extremely difficult.

5.1.2 Optimal Control Methods Using the Linearized Model

Steady-state Optimization of WECS
The control goal is formulated as maintaining the tip speed ratio at its optimal value, λ_{opt}. To this end, a rotational speed tracking system is conceived (Miller *et al.* 1997; De Broe *et al.* 1999), whose reference $\Omega_{h_{opt}}(t)$ depends on the instantaneous wind speed, $v(t)$. For a WECS with rigid drive train of ratio i Equation 4.2 gives

$$\Omega_{h_{opt}}(t) = i \cdot \Omega_{l_{opt}}(t) = i \cdot \frac{\lambda_{opt}}{R} \cdot v(t) \tag{5.4}$$

The tracking system is based upon a PI classical controller (see Figure 5.1), designed as follows. The cross-point of the mechanical characteristic and the ORC is identified; this point is placed on the falling zone of the torque characteristic. The turbine model, *i.e.*, the wind torque expression, is then linearized in the neighbourhood of this point, allowing the design of a PI (linear) controller (*e.g.*, using a pole-placement procedure for the closed-loop system). The instance of the corresponding WECS control structure is presented in Figure 5.2a. One can derive

the following remarks:
- large inertia of WECS generally does not allow the wind speed to be tracked within admissible limits of mechanical loads, but only in steady-state operation (slow wind speed variations);
- fast wind speed variations can lead to abruptly decreasing available power, even to modifying the power balance in the system, such that the induction machine works as a motor, whereas the turbine becomes a fan (Leithead 1990);
- parameters of the linearized model depend on the wind speed; therefore, choosing a PI control is even more suitable, provided its robustness to parametric variations.

Linear Quadratic (LQ) *Dynamic Optimization*
Tracking of the optimal rotational speed corresponding to a given wind speed induces variations of the electromagnetic torque, thus supplementary mechanical stress, which reduces the lifetime of the drive train mechanical parts. Furthermore, the turbine reliability is also decreased by involving supplementary maintenance costs and reducing the turbine availability.

The mechanical fatigue of the drive train can be reduced by imposing the electromagnetic torque variations, $\Delta\Gamma_G(t)$, to be minimized (Novak and Ekelund 1994; Ekelund 1997). Ekelund (1997) used the optimization criterion at Equation 4.11 complying with the minimization of the generator torque variations, which are responsible for the mechanical fatigue of the drive train.

Operation around optimality is ensured by minimizing only the first component from Equation 4.11, but allowing important torque variations (second term of the previous criterion). The positive coefficient α is introduced to adjust the trade-off between the two antagonistic demands; it confers flexibility to the control law in the following sense. If the wind turbulence is not significant, then the accent may fall on the energy efficiency of the WECS, and therefore α will take a large value. If, on the other hand, the wind turbulence is important, then, through a small value of α, the accent will fall on reducing the mechanical stress and increasing the life service of the WECS.

The optimal control input is obtained as state feedback, as shown in Figure 5.2b (DeCarlo *et al.* 1996):

$$u(t) = -\mathbf{K} \cdot \mathbf{x}(t) \qquad (5.5)$$

The gain vector **K** results from solving a Gaussian (stochastic) linear quadratic optimization problem (Athans and Falb 1966) (expressed by criterion at Equation 4.11), defined on the linearized model of the WECS (see Section 3.6), where the state vector **x** is composed of the LSS rotational speed variation and of either the wind speed variation (Ekelund 1997) or the wind torque variation (Munteanu *et al.* 2005).

The parameters of the dynamical system depend on the operating point's position on the turbine mechanical characteristic, which further depends on the average wind speed. Therefore, an adaptive control structure must be built in order to switch the gain vector, **K**, depending on the wind speed (see also Figure 4.24).

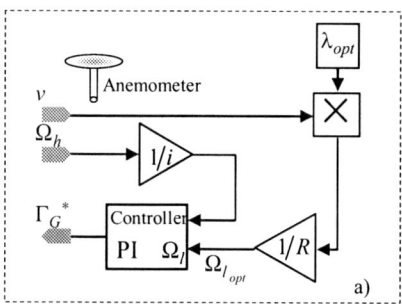 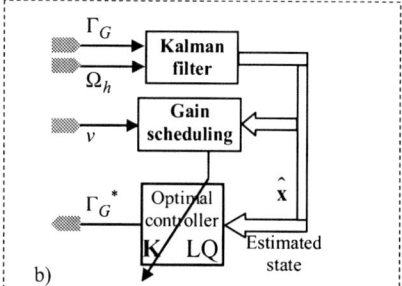

Figure 5.2. Linear control approaches for WECS: **a** steady-state optimization; **b** LQG optimization

QFT Control
An approach for the design of a robust controller can be based on the quantitative feedback theory (QFT) method (Horowitz 1993). Such a method proves its effectiveness for nonlinear systems with parameter uncertainty, using both the linearized and the nonlinear model. Its basic idea is to meet a set of prescribed performances by an iterative design procedure. Examples of successful application are, among others, flight control systems (Wu *et al.* 1998), elastic mechanical positioning systems, biotechnological processes (Skogestad *et al.* 1989) and hydraulic systems (Niksefat and Sepehri 2000). Regarding WECS, the QFT technique can be used for grid-connected (Torres and Garcia-Sanz 2004), as well as for autonomous (hybrid) wind power systems (Cutululis *et al.* 2006b), when an optimization goal is appropriately defined.

5.1.3 Concluding Remarks

The list of methods presented above is not exhaustive. The use of one or another from them depends on the envisaged control goal and on the available information about system parameters and feedback. Thus, the employed method will be more sophisticated as the control goal is more complex and feedback information is scarce.

Some of these methods are potentially more flexible and their drawbacks can be alleviated to a certain degree. Thus, the MPPT method has the advantage of employing very few parametric and feedback information in the controller construction. The PI control is the simplest but robust control technique, which uses a linearized model around a steady-state operating point. The sliding-mode control and the feedback linearization are appropriate and effective when wishing to use the nonlinear model. Requirements like robustness to parametric uncertainties can be met by using frequency-domain techniques, such as QFT robust control. Versions of these control techniques, used for energy maximization in the partial-load regime, will be presented later in this chapter. The LQG control allows the alleviation of mechanical stress, by imposing a trade-off between captured power from the wind and wind turbine reliability; this control technique will be extensively dealt with in Chapter 6 of this book. Each method is accompanied by case studies reporting MATLAB®/Simulink® simulation results.

5.2 Maximum Power Point Tracking (MPPT) Strategies

5.2.1 Problem Statement and Literature Review

The so-called MPPT is a very reliable and robust control method (Hilloowala and Sharaf 1994; Wang and Chang 1999; Schiemenz and Stiebler 2000, 2001; Tsoumas et al. 2003), which covers in fact an entire class of extremum search algorithms. Basically, the approach employs the hill-climbing method (Datta and Ranganathan 2003; Wang and Chang 2004; Tan and Islam 2004) for dynamically driving the operating point to the ORC, by using some searching (probing) signal in order to obtain gradient estimations of some (few) measurable variables.

Given that the WECS parameters (i.e., the optimal tip speed and maximum of the aerodynamic efficiency) are unknown, the MPPT algorithms generally aim at maintaining the optimal operating point by zeroing the value $\partial P_{wt}/\partial \Omega_l$, where P_{wt} is the total wind power and Ω_l is the low-speed shaft (LSS) rotational speed. Therefore, the control input, representing the wind turbine speed reference, depends on the operating point position and on its moving trend, expressed by the sign of $\partial P_{wt}/\partial \Omega_l$ (see Table 5.1 and Figure 5.3).

Table 5.1. MPPT-control logic

$\dfrac{d\Omega_l}{dt}\ \diagdown\ \dfrac{dP_{wt}}{dt}$	< 0	> 0
< 0	$\Omega_l \nearrow$ – case I	$\Omega_l \searrow$ – case II
> 0	$\Omega_l \searrow$ – case III	$\Omega_l \nearrow$ – case IV

Since the turbine power is not available for measurement, an estimation of its value, obtained based on the measured active power, is used for operating point localisation, leading to a somehow static approach. Also, for obtaining the LSS rotational speed, a measurement of the generator speed is used.

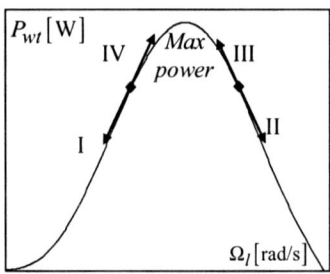

Figure 5.3. Decision cases for MPPT control on the static power curve of a WECS

Essentially, the MPPT algorithm (as stated in most of the related papers) consists of the steps listed below.

Algorithm 5.1. General form of the MPPT-control algorithm

#k-th step:
1. Measure rotational speed $\Omega_k \sim \Omega_l$, electrical power $P_k \sim P_{wt}$ and the wind speed $v_k = v$.
2. Depending on the wind speed, set the rotational speed variation step (equivalent to the searching speed), $\Delta\Omega_0$.
3. Estimate the power and speed gradients and deduce their signs: $A = \text{sign}\left((P_k - P_{k-1})/(t_k - t_{k-1})\right)$ and $B = \text{sign}\left((\Omega_k - \Omega_{k-1})/(t_k - t_{k-1})\right)$.
4. Deduce the rotational speed variation sign as $\text{sign}(C) = \overline{A \oplus B}$ (see Table 5.1), and obtain the speed variation value at step k: $\Delta\Omega_k = \text{sign}(C) \cdot \Delta\Omega_0$.
5. Obtain the control input (the HSS rotational speed reference) by integration, as $\Omega_k^* = \Omega_k + \Delta\Omega_k$; $\Omega_h^* = i \cdot \Omega_k^*$.
6. Rewind for $k \leftarrow k+1$.

Based on the operating point position on the power characteristic, the rotational speed is controlled in the sense of approaching the maximum power available. The search of the power maximum, combined with the wind speed variations and high turbine inertia, has some drawbacks, the main ones being the significant estimation errors and important high-frequency power fluctuations with negative influence on the system overall reliability.

The first drawback can be cancelled by using heuristic methods which need wind speed estimation, combined with power coefficient characteristic identification (Spée et al. 1994; Bhowmik and Spée 1998; Farret et al. 2000). As for the second drawback, one can use extensions of the MPPT algorithms, obtained by employing fuzzy control techniques (Hilloowala and Sharaf 1996; Simoes et al. 1997; Chen et al. 2000; Abo-Khalil et al. 2004), leading to extended and more flexible, but also quite context-dependent controllers. The rules base is built to maintain the operating point at a low value of $\partial P_{wt}/\partial \Omega_l$ and represents an extension of the control logic presented in Table 5.1. The step variation of the rotational speed, $\Delta\Omega_0$, is variable and depends on the operating point distance from the optimal operating point and on its movement direction and speed.

The *extremum seeking control* (ESC) method relies upon finding the extremum of some unimodal hard-to-model dynamics by exciting the plant with some sinusoidal probing signals (Åström and Wittenmark 1995; Krstič and Wang 2000; Ariyur and Krstič 2003). The basic functional diagram of this control method is given in Figure 5.4 (Krstič and Wang 2000). This figure shows that the controller performs a modulation/demodulation operation and that its output has a harmonic component called the *probing signal*. It usually contains a washout filter (for separating the high-frequency components of the plant output), a demodulator, a low pass filter and an integrator for obtaining the average component of the control input, and a summation with the probing signal.

Let us consider that the plant contains a dynamic described by a generic unimodal function denoted by $f(\lambda)$, having a maximum at λ_{opt}. Let us assume that the argument of function f has two components: an average one, $\bar{\lambda}$, and a harmonic probing component, of amplitude a (Figure 5.4).

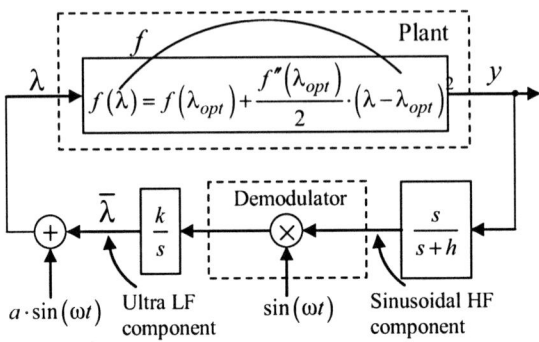

Figure 5.4. Explaining the general ESC principle

Consider then the Taylor series of this function around its maximum, $f(\lambda_{opt})$, with $\tilde{\lambda} = \lambda_{opt} - \bar{\lambda}$ denoting the optimum searching error. Given that the integrator constant, k, and the excitation amplitude, a, are positive and that $f''(\lambda_{opt})$ is negative, based on the diagram described above, one can deduce that the search error gradient is negative (Ariyur and Krstič 2003):

$$\dot{\tilde{\lambda}} = \frac{k \cdot a^2 \cdot f''(\lambda_{opt})}{4} \cdot \tilde{\lambda}, \qquad (5.6)$$

and, therefore, the searching process is convergent. One must note that the excitation frequency, ω, must be sufficiently large to ensure stability of the closed-loop system; the washout filter parameter, h, depends on this frequency (Ariyur and Krstič 2003).

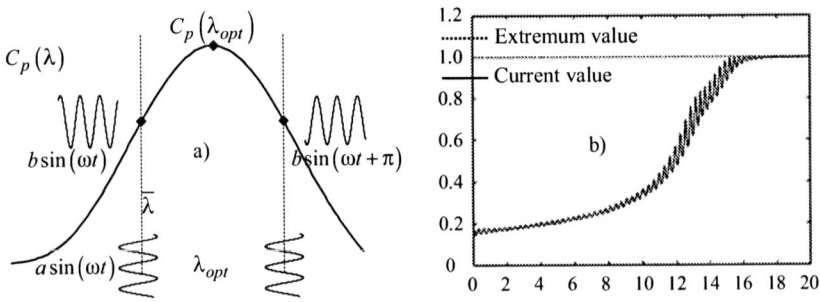

Figure 5.5. Explaining the maximum search idea applied to WECS

5.2 Maximum Power Point Tracking (MPPT) Strategies

Consider now the case of a wind turbine having an aerodynamic efficiency described by the power coefficient, C_p, as in Figure 5.5, and consider a hypothetic ω-frequency sinusoidal variation of the tip speed and a sufficiently small amplitude a. Depending on the position of the operating point (OP) on the slope of the aerodynamic efficiency curve, the $C_p(\lambda)$ variation will be sinusoidal, in phase with the λ variation for the ascending part and with a phase lag of π for the descending part (the intermodulation components for small a are neglected). Simple calculations give that the integrator input, $d\bar{\lambda}/dt$, will toggle its sign as the OP moves from a side to the other of the C_p optimum (see Figure 5.5a), and, assuming equal slopes of the $C_p - \lambda$ curve, its value – and consequently the search direction – varies as in Equation 5.7:

$$d\bar{\lambda}/dt = \pm k \cdot b \cdot \sin^2(\omega t) \qquad (5.7)$$

Therefore, the OP will move to the optimal position and the speed of convergence depends proportionally on k, a and $1/\omega$ (Ariyur and Krstič 2003).

This principle can be used in a WECS optimal control application. Namely, instead of feeding the system with a sinusoidal probing signal, one can use an already existing perturbation: the wind speed turbulence. In this case, the modulation process is naturally achieved by means of (nonharmonic) high-frequency wind variations. Let us also note that using the ESC in its standard form, with a sinusoid as probing signal, would have in the case of WECS the following drawbacks. Given that the plant is naturally excited by a random signal, namely the wind, it would be difficult to separate the answer to the probing sinusoid from the total output signal, which appears as random too. This can however be avoided by injecting a large magnitude signal and/or by filtering. But large signals could create reliability problems by inducing supplementary mechanical loads, whereas filters provide delayed feedback information.

A new method for driving the operating point to the energy maximum, based upon extremum seeking with the wind turbulence as search excitation (probing) signal, is next presented. Numerical simulation tests suggest that the new form of the MPPT algorithm thus obtained is satisfactory, provided that information about the system state is poor.

5.2.2 Wind Turbulence Used for MPPT

Modelling Assumptions
The method approached here is intended to be applied to various types of variable-speed WECS, having different aerodynamic, transmission and generation configurations.

Figure 5.6 presents the case of a SCIG-based WECS subjected to MPPT control. The approached system presents a higher level control loop superposed on the generator torque control: the HSS rotational speed control. This allows the direct control of the tip speed ratio, as required by the principle detailed in

Figure 5.4. Operating an induction machine by using vector control is a well-known and widely-employed technique (Bose 2001; Leonhard 2001), which supposes a PI controller being utilized in conjunction with the torque generator control (see Section 4.7.1). As the torque gradients must not exceed certain limits imposed by reliability requirements, the PI speed controller parameters are designed for narrow-frequency passband of the speed loop; moreover, the speed reference gradient is also limited.

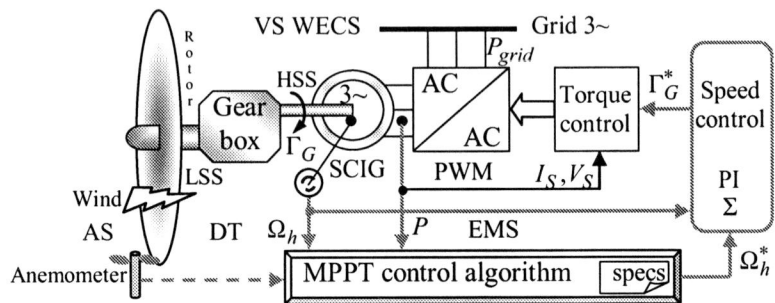

Figure 5.6. The MPPT-controlled SCIG-based WECS

At the general modelling frame (from Chapter 3) some supplementary assumptions are added, as listed bellow:
- measures of the wind speed, rotational speed and electrical power are available;
- the blade length, R, the transmission step, i, an estimation of the turbine's rotor inertia and the SCIG parameters are known, whereas the aerodynamic efficiency curve is totally unknown;
- the turbulence intensity varies in the usual range, *i.e.*, [0.12;0.18] (Burton *et al.* 2001); given that the considered WECS is of low power, the rotational sampling effect is negligible and the wind is characterized by its fixed point spectrum;
- a reliable estimation of the captured power can be obtained using the electrical power, $P = \eta \cdot P_{wt}$, since the power conversion efficiency, η, is considered constant in the concerned rotational speed range;
- time moments when control inputs are applied to the system are sufficiently rare compared to its dynamic (*i.e.*, a rotational speed reference, Ω_h^*, is practically instantaneously transmitted at the low-speed shaft, as compared with the variations of the low-frequency wind speed component, v_s).

Using the previous assumptions and remarks in the general modelling framework discussed in Chapter 3, the model of the variable-speed WECS can be sketched as in Figure 5.7.

5.2 Maximum Power Point Tracking (MPPT) Strategies

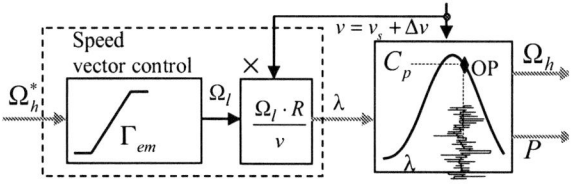

Figure 5.7. The WECS model used for MPPT

Available Feedback Information Processing
The stated scope of the control can be achieved by changing the tip speed ratio. From Equation 2.30 (in Section 2.3.2) the tip speed can be expressed in variations:

$$\Delta\lambda = \frac{R}{v} \cdot \Delta\Omega - \frac{\lambda}{v} \cdot \Delta v, \qquad (5.8)$$

therefore a variation of the tip speed depends on the rotational speed change, but also on the wind change. A variation of rotational speed will not be able to ensure by itself a targeted variation of the tip speed, since the wind speed is not controllable. The wind turbulences, together with rotational speed variations around a steady-state value, induce some tip speed variations around its average value and consequently some power coefficient variations, ΔC_p, which are non-harmonic, but have a bounded spectrum.

The tip speed ratio, $\lambda(t)$, can be computed using measures of the wind speed and rotational speed, as to its definition relation (Equation 2.30). Under the previously stated modelling assumptions, the power coefficient can be estimated from the electric power, $P(t)$, and the wind speed using the following relation:

$$C_p(t) = P(t) \big/ \left(0.5\pi \cdot \rho \cdot \eta \cdot R^2 \cdot v(t)^3\right), \qquad (5.9)$$

with η being the non-unitary efficiency of the mechanical to electrical power conversion.

Below, normalized variations of these two signals, in relation to their maximum values, will be used. Suppose now a Fourier decomposition of $\lambda(t)$ and $C_p(t)$ signals; each $\lambda(t)$ harmonic component will generate a response which is a part of $C_p(t)$. The method presented in the previous section can be adapted to this new situation by taking the components of the Fourier decomposition and then by composing/averaging their effects. The OP position on the C_p curve is then obtained by demodulation, which is possible by using one of the following approaches:

1. By using the Discrete Fourier Transform (DFT) for extracting the phase of each harmonic component of $\lambda(t)$ and $C_p(t)$ and then compute the phase lag of each component. An average of these values gives the angle – further denoted by $\theta(t)$ – containing the OP average position information. Values of $\theta(t)$ will get

closer to 0 if the OP is on the left (rising) slope of the C_p curve or closer to π if the OP is on the right (falling) one.

2. By convoluting $\lambda(t)$ and $C_p(t)$ signals, as the product of their harmonic components (obtained by DFT) is equivalent with a time convolution operation (Oppenheim *et al.* 1996).

The control method exposed in the following will use the first approach in obtaining the OP position information. Thus, instead of computing the *instantaneous* gradients – which is not an obvious task in dynamic conditions – the method used here is based upon deducing the *average phase shift*, denoted by θ, between the power coefficient and the tip speed, by computing the DFTs of the two signals on an appropriately chosen time window. Hence, zone 1) from Figure 5.8 will be characterized by $\theta<\pi/2$ and zone 2) by $\theta\geq\pi/2$.

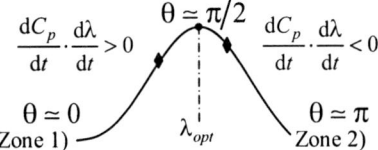

Figure 5.8. Feedback on the operating point position

One can note that the control is implemented using an integrator fed by the OP position information, which gives the direction of search, whereas the search step is the integration coefficient.

As the C_p curve is somehow flat, the OP must move smoothly; this action is not well illustrated by the binary information signal (θ(t) angle), which evolves abruptly. So, around the C_p maximum, the θ(t) signal is more suitably processed using a continuous function instead of a relay as in Relation 5.7. Even if the C_p curve is in general non-symmetric, the shape of the continuous function employed in the processing of θ(t) signal can in a first approximation be symmetric, because the C_p shape is unknown anyway. Therefore, a slight static error of the average tip speed *vs.* its optimal value is expected.

The system behaviour depends primarily on the turbulence wind speed evolution; meanwhile, the concerned measured information is updated at each sampling period, further denoted by T_S, which depends mainly on the turbine dynamics. The average position signal, θ(t), must therefore be computed, using the Fast Fourier Transform (FFT), within a time window of width $T_C = 2^n \cdot T_S$, where n is a positive integer resulting from a trade-off between suitably tracking the wind speed's variations and complying with the system inertia. It is obvious that a given system will not be ever made to track an unlimitedly turbulent wind.

Control Algorithm
The control law logic is based on the following idea. The control input tends to move slowly (as compared with the turbine's dynamics) the average OP to the top

of the aerodynamic efficiency curve, by regulating the average value of the tip speed computed on a relatively large time window. Therefore, we expect some large deviations of the instantaneous tip speed around its optimal value, λ_{opt}, whereas its average value will remain close to λ_{opt}.

The control structure, presented in Figure 5.9, has two main parts, namely an information processing block and a rotational speed reference generator.

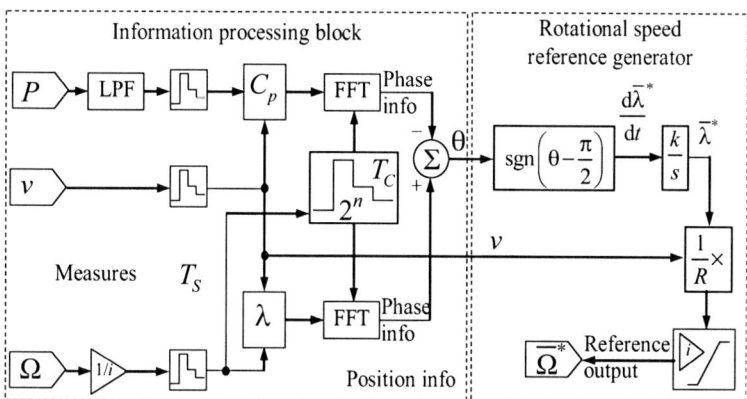

Figure 5.9. MPPT-control structure: feedback information processing and generation of the rotational speed reference

When the OP is placed on the positive slope of the C_p curve, the integrator will process a positive value, giving an increased tip speed reference. In contrast, when the OP is on the negative slope of the C_p curve, the tip speed reference must diminish. This can be achieved by integrating a nonlinear function of average phase information, $\text{sgn}(\theta - \pi/2)$.

On the left side of diagram from Figure 5.9, $C_p(t)$ and $\lambda(t)$ normalized signals are obtained using the electrical power, rotational speed and wind speed measurements. These signals – passed trough a Blackman window for avoiding the Gibbs effect – are feeding a FFT algorithm for obtaining their phase spectrum and the corresponding average position signal, $\theta(t)$. For information consistency, a sufficiently large window (compared to the turbulence dynamics) must be chosen.

As the OP position changes smoothly around the OOP (the slope of the aerodynamic efficiency is continuous), in the second part of the control scheme a preliminary processing of $\theta(t)$ around the optimal OP is necessary and can be made as Figure 5.10 suggests. Namely, instead of discontinuous feedback information in the form of the sign function a continuous approximation of slope m can be used. The result of this operation, applied to the integrator, produces a tip speed reference, renewed every T_C seconds after the computation of the phase shift, θ. The control structure outputs the rotational speed reference, obtained from the tip speed reference. For technical reasons concerning the generator control, low-pass filtering and rate-limiting operations are necessary.

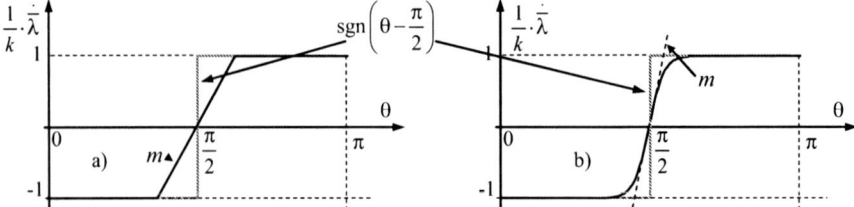

Figure 5.10. Feedback on the operating point position: **a** linear approximation; **b** hyperbolic tangent approximation of the sign function

Once the computing time T_C is established, for reasons concerning the turbulence dynamics, the key parameter is the *integrator coefficient k*, which results from T_C and the desired search speed. A good performance of this control structure can be judged by a sufficiently small standard deviation of the tip speed around its average value. The value of k can thus be chosen such that the variable-speed turbine compensates the tip speed standard deviation in the computing time T_C. Smaller values of m induce smoother behaviour around the optimality.

5.2.3 Case Study (2): Classical MPPT *vs.* MPPT with Wind Turbulence as Searching Signal

A low-power variable-speed fixed-pitch WECS has been used here as case study; its features can be found in Appendix A and Table A.2. This WECS has been subjected to both classical MPPT control and to the MPPT control using the wind turbulence as searching signal. Below some simulation results are discussed comparatively. Both sets of simulations have been done for a wind sequence having the average speed of about 8 m/s and a medium turbulence intensity of I=0.15 (Figure 5.11), obtained using the von Karman spectrum in the IEC standard.

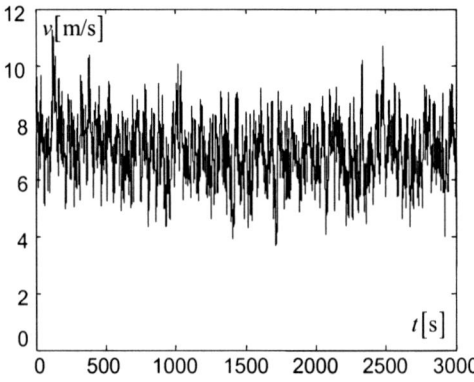

Figure 5.11. A 50-min wind speed sequence used for assessing the MPPT-control laws

As regards the use of the classical MPPT control, the rotational speed has been adjusted by steps of $\Delta\Omega_0 = 5$ rad/s every $T_s = 2$ s. Figure 5.12 shows the tracking performance of this algorithm as the wind speed varies, *i.e.*, tracking of the rotational speed reference (Figure 5.12a) and of the ORC (Figure 5.12b).

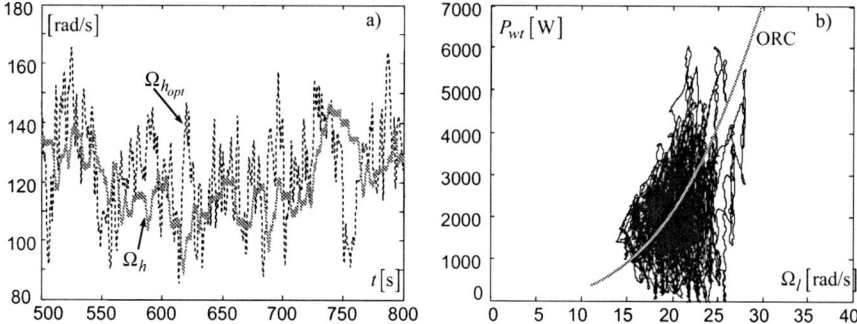

Figure 5.12. Performance of the classical MPPT control: **a** tracking the optimal rotational speed; **b** achieving energy optimization by tracking the ORC

Figure 5.13a presents a histogram of tip speed values in the same time interval, showing that values around the optimal one ($\lambda_{opt} = 7$) appear the most often. The tip speed values have a standard deviation of $\sigma_\lambda = 1.31$ around a mean of $\overline{\lambda} = 7.32$. Figure 5.13b displays the corresponding evolution of the power coefficient.

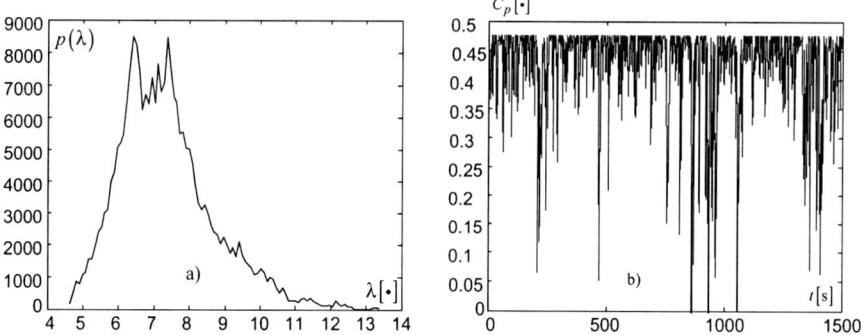

Figure 5.13. Maximizing the energy efficiency by classical MPPT control: **a** histogram of tip speed values as the wind speed varies; **b** corresponding evolution of the power coefficient

An improvement of the simulation results may be noted when applying the MPPT-control method presented above, which uses the wind turbulence as searching signal, to the same WECS. The turbine has variable-speed capability, being equipped with a speed controller based on a vector control structure. The tests concern only the partial-load region for medium wind turbulence and no wind gusts.

The sample time has been chosen here as $T_S = 0.1$ s and the FFT has been performed on a 256-element sequence; therefore the time for computing the average lag between the tip speed and the power coefficient, θ, $T_C = 256 \cdot T_S$, has been about 25 s. Parameters of the PI controller are $K_p = 1$, $K_I = 4$.

Figure 5.14 presents the change in the θ angle as the tip speed increases. In order to obtain this result, the rotational speed is imposed to vary in ramp between 20 rad/s and 250 rad/s, while the wind speed has an average component of 8 m/s and an indispensable turbulence component with medium intensity and constant statistical parameters.

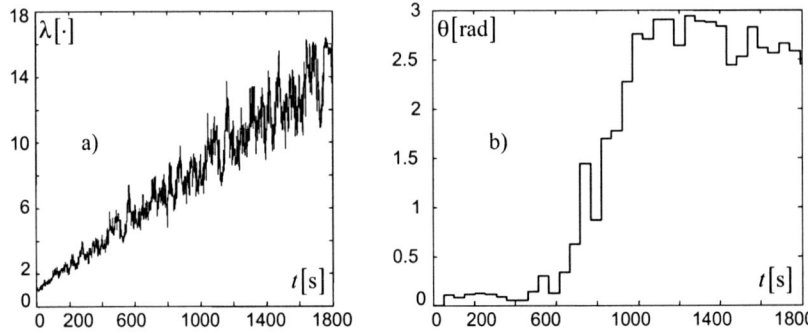

Figure 5.14. Information pre-processing for MPPT with wind turbulence as searching signal: **a** the considered tip speed variation; **b** the corresponding position signal, θ

The tip speed evolution over 30 min is presented in Figure 5.14a. The corresponding evolution of the θ angle can be seen in Figure 5.14b, confirming the possibility of deciding the operating point's position relative to the optimal operating point, either at the left side if $\theta < \pi/2$ or at the right side if $\theta > \pi/2$.

For the following two figures, the controller gain has been chosen as $k=0.02$. Figure 5.15a presents how the actual rotational speed of the turbine differs from the optimal speed imposed as reference.

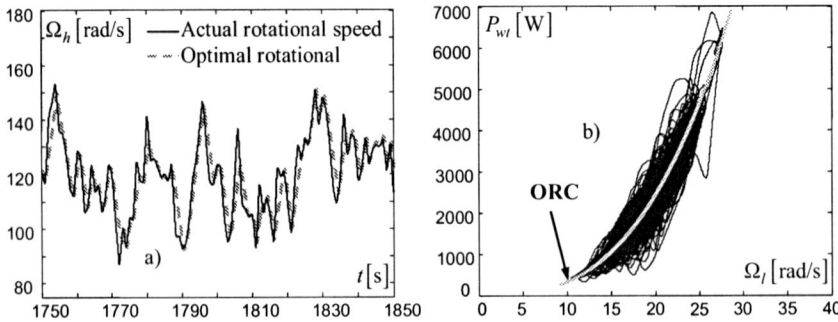

Figure 5.15. Performance of the MPPT control with wind turbulence as searching signal: **a** tracking the optimal rotational speed; **b** achieving energy optimization by tracking the ORC

In Figure 5.15b one could remark the operating point distribution around ORC (in speed-power plane). Both figures show a better performance of the control law as compared with the classical MPPT version (see Figure 5.12).

Time evolutions of some variables illustrating the quality of tracking the maximum power point are presented in Figure 5.16.

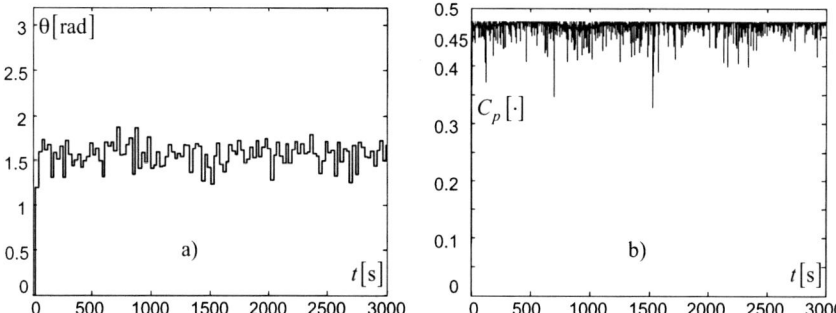

Figure 5.16. Energy efficiency maximization in case of MPPT control with wind turbulence as searching signal: **a** evolution of the feedback signal as the wind speed varies; **b** corresponding evolution of the power coefficient

The oscillations of the signal $\theta(t)$ around $\pi/2$ describe the excursion of the operating point between the stall and the normal operating region, as suggested by Figure 5.16a. In Figure 5.16b the time evolution of the power coefficient is presented, corresponding to a tip speed average value of $\bar{\lambda} = 6.75$ and a standard deviation of $\sigma_\lambda = 0.36$, smaller than in the case of the classical MPPT control.

Simulations with different values of the control parameter, k, in the range from 0.005 to 0.5, have been also performed. Figure 5.17a shows how the tip speed's density of probability is changing.

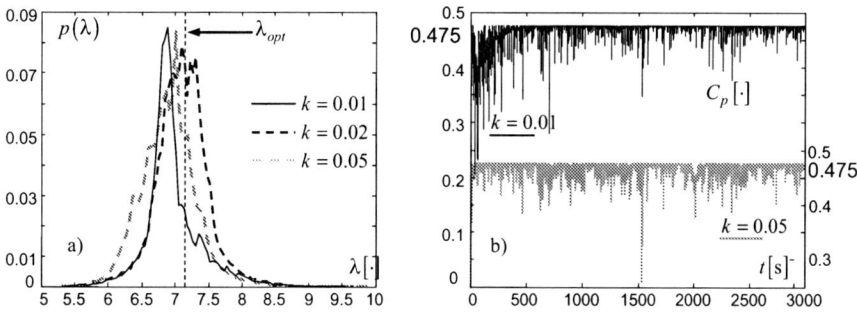

Figure 5.17. Influence of the integration parameter, k, on the quality of MPPT with wind turbulence as searching signal

For small k (small searching speed), one obtains smaller standard deviations, but the average value differs from the optimal one, which denotes poor energy efficiency. For large k (large searching speed), the standard deviation becomes large even if the mean of the tip speed remains close to the optimal value, meaning poor efficiency too. Figure 5.17b suggests that, once the system is in steady-state regime, a small value of k can produce better performance. These remarks suggest the existence of a range of k values for which closeness to optimality can be successfully designed: a good tracking of the maximum power along with a sufficiently small variance.

Details about the MATLAB®/Simulink® implementation of this case study can be found in the folder case_study_2 from the software material.

5.2.4 Conclusion

As MPPT methods prove their efficiency when the plant is not sufficiently and/or precisely known, their improvement concerns mainly the measure information processing. Classical MPPT also has the drawback of requiring real-time computation of gradients. An idea of improvement was that, based on measurements of the wind speed, the electrical (active) power and the rotational speed, one can compute the power coefficient, $C_p(t)$, and the tip speed, $\lambda(t)$, normalized signals over a sufficiently large time window. If the operating point on the $C_p - \lambda$ characteristic is on the stalling (ascending) part of the aerodynamic efficiency curve, the two signals will be in phase, otherwise (on the descending slope), their phase lag will be about π. To decide the position of the actual operating point in relation to the optimal one, a phase discrimination is performed, by computing the FFT of the two signals, extracting the phase information and computing the average phase lag, $\theta(t)$. The rotational speed reference is given by the output of an integrator whose gain represents the optimum searching speed; its sign is given by signal $\theta(t)$.

As suggested by simulations, the above improved MPPT method has shown better control performances as compared to the classical MPPT, that can be obtained with a reasonable control effort. The control input's variations are of low-frequency, adding small mechanical stress.

The most valuable feature of the presented method appears to be the use of minimal knowledge about the system's state and parameters. The control algorithm has been designed while neglecting the system dynamics and this has resulted in a nonzero steady-state error. Indeed, the method cannot ensure an average value of the tip speed equal with the optimal one, as the plant dynamics introduce an unknown phase lag. The method becomes more effective as the turbulence level increases. As this latter increases with average wind speed, the control efficiency is expected to increase at large wind speeds.

An issue to address further is to get the operating point's position by convoluting measurements of the active power and rotational speed; thus, the wind speed measure would no longer be necessary. This issue is especially interesting for high power WECS, where the wind speed varies across the swept area. The influence of the plant dynamics on the average phase lag signal and on ensuring the searching convergence should also be considered.

5.3 PI Control

5.3.1 Problem Statement

The classical PI (proportional integral) control is widely used in industry applications, owing its popularity to some key features (Åström and Hägglund 1995). Its design procedure is simple enough, requiring little feedback information and giving rise to solutions easy to implement, with intrinsic robustness properties and which can be employed over most plants having smooth models, in conjunction with other control and modelling techniques such as linearization, gain scheduling, *etc.*

As stated before, for fixed-pitch turbines operating in partial load, maximum energy capture available in the wind can be achieved if the turbine rotor operates on the ORC. Equivalently, the tip speed ratio must be made to equal the optimal value. This regime can be obtained by tracking some target variables: the optimal rotational speed, depending proportionally on the wind speed, or the optimal rotor power, which depends proportionally on the rotational speed cubed (Relations 4.2 and 4.4). Following the same idea, one can impose the wind torque proportionally with the rotational speed squared:

$$\Gamma_{wt_{opt}} = \Gamma_{ref} = K \cdot \Omega_l^2, \tag{5.10}$$

with K defined by Relation 4.5.

The use of the controller can easily be extended in the full-load region by imposing a constant torque value as target. Some elements for PI controlling the generator torque are given in Bossanyi (2000) and Muljadi *et al.* (2000). One can envisage three kinds of control loops for tracking the ORC, as follows.

Based on the wind velocity, rotational speed measurement and an inner torque control loop, a tip speed ratio loop can be built. The PI controller zeroes the difference between the target and the measured rotational speed and imposes the generator torque reference (Figure 5.2a). One can expect large torque variations, as the torque demand varies rapidly in this configuration.

Based only on the rotational speed feedback, a torque control loop can be built using as reference Expression 5.10 (see also Figure 5.1d) which is often called the $K \cdot \Omega^2$ law (Connor and Leithead 1993; Pierce 1999). For the induction machine, the PI control of the generator torque can be found in the vector control scheme ("i_{Sq} controller" – see Figure 4.29). If a PMSG is considered, the torque is controlled directly by manipulating the current in the DC circuit. An advantage of this control structure is the increased mechanical compliance of the system.

An active power loop can also be built using again the measured rotational speed, the active power and the inner torque control loop. The target power results from Equation 4.4 and the controller output represents the torque demand. By zeroing the power error, the operating point moves to the maximum power point (Muljadi *et al.* 2000; Burton *et al.* 2001).

5.3.2 Controller Design

This section aims at giving some guidelines to the classical/PI design of wind turbine control in the partial-load region, for the three above-mentioned cases.

Torque Loop
The control structure is formed by the reference generator in Figure 5.1d – upper part and the vector control scheme in Figure 4.29. The torque target value is given by the relation below:

$$\Gamma_{ref} = \frac{K}{i} \cdot \left(\frac{\Omega_h}{i}\right)^2 \cdot \eta, \qquad (5.11)$$

where K is given by Equation 4.5 and η denotes here the total mechanical efficiency of the power train.

As the rotor speed varies quite slowly, no special requirements are needed for the PI torque controller tuning. One expects both low-frequency and amplitude torque variations corresponding to the wind speed variations and therefore quite poor tracking of the ORC.

Speed Loop
The controller design is based upon the WECS linearized model (Equation 3.75), where the steady-state (linearization) point corresponds to maximum energy efficiency for the wind speed in the middle of its variation range (*i.e.*, about 7 m/s).

The simplified closed-loop structure is shown in Figure 5.18, where the power train model is essentially the one given by Equation 3.75, having as main time constant $T_{pt} = 1/K_2$ and as gain $K_{pt} = K_1/K_2$.

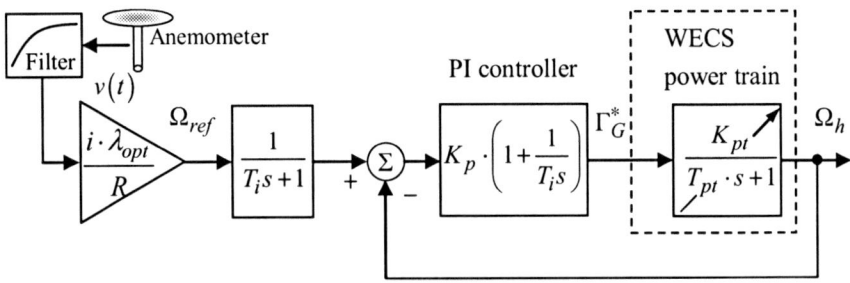

Figure 5.18. WECS PI control structure: speed loop case

The PI tuning follows the pole-placement procedure (Åström and Hägglund 1995). The closed-loop system with PI controller will exhibit a two-pole-one-zero dynamic. The second-order dynamic is imposed by choosing the natural frequency, ω_n, and the damping factor, ζ which further gives the controller parameters:

$$\begin{cases} T_i = \dfrac{2\zeta}{\omega_n} - \dfrac{1}{\omega_n^2 \cdot T_{pt}} \\ K_p = \dfrac{T_i \cdot T_{pt}}{K_{pt}} \cdot \omega_n^2 \end{cases} \quad (5.12)$$

A high K_p will thereby ensure better tracking performance; however, one must take account of control effort (torque) limitations, so values of K_p must also be limited. The zero effect, of increasing the overshoot, is compensated by first-order filtering the reference signal (Figure 5.18).

Although the steady-state speed error is zero, there will always be nonzero dynamical errors due to tracking the significantly variable reference signal, Ω_{ref}.

One must note that the imposed closed-loop performances are guaranteed for the chosen operating point. As to Section 3.6.1, both the gain and the time constant of the torque controlled system around a certain steady-state operating point depend on that operating point (through wind velocity and rotational speed). In conclusion, the dynamic performances of the tracking system also vary upon the operating point.

Power Loop

Here the input of the plant is the electromagnetic torque, its output being the generator active power, P. The controller design is based on the parameterization of the system response at step changes in generator torque for a given wind speed (Figure 5.19a). In the first moment, the negative torque step results in a step power change, the rotational speed being the initial one. Then, the difference between the wind torque and the electromagnetic torque produces the rotational speed increasing, such that the power increases according to the speed dynamics until it reaches the new steady-state value. One can note that the plant dynamics in the case of the power control loop sensibly differ from those corresponding to the speed control loop case (see Figure 5.18).

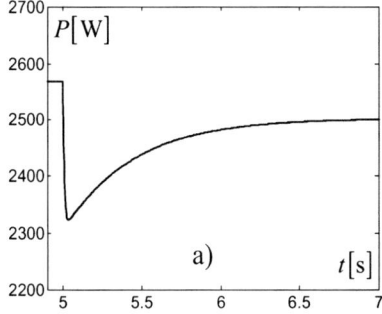
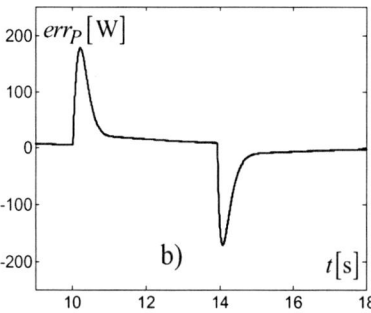

Figure 5.19. **a** WECS step response from Γ_G to P. **b** The power tracking error for step changes in wind speed

One can deduce that the dynamic behaviour is well enough captured by a two-pole-one-zero transfer function, as follows:

$$H_{pt}(s) = \frac{K_{pt} \cdot (T_z s + 1)}{(T_\Sigma s + 1) \cdot (T_{pt} s + 1)}, \quad (5.13)$$

with T_{pt} and T_Σ being the main and respectively the parasitic time constant. The closed-loop structure is shown in Figure 5.20.

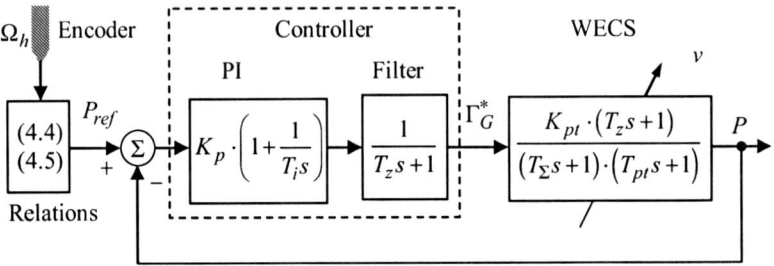

Figure 5.20. WECS PI control structure: power loop case

The controller results according to the following procedure. The main time constant is compensated by taking $T_i = T_{pt}$, whereas the plant zero is compensated by a first-order filter having T_z as time constant, embedded in the controller. Therefore, in this case the controller will exert a filtered PI action, such that the closed-loop system will have a second-order dynamic. Then, one suitably imposes this dynamic, T_{cloop}, and computes the controller gain:

$$K_p = \frac{T_{pt} \cdot T_\Sigma}{K_{pt} \cdot T_{cloop}^2} \quad (5.14)$$

High K_p values ensuring good tracking performance must nevertheless be kept under a certain limit for reliability reasons. For reasonable high values of K_p one can obtain a power error response at step changes in wind speed, err_P, as in Figure 5.19b.

As in the previous case, the imposed closed-loop performances are guaranteed for the chosen operating point. Because the plant parameters vary, the closed-loop dynamic performances are expected to change with the operating point.

5.3.3 Case Study (3): 2 MW WECS Optimal Control by PI Speed Control

A high-power variable-speed fixed-pitch WECS – with features given in Appendix A and Table A.7 – has been used as case study in order to illustrate the design of the PI speed controller.

The system parameters have been identified based on the step response from

Γ_G to Ω_h under constant wind speed at 8 m/s; thus, the gain is $K_{pt} = 0.012$ and the time constant is $T_{pt} = 10.5$ s.

A smaller time constant for the closed-loop system has been imposed, namely $0.24 \cdot T_{pt}$, which gives $\omega_n = 0.4$ rad/s. If the damping factor is chosen as $\zeta \approx 0.7$, the controller parameters results as to Relation 5.12: $T_i = 3$ s and $K_p \approx 412$ – this controller will be referred later as C_1. Figure 5.21a presents the corresponding generator speed error evolution in response to a step change in wind velocity (from 7 m/s to 8 m/s), whereas the aerodynamic efficiency evolves as can be seen in Figure 5.21b.

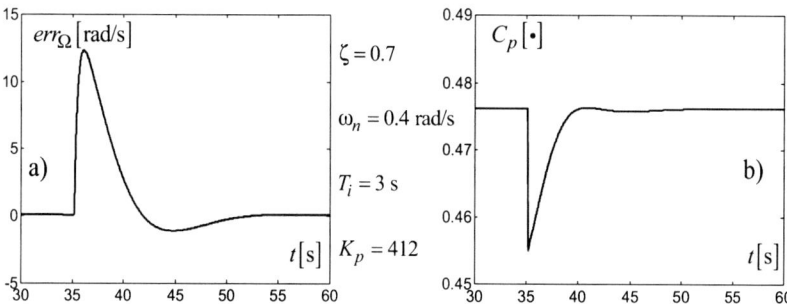

Figure 5.21. Closed-loop system response when switching wind speed from 7 m/s to 8 m/s: **a** speed tracking error; **b** variation of C_p

The closed-loop system behaviour has been simulated under a realistic wind profile, namely for a wind sequence having the average speed of about 7.5 m/s and a medium turbulence intensity of $I=0.15$, obtained using the von Karman spectrum in the IEC standard. Figure 5.22 presents the closed-loop system performances in tracking the ORC (Figure 5.22a) and the corresponding difference from the optimal rotational speed (as to Equation 4.2 – Figure 5.22b).

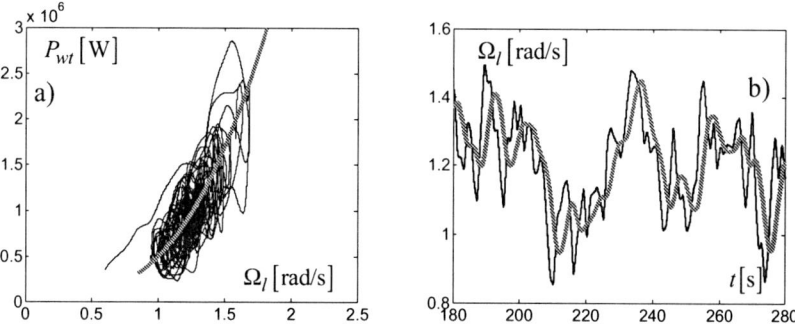

Figure 5.22. Performance of PI speed control system: **a** tracking of the ORC; **b** tracking of the optimal speed reference (reference – *thin line*; actual speed – *thick line*)

Figure 5.23 shows the change in performance and in the control effort when the control parameters are switched from the ones corresponding to C_1 to another set, denoted by C_2: $T_i = 1.88$ s and $K_p \approx 733$. The latter has been obtained for $\omega_n = 0.66$ rad/s and $\zeta \approx 0.7$ using the same computation frame. One can note that a greater control effort (Figure 5.23a) does not produce significant improvement in control performance (Figure 5.23b).

If imposing very high ω_n with respect to the plant dynamic, the power variations become unacceptable; moreover, the electromagnetic torque can take positive values, thus leading to abnormal regimes of the electrical machine.

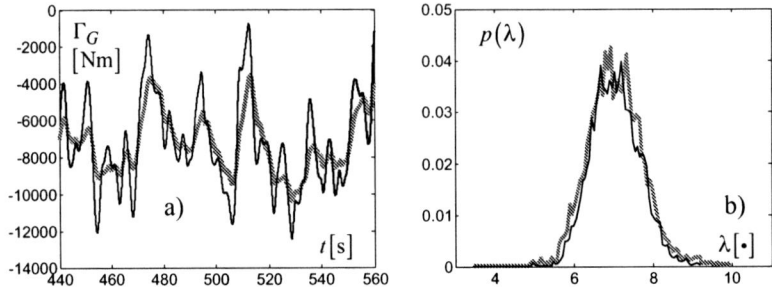

Figure 5.23. Comparison between two PI speed controllers: C_1 – *thin line*, C_2 – *thick line*

In order to find out details about the MATLAB®/Simulink® implementation of this case study, the reader can consult the folder case_study_3 within the software material.

5.3.4 Case Study (4): 6 kW WECS Optimal Control by PI Power Control

In order to illustrate the PI power controller design, a low-power variable-speed fixed-pitch WECS – with features given in Appendix A and Table A.2 – has been used as case study.

The system parameters have been identified based on the step response (see Figure 5.19a) when input Γ_G switched from –18 Nm to –20 Nm under constant wind speed of 8 m/s. The following set of parameters has been obtained: $K_{pt} = 35$, $T_{pt} = 0.36$ s, $T_\Sigma = 0.01$ s, $T_z = 1.42$ s. By imposing $T_{cloop} \approx 0.1 \cdot T_{pt}$ results $K_p \approx 0.1$, according to Equation 5.14. For this K_p the power error evolution at step changes in wind speed (from 7 m/s to 8 m/s and back) looks like in Figure 5.19b. K_p ranges between 0.05 and 5, being limited by an unstable behaviour for small values and by control effort magnitude for high values. The closed-loop response can exhibit oscillations for higher values of K_p.

The closed-loop system behaviour has also been simulated under the same wind velocity sequence as in the previous case study. Figure 5.24 contains the main simulation results showing the control law performance.

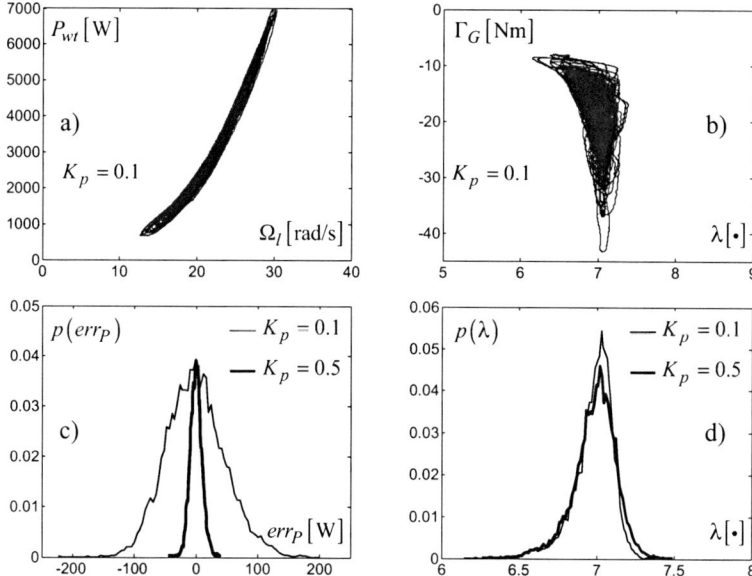

Figure 5.24. Performance of PI control system in the power loop case

The variance of the operating point around ORC is satisfactory (see Figure 5.24a). The closed-loop performs better in high than in low winds, as shown in Figure 5.24b. The power tracking quality increases as the K_p value increases, as reflected in Figure 5.24c. However, even if the power error reduces when increasing K_p, the performance in ensuring $\lambda_{opt} = 7$ does not increase accordingly (Figure 5.24d). This is because the optimisation is indirectly achieved, based on measuring the rotational speed and on tracking the corresponding active power. One can conclude that the choice of K_p is not critical in fact, high values will eventually induce supplementary mechanical stress without greatly improving the aerodynamic efficiency.

The files containing the MATLAB®/Simulink® implementation of this case study are grouped together in the folder `case_study_4` from the software material.

5.4 On–Off Control

5.4.1 Controller Design

This approach supposes that the WECS reacts sufficiently fast to variation of the low-frequency wind speed; this actually happens in the case of low-power WECS. Thus, for ensuring the optimal energy conversion it is sufficient to feed the electrical generator with the torque control value corresponding to the steady-state operating point placed on the ORC. To this end, an on-off-controller-based structure can be used in order to zeroing the difference $\sigma(t) = \lambda_{opt} - \overline{\lambda}(t)$, where $\overline{\lambda}(t)$ is given by the low-frequency component of the wind speed, v_s:

$$\overline{\lambda}(t) = \frac{R \cdot \Omega_l}{v_s} \tag{5.15}$$

The choice of such a structure (Figure 5.25 – Munteanu et al. 2004b, 2006c) is justified by its robustness to the parametric uncertainties inherent to any WECS.

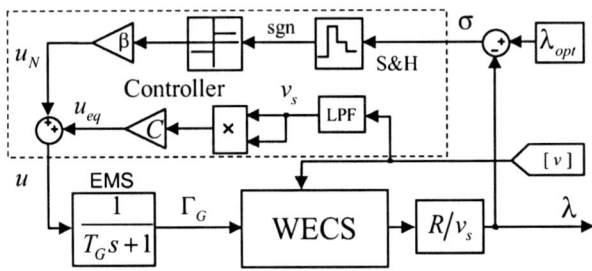

Figure 5.25. Block diagram of on-off-controller-based energy efficiency optimization loop

The control law associated to the diagram in Figure 5.25 provides the steady-state torque reference, u, which has two components:

$$u = u_{eq} + u_N, \tag{5.16}$$

where the *equivalent* control, u_{eq}, is a smooth component, corresponding to the optimal operating point (at λ_{opt}), and depends proportionally on the low-frequency wind speed squared, v_s^2:

$$u_{eq} = 0.5 \cdot \pi \cdot \rho \cdot R^3 \cdot v_s^2 \cdot \frac{C_p(\lambda_{opt})}{i \cdot \lambda_{opt}} = C \cdot v_s^2 \tag{5.17}$$

As it is imposed that the control loop operate in self-oscillations, u_N is an alternate, high-frequency component, which switches between two values, $+\beta$ and $-\beta$, $\beta > 0$:

$$u_N = \beta \cdot \text{sgn}(\sigma(t)) \tag{5.18}$$

Component u_{eq} must drive the system to the optimal operating point, whereas u_N has the role of stabilising the system behaviour around this point, once reached. The control input has in this case a large spectrum, such that the time constant, T_G, of the electromagnetic subsystem (EMS) cannot be neglected any more. The zero-order sample-and-hold (S&H in Figure 5.25) has been introduced to limit the loop switching frequency. If this frequency is too large, the EMS dynamic is too fast, such that the control loop becomes inefficient. Moreover, S&H grows the system order, allowing the system to be stabilised by self-oscillations (Munteanu *et al.* 2004b).

Self-oscillations Analysis
The self-oscillating regime is analysed using the describing function method. This requires the torque characteristics to be linearized for a given value of the low-frequency wind speed, v_s, around a typical operating point, that is, placed on the falling region. The modelling approach is the one detailed in Section 3.6.1, leading at the transfer function from Equation 3.75.

For analysis purpose, the zero-order S&H element is approximated as a first-order low-pass filter with time constant $T_1 \approx T_s/2$, where T_s is the sampling period of the S&H, determined by fitting the step response in a least squares sense. The analysis is performed on the loop branch governed by u_N. In Figure 5.26a the nonlinear part consists of an on-off relay ("sgn" block), whereas the linear part contains the series connection of all the other blocks. Diagram from Figure 5.26a is usually represented as in Figure 5.26b (Voicu 1986).

Let $N(A)$ be the describing function of the on-off relay (Voicu, 1986):

$$N(A) = \frac{4\beta}{\pi A}, \tag{5.19}$$

where A is the amplitude of the relay input.

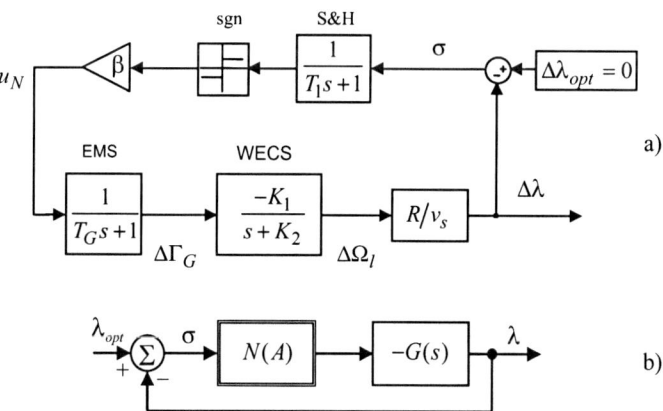

Figure 5.26. a Closed-loop system with on-off controller used for self-oscillations analysis. b Identifying the linear and nonlinear parts of the self-oscillating system from a

The transfer function of the linear part from Figure 5.26b is

$$G(s) = \frac{K_1 R}{v_s \cdot (T_G s + 1)(T_1 s + 1)(s + K_2)} \tag{5.20}$$

A closed loop containing a non-symmetric nonlinearity is likely to self-oscillate (corresponding to some limit cycles in the state space) depending on whether the harmonic balance equation

$$N(A) \cdot G(j\omega) + 1 = 0 \tag{5.21}$$

has solutions or not (Voicu 1986; Teel *et al.* 1996). In this case, Equation 5.21 has a unique solution, deduced by combining Equations 5.19 and 5.20:

$$\begin{cases} A_0 = \dfrac{4\beta R \cdot K_1 T_G T_1}{\pi v_s \cdot \left[(T_G + T_1) + T_G T_1 (T_g + T_1) K_2^2 + (T_G + T_1)^2 K_2 \right]} \\ \omega_0 = \sqrt{\dfrac{1 + (T_G + T_1) \cdot K_2}{T_G T_1}} \end{cases} \tag{5.22}$$

which represent the self-oscillations' amplitude and respectively frequency of the nonlinear part (*i.e.*, relay) input signal. Therefore, the amplitude and frequency of the tip speed $\bar{\lambda}(t)$ are

$$A_{0_\lambda} = A_0 \cdot \sqrt{\omega_0^2 T_1^2 + 1}, \quad \omega_{0_\lambda} = \omega_0 \tag{5.23}$$

Let $N_i(A) = -1/N(A)$ be the inverse negative describing function. Figure 5.27 identifies the point of self-oscillation in the complex plane as the cross-point of the linear part's Nyquist diagram and the nonlinear part's inverse negative describing function.

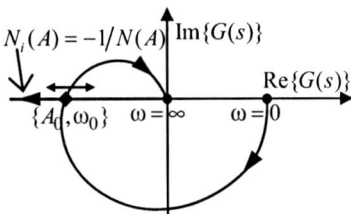

Figure 5.27. Point of self-oscillation of the closed-loop system in the complex plane

For any operating point (any v_s fixed), the closed-loop system from Figure 5.26a with parameters given by Relation 5.22 has stable self-oscillations. This claim is justified by using the Loeb's rule (Voicu 1986), according to which (A_0, ω_0) is point of stable self-oscillations if and only if

$$S_0 = \left(\dfrac{dG_R}{d\omega} \right)_{\omega_0} \cdot \left(\dfrac{dN_{i_I}}{dA} \right)_{A_0} - \left(\dfrac{dG_I}{d\omega} \right)_{\omega_0} \cdot \left(\dfrac{dN_{i_R}}{dA} \right)_{A_0} > 0, \tag{5.24}$$

where $G(j\omega) = G_R(\omega) + jG_I(\omega)$ and $N_i(A) = N_{i_R}(A) + jN_{i_I}(A)$. Simple calculation gives that $dN_{i_I}/dA = 0$ for any A, $dN_{i_R}/dA = -\pi/(4\beta)$ and

$(dG_I/d\omega)_{\omega_0} > 0$; consequently, Relation 5.24 is true.

One should note that values given by Relation 5.22 depend on changing the operating point, due to wind speed changing, v_s, but also on changing parameter K_2, which depends on v_s. This means that the position of point $\{A_0, \omega_0\}$ in the complex plane changes. According to the above discussion, each such point is characterized by stable oscillations.

Robustness to Parametric Uncertainties

From Equation 5.17 one can deduce that, for a given v_s, the equivalent control input is the optimal wind torque referred to the high-speed shaft:

$$u_{eq} = \Gamma_{wt}(\Omega_{l\,opt}, v_s)/i = \Gamma_{wt_{opt}}/i,$$

where $\Omega_{l\,opt} = \lambda_{opt} \cdot v_s/R$. In most cases, λ_{opt} is known with sufficient accuracy, but when used for computation of the optimal wind torque value, $\Gamma_{wt_{opt}}$, it is affected by uncertainties. Let u_{eq_c} be the computed value of the equivalent control and $u_{eq_{nec}} \equiv u_{eq_{opt}}$ be its necessary value, given by the real value of $\Gamma_{wt_{opt}}$.

A necessary condition for self-oscillations, like those in Figure 5.28, is that difference $\sigma = \lambda_{opt} - \overline{\lambda}$ changes its sign due to applying the switched control u_N. If $u_{eq_c} = u_{eq_{nec}}$, then σ and u_N change their signs for any β (Equation 5.18). When u_{eq_c} significantly differs from the necessary value $u_{eq_{nec}}$, there are values of coefficient β for which σ never changes its sign.

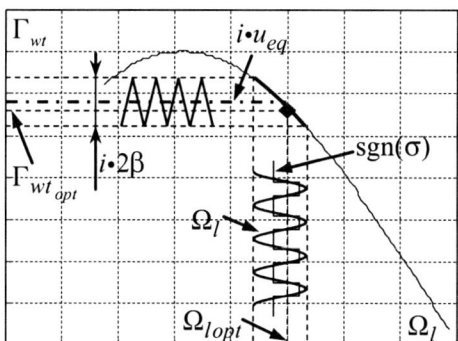

Figure 5.28. Self-oscillations around a steady-state operating point

Let $\Delta u_{eq} = u_{eq_{nec}} - u_{eq_c}$. Ideally, this difference is small enough (the control law parameters are sufficiently precise) such that σ changes its sign for a small enough value of β. Taking into account the filtering effect of EMS (first-order dynamic), a sufficient condition for self-oscillations to take place may be

$$\frac{\beta}{\omega_0 T} > \Delta u_{eq}, \qquad (5.25)$$

which provides a constraint for choosing β, given an estimated value of Δu_{eq}. If the control law parameters are imprecise (that is, Δu_{eq} is significantly large), then the control goal can be reached by increasing β, at the cost of a stronger mechanical stress.

In order to alleviate the torque variations induced by the alternate control input, in Munteanu et al. (2006c) a modified control law is proposed, based also upon Equation 5.16. A controller ensuring robustness to parametric uncertainties is thus obtained, which can also limit the torque variations due to the switched control input component. The improved controller block diagram is presented in Figure 5.29.

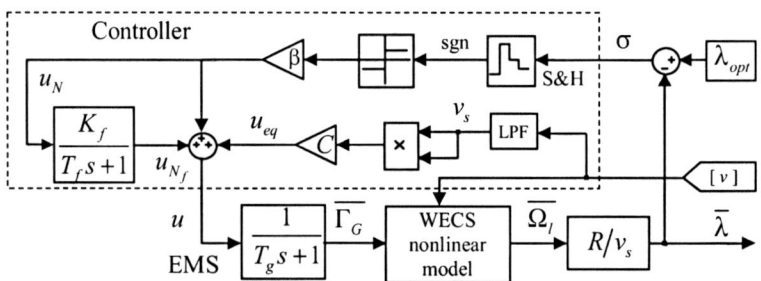

Figure 5.29. Improved on-off control law for energy optimization

The modified control law, u_{mdf}, contains an additional low-frequency component, denoted by u_{N_f}, obtained by filtering the alternate component, u_N (Figure 5.29):

$$u_{mdf} = u_{eq} + u_N + u_{N_f}, \qquad (5.26)$$

which thus is proportional to the average value of u_N. It is this component that aims at zeroing the difference Δu_{eq}. The filtering time constant, T_f, results by imposing a maximum admissible ripple for u_{N_f}; as regards the gain K_f, a sufficient value may be $K_f = \Delta u_{eq_{max}} \cdot \omega_0 T / \beta$, where $\Delta u_{eq_{max}}$ represents the maximum expected value of the difference Δu_{eq}.

5.4.2 Case Study (5)

The proposed on-off control law has been tested by numerical simulation for the low-power fixed-pitch HAWT-based WECS whose parameters are given in Appendix A and Table A.2. The control specifications are the following:

- the S&H element has $T_s = 0.01$ s, therefore $T_1 = 0.0054$ s;
- the amplitude of the alternate component u_N is $\beta = 3$ Nm;
- $C \approx 0.25$ (multiplying factor of the wind speed squared) for the considered WECS;
- time constant of the averaging filter used for the modified control law is 5 s, its gain is 20.

Figure 5.30a suggests the control goal being accomplished. The tip speed has small amplitude oscillations around its optimal value. The same figure presents the total and low-frequency wind speed evolutions used in simulations (one can remark that $v_s(t)$ varies around an hourly average of 7 m/s).

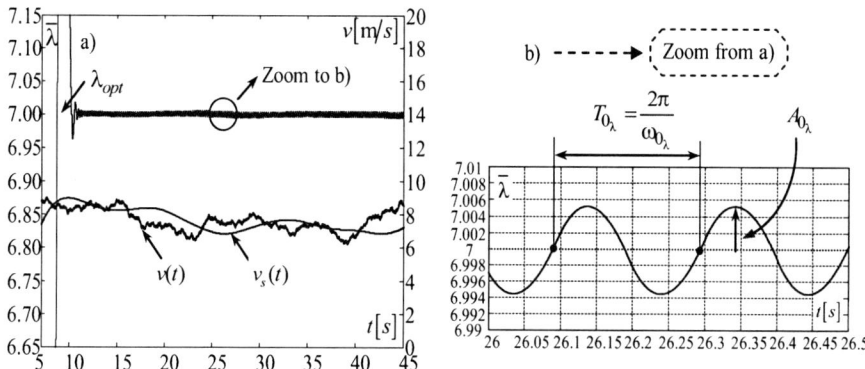

Figure 5.30. a Wind speed and tip speed evolutions. **b** Small amplitude tip speed oscillations (relative amplitude under 1‰)

Figure 5.30b is a zoom of Figure 5.30a, showing the tip speed oscillations, of amplitude $A_{0\lambda} = 0.005$ and period $T_{0\lambda} = 0.193$ s; these values meet Expressions 5.22 and 5.23 for $v_s = 7$ m/s. Sufficiently small amplitude oscillations have been obtained for the entire speed range (partial load).

Figure 5.31 illustrates the self-oscillations starting process, corresponding to a limit cycle in the (Ω_h, Γ_G) plane for a constant low-frequency wind speed $v_s = 7$ m/s (Γ_G and $\Omega_l = \Omega_h/i$ are state variables of the linearized system $G(s)$ – Relation 5.20).

From Figure 5.32 one can conclude the robustified control law (Equation 5.26) efficiency vs. the initial one (Equation 5.16), under the same uncertainty conditions, Δu_{eq}. The improved control law allows the same results to be obtained, but employing a smaller value of β, which means smaller amplitude and frequency of the torque ripple, thus reduced mechanical fatigue. Robustness properties are maintained.

For details about the MATLAB®/Simulink® implementation of this case study, the reader is referred to the folder `case_study_5` included in the software material.

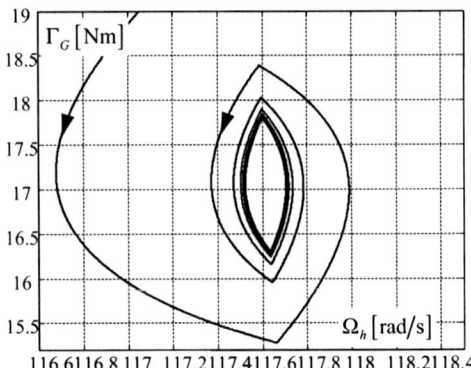

Figure 5.31. Limit cycle in the (Ω_h, Γ_G) plane

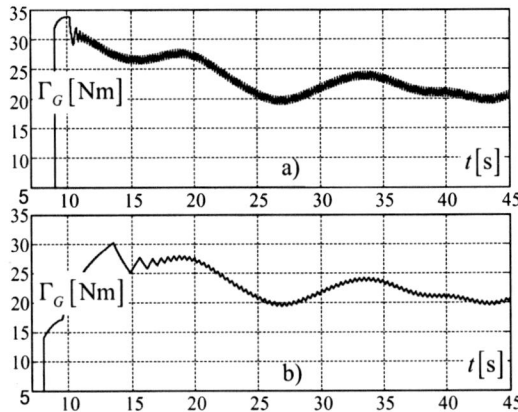

Figure 5.32. Electromagnetic torque variations when using the initial control law (**a**) *vs.* using the modified control law (**b**)

5.5 Sliding-mode Control

Variable structure control (VSC) has proven particularly suitable to variable-speed WECS control, being easy to implement by the already existing electronics. Its main drawback is an increase of mechanical stress due to chattering, which can however be alleviated by various methods.

This section proposes a sliding-mode approach for tracking the energetic optimum of a variable-speed doubly-fed-induction-generator-based WECS. The sliding surface is systematically derived from imposing a desired reduced-order dynamics and allows the turbine operation more or less close to the ORC, according to an imposed trade-off between the torque (control input) ripple and the optimum tracking. In this way, by torque controlling the generator, a multipurpose (energy-reliability) optimization is actually achieved.

5.5.1 Modelling

Figure 5.33 presents the block diagram of the considered rigid-drive-train DFIG-based WECS. The variable-speed regime is achieved by torque control (vector-control scheme with first-order global dynamic, T_G – Peña *et al.* 2000).

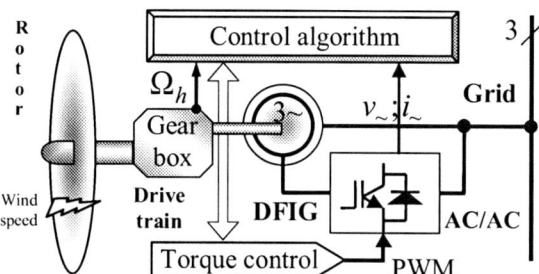

Figure 5.33. Rigid-drive-train DFIG-based WECS operating at variable-speed by torque control

Modelling is made under the assumption of the single-mass model (Wilkie *et al.* 1990) and of the limited frequency domain of the wind speed. The Van der Hoven wind speed model holds (a low-frequency wind speed component and a turbulence component). The power coefficient curve is considered known and the structural dynamics are negligible. The electric generator is ideal, *i.e.*, it has constant parameters. Power losses induce a constant efficiency for the whole wind speed domain.

Starting from the motion equation of the high-speed shaft (Wilkie *et al.* 1990), the system equations are written as

$$\begin{cases} \dot{\Omega}_h(t) = \Gamma_{wt}(i \cdot \Omega_h, v)/(i \cdot J_t) - \Gamma_G/J_t \\ \dot{\Gamma}_G(t) = -\Gamma_G/T_G + u/T_G \end{cases}, \quad (5.27)$$

with J_t being the inertia of the high-speed shaft (equal to J_{wt}/i^2, where J_{wt} is the rotor's inertia) and u being the electromagnetic torque reference, $u = \Gamma_G^*$.

5.5.2 Energy Optimization with Mechanical Loads Alleviation

Sliding Surface
The goal is to find a sliding surface allowing the turbine to operate more or less close to the ORC, implicitly requiring an antagonistic sizing of the control effort. This requires the sought surface to depend on the wind speed; its image in the (Ω_l, P_{wt}) plane must have a nonempty intersection with the ORC for each value of the wind speed and also an adjustable slope for tuning the sliding-mode dynamics. State equations (Equation 5.27) may be written in the form

$$\dot{x} = f(x,t) + B(x,t) \cdot u,$$

where $x = [\Omega_h \ \Gamma_G]^T$, $f(x,t) = [\Gamma_{wt}(\Omega_h/i,v)/(J_t \cdot i) - \Gamma_G/J_t \ \ -\Gamma_G/T_G]^T$ is nonlinear because of the wind torque, $\Gamma_{wt}(\Omega_h/i,v)$, and $B(x,t) = [0 \ \ 1/T_G]^T$. The state equation is already in the regular form (DeCarlo et al. 1996). Therefore, the reduced-order dynamics are expressed by the first equation from below:

$$\begin{bmatrix} \dot{\Omega}_h \\ \dot{\Gamma}_G \end{bmatrix} = \begin{bmatrix} \Gamma_{wt}(\Omega_h/i,v)/(i \cdot J_t) - \Gamma_G/J_t \\ -\Gamma_G/T_G \end{bmatrix} + \begin{bmatrix} 0 \\ 1/T_G \end{bmatrix} \cdot u \quad (5.28)$$

The sliding-mode dynamics (that is, on the sliding surface) may be imposed as equivalent to some linear ones:

$$\dot{\Omega}_h = (\Gamma_{wt}(\Omega_h/i,v)/i - \Gamma_G)/J_t \equiv [a_1 \ a_2] \cdot [\Omega_h \ \Gamma_G]^T, \quad (5.29)$$

where a_1 and a_2 correspond to the first-order dynamics on the sliding surface. A first form of the sliding surface results after some algebra (see Appendix B.1) as

$$\sigma = a_1 \cdot J_t \cdot \Omega_h + \Gamma_G \cdot (1 + a_2 \cdot J_t) - \Gamma_{wt}/i \quad (5.30)$$

Parameter a_1 represents the time constant of the sliding-mode dynamics. The steady-state regime is imposed by choosing parameter a_2; in this way, the equilibrium on the sliding surface may be characterized. In our case, provided that the energetic optimization is of interest, the equilibrium point is set to *the optimal one* (that is, on the ORC). One can note that the form at Equation 5.30 of the sliding surface depends upon the wind torque, which is practically impossible to measure. From the motion equation it holds that $\Gamma_{wt}/i - \Gamma_G = J_t \cdot \dot{\Omega}_h$; therefore, one obtains an equivalent form of the sliding surface:

$$\sigma = a_1 J_t \Omega_h + \Gamma_G a_2 J_t - \underbrace{(\Gamma_{wt}/i - \Gamma_G)}_{J_t \cdot \dot{\Omega}_h} = a_1 J_t \Omega_h + a_2 J_t \Gamma_G - J_t \dot{\Omega}_h, \quad (5.31)$$

showing that the commutation surface depends on the derivative of a state variable, which is an inconvenience for the real-time implementation. This derivative can be estimated by using a suitably chosen first-order high-pass filter, with $\dfrac{s}{T_f s + 1}$ as transfer function.

Sliding-mode Control Law
The two components of the sliding-mode control law: the equivalent control input, u_{eq}, and the on-off component, u_N (DeCarlo et al. 1996), must now be computed.

The equivalent control input is obtained as detailed in Appendix B.1 (Munteanu et al. 2006b):

$$u_{eq} = \Gamma_G - \frac{T_G}{1+a_2 J_t} \cdot (a_1 J_t \Omega_h + a_2 J_t \Gamma_G) \cdot (a_1 - A(\lambda, v)), \quad (5.32)$$

where

$$A(\lambda, v) = \left(K \cdot v \cdot R^2\right)/i^2 \cdot \left(C_p'(\lambda) \cdot \lambda - C_p(\lambda)\right)/\lambda^2,$$

with $K = 0.5 \cdot \pi \cdot \rho \cdot R^2$ considered as invariant and $C_p'(\lambda)$ being the derivative of the power coefficient in relation to λ.

The control law searched for has torque dimensions, as Expression 5.32 shows. Parameter a_1 results from imposing the convergence speed to the sliding-mode regime: $a_1 = -1/T_{sm}$, where T_{sm} is a time constant. The value of a_2 results from imposing the "target" steady-state, namely the optimal operating point (OOP), corresponding to λ_{opt}. Hence, from $\dot{\Omega}_h = a_1 \cdot \Omega_{h\,opt} + a_2 \cdot \Gamma_{G\,opt} = 0$ one gets:

$$a_2 = -a_1 \cdot \Omega_{h\,opt} / \Gamma_{G\,opt}$$

Once entering the sliding mode, a_1 is the inverse of the time constant and a_2 is the (static) gain. By adopting a dynamic modification of a_2, namely by using the following expression:

$$a_2 = -a_1 \cdot \frac{\Omega_{h\,opt}}{\Gamma_{G\,opt}\left(1 + k \cdot (\Omega_h - \Omega_{h\,opt})/\Omega_{h\,opt}\right)}, \quad (5.33)$$

with $k \geq 0$, the operating point variations around the OOP can be reduced. Expression 5.33 is not valuable but around the OOP, otherwise parameter a_2 can take sufficiently large values such that the system leaves the normal operating regime. The larger the value of k is, the more quickly the optimal steady-state, $(\Omega_{h\,opt}, \Gamma_{G\,opt})$, is reached, and therefore the ORC tracking has a better quality (numerical and real-time simulation results for the case study in the next section sustain this idea). The system can be thus forced to follow more accurately the control goal, *i.e.*, the energetic optimization, but the control input variations are more significant, affecting the reliability. The conclusion is that parameter k adjusts the control effort and can be used to design a desired energy-reliability trade-off. Accordingly, as k increases, the slope of sliding surface's image in the rotational speed – power plane also increases; in this way, ORC is tracked more closely. Because $(\Omega_{h\,opt}, \Gamma_{G\,opt})$ pair depends on the wind speed, the sliding surface changes with the wind speed.

The on-off component of the sliding control law, u_N, is obtained by choosing as Lyapunov (energy) function the square of the obtained sliding surface:

146   5 Design Methods for Optimal Control with Energy Efficiency Criterion

$$u_N = -\alpha \cdot \text{sgn}_h(\sigma), \quad (5.34)$$

where $\text{sgn}_h(\cdot)$ is a hysteretic sign function of width h. Finally, the total sliding-mode control law is the sum of the equivalent component and on-off component:

$$u = u_{eq} + u_N \quad (5.35)$$

5.5.3 Case Study (6)

MATLAB®/Simulink® numerical simulations have been performed in order to assess the sliding-mode optimization approach detailed above. The low-power fixed-pitch HAWT-based WECS whose parameters are given in Appendix A and Table A.2 has been chosen for illustration.

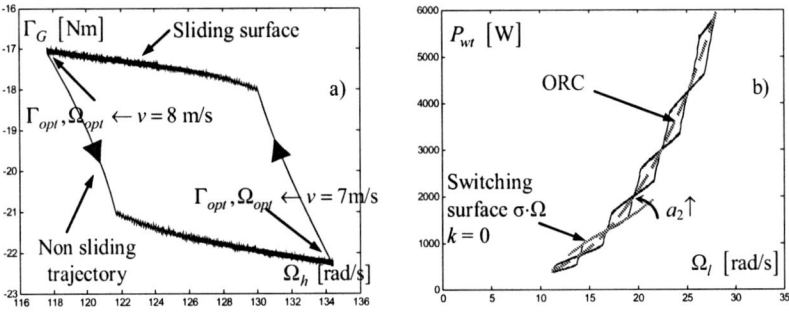

Figure 5.34. WECS optimised by sliding-mode control: response to step changes in the wind speed for $k=0$: **a** state space evolution; **b** wind power evolution vs. the ORC and the image of the sliding surface in the rotational speed–power plane

Figure 5.34 presents simulation results concerning the closed-loop evolution in response to step changes in the wind speed for trade-off parameter $k=0$. Figure 5.34a shows that the operating point is attracted to the new sliding surface and then evolves in sliding mode to the optimal one. Figure 5.34b shows the relative position of the ORC vs. the sliding surface's image in the ORC plane for different wind speeds.

Simulations detailed in Figure 5.35 have been done for a 5-min time horizon, using a pseudo-random sequence of wind speed with medium turbulence intensity of $I=0.17$, obtained by using the von Karman spectrum in the IEC standard, as shown in Figure 5.35a. Parameter k of the control law has been set at 5. Figure 5.35b,c shows the control law efficiency in maintaining the optimal conversion regime. This efficiency increases when the wind speed increases (Figure 5.35d), suggesting that an adaptive adjusting law of k parameter might be useful.

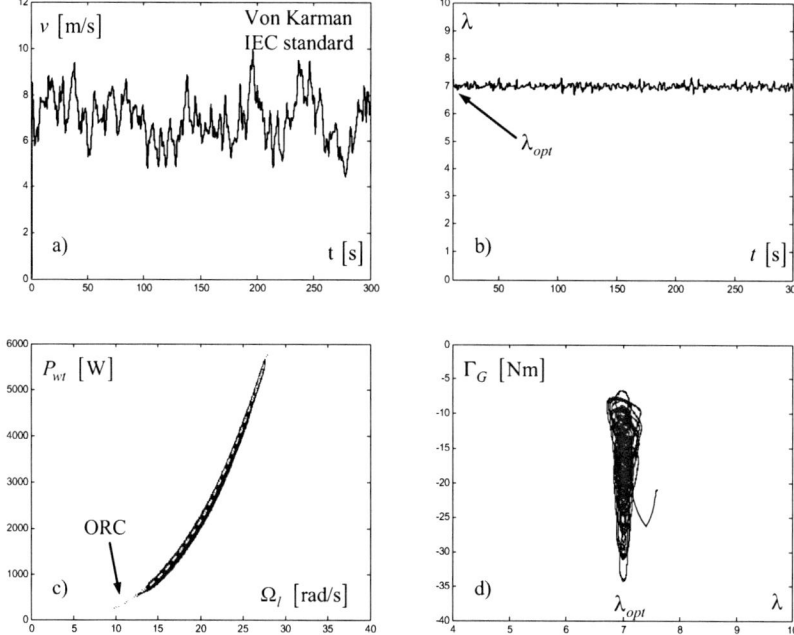

Figure 5.35. Performance of the sliding-mode control with $k=5$ and $\alpha = 0.3$: **a** wind sequence used; **b** tip speed evolution; **c** ORC tracking; **d** electromagnetic torque versus tip speed

Details about the numerical implementation of this case study can be found in the folder case_study_6 included in the software material.

5.5.4 Real-time Simulation Results

The initial form of the sliding-mode control law – Equations 5.32 and 5.34 replaced in Equation 5.35 – has been tested on a physical simulator designed according to HILS methodology as described in detail in Chapter 7. The initial control law had to suffer some changes in order to be real-time tested because of some drawbacks. The latter and the changes required in order to alleviate them are presented below.

The optimal operating point's coordinates rendered to the high-speed shaft, $(\Omega_{opt}, \Gamma_{opt})$, must be computed using the turbine parameters, the mechanical transmission and the electrical generator efficiency. This can be surpassed by using a multiplying correction applied to Γ_{opt}, denoted by η_{em}, which is an estimation of the electromechanical efficiency of the wind power system, to which the integral of the tip speed ratio steady-state error, $\lambda - \lambda_{opt}$, can be added. η_{em} has a sigmoid variation with the rotational speed, as established by measurements. This compensation concerns the equivalent control input, originally given by Equation 5.32, where parameter a_2 is replaced by $a_2 = -a_1 \cdot \eta_{em} \cdot \Omega_{opt}/\Gamma_{opt}$.

Some drawbacks concern the necessity of relying upon the functional and constructive parameters when computing the control law. First, the values of parameter $A(\lambda, v)$ around the OOP must be estimated. If the operating point is sufficiently close to the ORC, then one can obtain a consistent estimation of these values. Second, since the control input is based on a measure of the low-speed shaft's rotational speed gradient, a sufficiently smooth measure of the rotational speed is necessary. One can note that generally the control law can have a smaller dependence on the constructive parameters if increasing the value of parameter α of its on-off component, u_N. This consequently will increase the sliding surface's attractiveness, so the robustness, but also the chattering.

Because the effectiveness of the proposed optimal control law is not the same for all the operating range, it might be possible to find an adaptation law for parameter k, depending on the wind speed. It is this parameter that represents, in fact, the main degree of freedom of the sliding-mode control law.

Finally, the use of a relay function for obtaining the alternate control input induces unacceptable generator torque/current variations. Use of a continuous hysteretic sigmoid (*e.g.*, hyperbolic tangent) function is a well-known method for alleviating the supplementary fatigue due to electromagnetic torque and current oscillations without greatly affecting the control law robustness.

Real-time simulations have been performed for the control law modified according to the previous comments; some results are discussed next.

The overall performance of the real-time simulated closed-loop WECS can be judged by means of the set of ControlDesk® captures from Figure 5.36, illustrating the evolution of the main variables when an energy-reliability optimization by variable speed is implemented, as previously proposed. The trade-off coefficient has been set to a medium value, $k=5$, and the amplitude of u_N is $\alpha=0.3$.

The 2-min pseudorandom wind speed sequence in Figure 5.36a has medium turbulence and was obtained using the von Karman spectrum in the IEC standard. The tracking precision of the optimal conversion regime can be estimated from the tip speed temporal evolution (Figure 5.36b) and from the power coefficient evolution (Figure 5.36c). Figure 5.36d shows a non-uniform effectiveness of the control law along the concerned operating range. The ORC tracking precision in the rotational speed – power plane is given in Figure 5.36e. The evolution of the active power fed into the electrical grid is given in Figure 5.36f, thus illustrating the final conversion result.

Finally, Figure 5.37 shows the influence of the trade-off parameter, k, on accomplishing the control goal. Indeed, the ORC tracking precision increases with the value of k (Figure 5.37d–f). But the control input (electromagnetic torque) variations increase also (Figure 5.37a–c), and so does the mechanical stress involved.

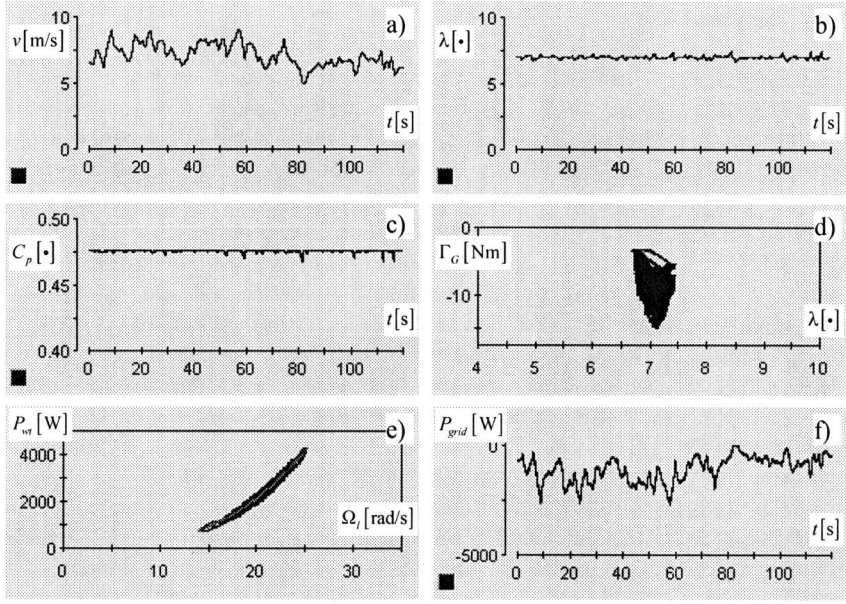

Figure 5.36. ControlDesk® captures to illustrate the overall behaviour of the real-time simulated controlled WECS (sliding-mode control law with $k=12$, $\alpha=0.3$, sigmoid hysteresis)

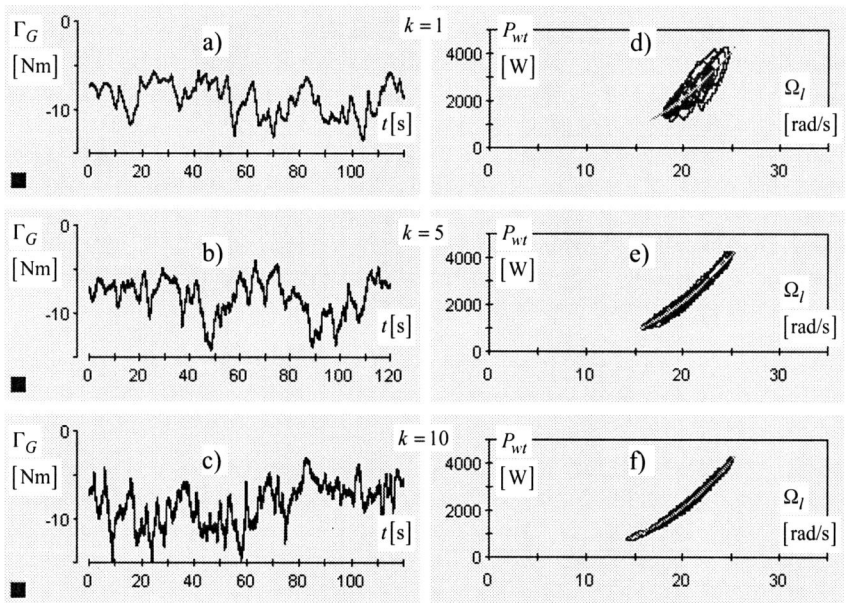

Figure 5.37. WECS optimised by sliding-mode control: real-time simulation results, showing the energy-control effort trade-off as k parameter increases

5.5.5 Conclusion

The sliding-mode control law presented in this section aims at maximizing the wind harvested power, while limiting the electromagnetic torque variations. Provided that ensuring the optimal tip speed ratio implies strong torque variations, the choice of an appropriate sliding surface is not a trivial task. Such a surface, having a non-empty intersection with the ORC, variant with the wind speed and depending on a desired energy-reliability trade-off, has been systematically found. The possibility of driving the operating point in a conveniently sized neighbourhood of the ORC is thus ensured by imposing some desired reduced-order dynamics, implicitly allowing the generator torque variations limitation in the high-frequency range. The drive train mechanical fatigue induced by the generator control is alleviated with positive influence on the WECS overall reliability. The possibility of adjusting the trade-off coefficient, k, confers flexibility to the WECS, so that the wind energy conversion efficiency be significantly increased when the particular conditions of the site allow it (*i.e.*, the mechanical stress induced by the turbulences is not important).

Real-time simulation has allowed validation of the proposed control law under various wind regimes, irrespective of the actual meteorological context. The theoretical form of the control law had to be modified such as to comply with some practical limitations and modelling uncertainties. Even if not controlled, the chattering frequency is sufficiently high *vs.* the fastest dynamic of the system.

Two directions deserve to be further investigated, concerning both completions as well as extensions of this framework.

Among completions, a study of the closed-loop system's robustness and the design of a sliding-mode control law for harvested power regulation would be interesting to investigate. Also, a quantitative expression of the achievable trade-off must be sought; that is how much is the stress reduced when giving up the tracking performance at a certain degree.

As regards extensions, to study the effectiveness of the proposed approach for large wind power systems is a challenging issue. Time constants are larger and dynamics are more complex, *e.g.*, new phenomena induced by a very large blade appear (oscillations, vibrations, tower shadow effect, *etc.*) and need proper modelling. A thorough fatigue analysis is also necessary to clarify the quantitative impact of the system variables variations over its reliability.

5.6 Feedback Linearization Control

Because WECS are highly nonlinear systems, but with smooth nonlinearities, a possible optimal control design solution can be the feedback linearization control (Isidori 1989). Being suitable to electrical drive control (Lee *et al.* 2000), this approach has been applied to grid synchronous-generator-based energy conversion systems (Chapman *et al.* 1993; Savaresi 1999; Akhrif *et al.* 1999; Wang *et al.* 1993).

This section presents the feedback linearization design of a robust control system for an autonomous hybrid wind power system.

5.6.1 WECS Modelling

The autonomous wind power systems are isolated power producing systems connected to a local grid. In order to assure an uninterrupted energy supply, such isolated systems also need other electrical energy sources such as diesel generators, batteries and/or photovoltaic panels. A wind power system combined with any of these sources (standard or renewable) form the so-called hybrid wind power systems (HWPS). The HWPS considered in this section consist of a variable-speed wind turbine, driving a permanent-magnet synchronous generator (PMSG) connected to the local grid through power converters (see Figure 5.38).

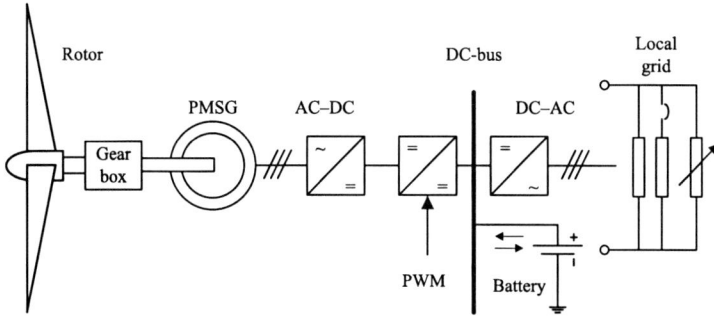

Figure 5.38. The analysed PMSG-based variable-speed WECS – reprinted (with modifications) from Wind Energy, 9/5, Cutululis NA, Ceangă E, Hansen AD, Sørensen P, Robust multi-model control of an autonomous wind power system, 399-419, Copyright (2006), with permission from John Wiley and Sons Ltd

In order to assure the proper operation of the PMSG-based WECS, there are several control loops. The local control loops are the control loop regarding grid connection through power electronics and the control loop that assures the optimum wind energy extraction. The latter controls the shaft speed according to the wind speed in order to assure optimum wind energy extraction, by means of an inner current control loop acting on the DC-link. This will result in modifying the equivalent resistance at the generator's terminal. The global control acts on the load/unload regimes of the battery in such manner that the active power produced by the wind turbine meets the active power required by the load.

The focus in this section is on the local control loop that assures the optimum wind energy extraction and assumes that the other control loops are present and working. The power electronics dynamic is significantly more rapid than the PMSG-based WECS dynamic and therefore neglected. The system presented in Figure 5.39 represents the equivalent subsystem that assures optimum wind energy conversion. An equivalent load defined by a constant inductance, L_s, and a variable resistance, R_s, being the control variable, replace the power electronics. The modelling details of such a system are given in Section 3.6.2.

A frequent difficulty involved in this approach is the synthesis computational complexity. In the WECS particular case, the wind torque coefficient variation on the tip speed must be modelled by a high-order polynomial function, in order to capture all operating regimes (including the starting one). One must assume a simplified expression to use for feedback linearization (*e.g.*, capture only the steady-state regime), otherwise, the design procedure becomes extremely difficult. Thus, the torque coefficient is modelled as a second-order instead of a sixth-order polynomial (see Equation 3.25), excluding the starting regime (Figure 5.40).

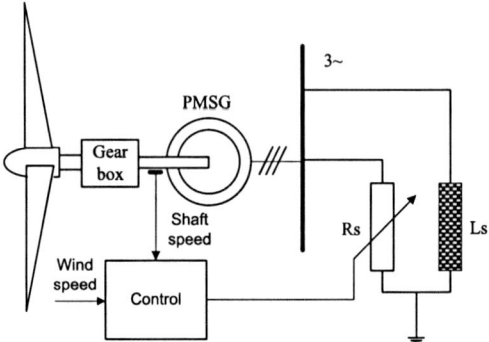

Figure 5.39. The equivalent PMSG-based WECS – reprinted (with modifications) from Wind Energy, 9/5, Cutululis NA, Ceangă E, Hansen AD, Sørensen P, Robust multi-model control of an autonomous wind power system, 399-419, Copyright (2006), with permission from John Wiley and Sons Ltd

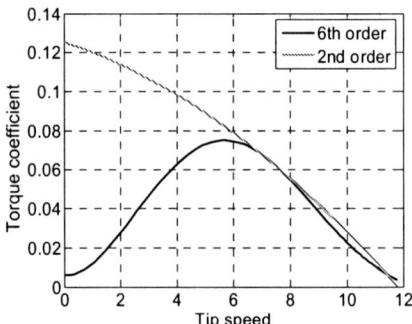

Figure 5.40. Sixth- *vs.* second-order torque coefficient

5.6.2 Controller Design

The aim is to maximize the power extraction when the WECS operates in the partial-load regime. This is accomplished by controlling the shaft speed. The PMSG-based WECS model, using Equations 3.58 and 3.30, is written in the form

$$\begin{cases} \dot{\mathbf{x}} = f(\mathbf{x}) + g(\mathbf{x})u \\ y = h(\mathbf{x}) \end{cases}, \tag{5.36}$$

where

$$\mathbf{x} = \begin{bmatrix} i_d & i_q & \Omega_h \end{bmatrix}^T$$

$$f(\mathbf{x}) = \begin{bmatrix} f_1 \\ f_2 \\ f_3 \end{bmatrix} = \begin{bmatrix} \dfrac{1}{L_d + L_s}\left(-Rx_1 + p\left(L_q - L_s\right)x_2 x_3\right) \\ \dfrac{1}{L_q + L_s}\left(-Rx_2 - p\left(L_d + L_s\right)x_1 x_3 + p\Phi_m x_3\right) \\ \dfrac{1}{J}\left(d_1 v^2 + d_2 v x_3 + d_3 x_3^2 - p\Phi_m x_2\right) \end{bmatrix}$$

$$g(\mathbf{x}) = \begin{bmatrix} g_1 & g_2 & g_3 \end{bmatrix}^T = \begin{bmatrix} -\dfrac{1}{L_d + L_a} x_1 & -\dfrac{1}{L_q + L_s} x_2 & 0 \end{bmatrix}^T \quad (5.37)$$

$$u = R_s$$

$$h(\mathbf{x}) = x_3 = \Omega_h$$

In order to determine the relative degree of the system (Isidori 1989), we calculate the Lie derivatives (see Appendix B.2):

$$\begin{cases} L_f h(\mathbf{x}) = d_1 v^2 + d_2 v x_3 + d_3 x_3^2 - d_4 x_2 \\ L_g L_f h(\mathbf{x}) = -d_4 a_3 x_2 \neq 0 \end{cases} \quad (5.38)$$

Since $L_g L_f^n h(\mathbf{x}) \neq 0$, with $n = 1$, the relative degree of the system is $r = n + 1 = 2$. This means that only a *partial* linearization is possible. The linearization affects only the system dynamics that are responsible for the input-output mapping, while the rest of the dynamics are internal and they do not influence the input-output mapping.

In order to bring the system in the *normal form* (see Appendix B.2), a coordinate transform, fulfilling the diffeomorphism condition, must be found:

$$\frac{\partial z_3}{\partial x_1} g_1 + \frac{\partial z_3}{\partial x_2} g_2 + \frac{\partial z_3}{\partial x_3} = \frac{\partial z_3}{\partial x_1} a_3 x_1 + \frac{\partial z_3}{\partial x_2} a_3 x_2 = 0, \quad (5.39)$$

where $a_3 = -1/(L_d + L_s)$. The condition is fulfilled for $z_3 = a_3 x_1 / x_2$. The coordinate transform that leads to a partial linearization of the system is

$$\mathbf{z} = \Phi(x_1, x_2, x_3) = \begin{bmatrix} \phi_1(x_1, x_2, x_3) \\ \phi_2(x_1, x_2, x_3) \\ \phi_3(x_1, x_2, x_3) \end{bmatrix} = \begin{bmatrix} x_3 \\ d_1 v^2 + d_2 v x_3 + d_3 x_3^2 - d_4 x_2 \\ a_3 \dfrac{x_1}{x_2} \end{bmatrix} \quad (5.40)$$

In order to be able to perform the inverse transform (Isidori 1989), $\Phi(x_1, x_2, x_3)$ should not be singular. This is verified using the Symbolic Toolbox in MATLAB®. The direct coordinates transform is

$$\begin{cases} z_1 = h(\mathbf{x}) = x_3 \\ z_2 = L_f h(\mathbf{x}) = d_1 v^2 + d_2 v x_3 + d_3 x_3^2 - d_4 x_2 \\ z_3 = a_3 x_1 / x_2 \end{cases} \quad (5.41)$$

and the inverse coordinates transform is

$$\begin{cases} x_1 = a_3 z_3 \cdot \dfrac{d_1 v^2 + d_2 v \cdot z_1 + d_3 x \cdot z_1^2 - z_2}{d_4} \\ x_2 = \dfrac{d_1 v^2 + d_2 v \cdot z_1 + d_3 \cdot z_1^2 - z_2}{d_4} \\ x_3 = z_1 \end{cases}, \quad (5.42)$$

with

$$A = d_1 v^2 + d_2 v z_1 + d_3 z_1^2 - z_2 \quad (5.43)$$

The control input will be (Isidori 1989)

$$u = \dfrac{1}{L_g L_f h(\mathbf{x})} \left(-L_f^2 h(\mathbf{x}) + u_v \right), \quad (5.44)$$

where

$$\begin{cases} L_f^2 h(\mathbf{x}) = -d_4 \cdot f_2 + (d_2 v + 2 d_3 x_3) \cdot f_3 \\ L_g L_f h(\mathbf{x}) = -d_4 a_3 x_2 \end{cases} \quad (5.45)$$

and are the ones in Equation 5.38, with f_2 and f_3 being components of function f from Equation 5.37. The control input has a state feedback component – the Lie derivates $L_f^2 h(\mathbf{x})$ and $L_g L_f h(\mathbf{x})$ – and a component, u_v, which imposes the dynamic of the linear input-output mapping. The latter is chosen to be a state-feedback control (Ceangă et al. 2001), as shown in Figure 5.41. In order to ensure zero error in steady-state regime, an integrator was added.

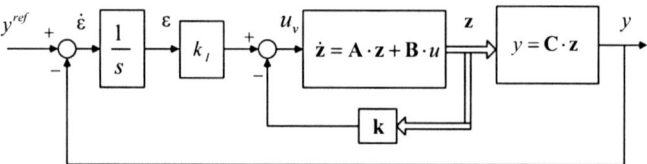

Figure 5.41. State-feedback control

5.6 Feedback Linearization Control

The linear model is

$$\begin{cases} \begin{bmatrix} \dot{z}_1 \\ \dot{z}_2 \end{bmatrix} = \begin{bmatrix} 0 & 1 \\ 0 & 0 \end{bmatrix} \cdot \begin{bmatrix} z_1 \\ z_2 \end{bmatrix} + \begin{bmatrix} 0 \\ 1 \end{bmatrix} \cdot u \\ y = \begin{bmatrix} 1 & 0 \end{bmatrix} \cdot \begin{bmatrix} z_1 \\ z_2 \end{bmatrix} \end{cases} \quad (5.46)$$

and the control input u_v is calculated using pole allocation technique:

$$u = -\underbrace{\begin{bmatrix} k_1 & k_2 \end{bmatrix}}_{k} \cdot \begin{bmatrix} z_1 \\ z_2 \end{bmatrix} + k_I \cdot \varepsilon \quad (5.47)$$

with

$$\varepsilon = y^{ref} - y = y^{ref} - \begin{bmatrix} 1 & 0 \end{bmatrix} \cdot \begin{bmatrix} z_1 \\ z_2 \end{bmatrix} \quad (5.48)$$

Defining the extended state vector $\hat{z} = \begin{bmatrix} z_1 & z_2 & \varepsilon \end{bmatrix}^T$, the linear system is

$$\begin{bmatrix} \dot{z}_1 \\ \dot{z}_2 \\ \dot{\varepsilon} \end{bmatrix} = \begin{bmatrix} 0 & 1 & 0 \\ 0 & 0 & 0 \\ -1 & 0 & 0 \end{bmatrix} \cdot \begin{bmatrix} z_1 \\ z_2 \\ \varepsilon \end{bmatrix} + \begin{bmatrix} 0 \\ 1 \\ 0 \end{bmatrix} \cdot u + \begin{bmatrix} 0 \\ 0 \\ 1 \end{bmatrix} \cdot y^{ref} \quad (5.49)$$

The control input, u_v, is obtained as

$$u_v = -\begin{bmatrix} k_1 & k_2 & -k_I \end{bmatrix} \cdot \begin{bmatrix} z_1 \\ z_2 \\ \varepsilon \end{bmatrix}, \quad (5.50)$$

thus the closed-loop system is described by

$$\begin{bmatrix} \dot{z}_1 & \dot{z}_2 & \dot{\varepsilon} \end{bmatrix}^T = \left(\begin{bmatrix} 0 & 1 & 0 \\ 0 & 0 & 0 \\ -1 & 0 & 0 \end{bmatrix} - \begin{bmatrix} 0 \\ 1 \\ 0 \end{bmatrix} \cdot \begin{bmatrix} k_1 & k_2 & -k_I \end{bmatrix} \right) \cdot \begin{bmatrix} z_1 \\ z_2 \\ \varepsilon \end{bmatrix} + \begin{bmatrix} 0 \\ 0 \\ 1 \end{bmatrix} y^{ref} \quad (5.51)$$

As mentioned above, k_1, k_2 and k_I are calculated using a pole-placement technique. Thus, a dominant pair of poles was imposed, defined by the cut-off frequency $\omega_0 = 20$ rad/s and the damping factor $\zeta = 0.9$, resulting in

$$k_1 = 4000; \; k_2 = 136; \; k_I = 40000$$

5.6.3 Case Study (7)

The performances of the feedback linearization control of the PMSG-based WECS described in Appendix A and Table A.3, were assessed through simulations in MATLAB®/Simulink®.

The block diagram of the maximum power extraction using feedback linearization control is presented in Figure 5.42.

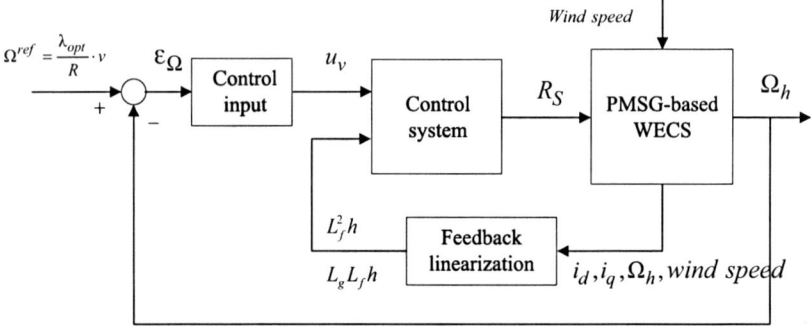

Figure 5.42. Feedback linearization control of PMSG-based WECS

The description of the block diagram is:
- *PMSG-based WECS* – the nonlinear model given in Equation 5.37;
- *feedback linearization* – coordinates transform, bringing the system in the new coordinates $\dot{z}_1, \dot{z}_2, \dot{z}_3$, inverse coordinates transform and Lie derivates $L_f^2 h(\mathbf{x})$ and $L_g L_f h(\mathbf{x})$ for the control system are computed in this block;
- *control input* – the control input for the linear system, see Figure 5.41 and Equation 5.50, is computed here;
- *control system* – the total control input applied to the WECS is computed here according to Equation 5.44.

In order to assess the proposed control structure, the first simulation used a deterministic speed reference covering in steps the operating regimes corresponding to wind speeds from 3 to 11 m/s. The generator speed vs. the reference optimal speed is shown in Figure 5.43. The controller manages to ensure the speed tracking.

The results of a simulation with a wind sequence having the average speed of about 7 m/s and a medium turbulence intensity of I=0.15 (Figure 5.44), obtained using the von Karman spectrum in the IEC standard, are shown in Figure 5.45, where the evolution of the power coefficient vs. the optimal power coefficient (Figure 5.45a) and the evolution of the tip speed (Figure 5.45b) are presented. The proposed feedback linearization controller manages to ensure maximum power extraction for a wind profile that varies between 5 and 10 m/s.

The MATLAB®/Simulink® implementation files of this case study can be found in the folder `case_study_7` from the software material.

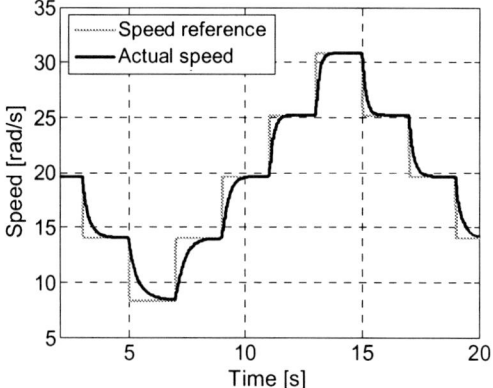

Figure 5.43. Shaft speed *vs.* reference speed

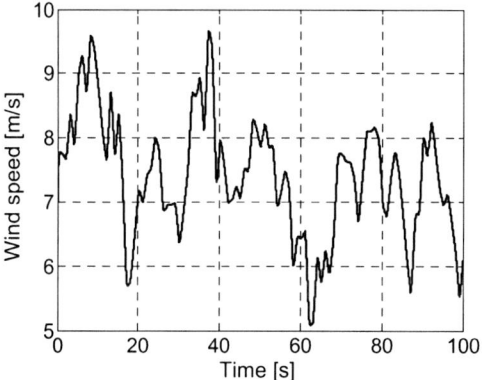

Figure 5.44. Wind profile

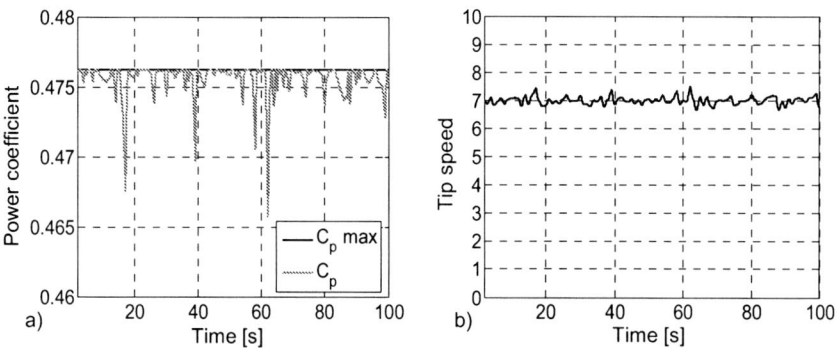

Figure 5.45. Power coefficient (**a**) and tip speed (**b**) evolution

158   5 Design Methods for Optimal Control with Energy Efficiency Criterion

5.7 QFT Robust Control

Robust control is "the control of unknown plants with unknown dynamics subject to unknown disturbances" (Chandrasekharan 1996). From this statement, it is clear that one of the key issues of robust control systems is uncertainty and how the control system deals with it.

An approach for the design of a robust controller can be based on the quantitative feedback theory (QFT) method (Horowitz 1993). The QFT is one of the robust control design methods used to achieve a given set of performance and stability criteria over a specific range of plant parameter uncertainty. The associated design procedure is a step-by-step one, whose most important step is the loop-shaping in frequency domain by manipulating the poles and zeros of the nominal loop transfer function. This section presents the design of a robust control system for an autonomous hybrid wind power system using the QFT method.

5.7.1 WECS Modelling

The WECS considered for illustrating the QFT design method is the same as in the feedback linearization control presented in Section 5.6. The use of the QFT design method for this system arises from the conclusion of the zero-pole distribution analysis, presented in Section 3.6.2. As observed from Figure 3.30, the system exhibits second-order dynamics whose cut-off frequency and damping factor strongly depend on the position of the operating point on the ORC. Thus, the system is nonlinear and the design of a robust controller for it is not a trivial task.

5.7.2 QFT-based Control Design

The QFT design technique is based on the frequency-domain shaping of the *open-loop* transfer function (Horowitz 1993). The uncertainties in the plant gain and phase are represented as a template on the Nichols chart and used to define regions in the frequency domain where the open-loop frequency response must lie in order to satisfy the performance and stability specifications.

The QFT design method employs a two-degree freedom control structure, using a compensator, $C(s)$, and a prefilter, $F(s)$, as shown in Figure 5.46.

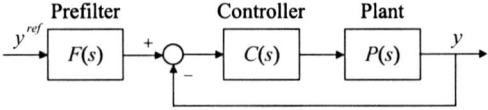

Figure 5.46. Two degree of freedom control structure

The QFT design method implies a number of steps to be accomplished. A brief description of these steps is given in the following (Wu *et al.* 1998).

1. *System formulation*. The plant is represented by a series of linear time invariant (LTI) transfer functions in various points of the operating range. The operating range is chosen according to the control objectives.

2. *Design specifications.* The performance specifications impose the desired dynamic and steady-state performance of the closed-loop system. The tracking specification defines the acceptable variations domain of the closed-loop tracking response due to parameter uncertainty and disturbances and it is usually defined in the time domain

$$y_L(t) \le y(t) \le y_U(t), \quad (5.52)$$

where $y(t)$ denotes the system tracking step response; $y_L(t)$ and $y_U(t)$ denote the lower and respectively the upper tracking bounds for step response. Since QFT is a frequency-domain design method, the time-domain tracking specifications must be transformed into the frequency domain.

The robust stability specification assures the closed-loop stability, regardless of the plant parameters variation in the considered uncertainty region and it is defined as follows:

$$\left| \frac{C(j\omega)P(j\omega)}{1+C(j\omega)P(j\omega)} \right| \le \alpha_B, \quad (5.53)$$

where α_B is usually chosen such that $\alpha_B < 2\text{dB}$.

3. *Choosing the frequency array.* The frequency array – chosen prior to the detailed design stage – consists of different frequency points at which the templates and the various bounds are computed. There are no strict criteria for choosing those frequencies, also known as *trial frequencies*.

4. *Plant uncertainty and template computation.* The plant uncertainty can be either parametric (structured) uncertainty or non-parametric (unstructured), or it can be a mix of both.

5. *Selection of the nominal plant model.* One of the representative functions defined in the uncertainty step must be considered the nominal plant model.

6. *Generation and integration of bounds.* The bounds are generated using the design specifications from step 2 and the templates from step 4. The open-loop transfer function of the system in Figure 5.46 is

$$L(s) = C(s)P(s) \quad (5.54)$$

There are different bounds: robust stability bounds, tracking bounds, ultra-high-frequency bounds and disturbance bounds. The transfer function $L(j\omega)$ must lie on or above the bounds at each of the trial frequencies. If more than one type of bounds is used, then the bounds are integrated together and the condition must be fulfilled for the resulting composite bound.

7. *Loop shaping.* The controller is designed in a very transparent and interactive way, *via* a loop-shaping process in the Nichols plane. The (composite) bounds, for each trail frequency, and the nominal open-loop transfer function are plotted together. The design itself is performed by adding gains, poles and zeros to the nominal plant frequency response in such a way that the boundaries are satisfied at each frequency. The resulted controller is the aggregate of these added gains, poles and zeros.

8. *Prefilter design.* The aim is to ensure that the closed-loop response of the system satisfies the requirements on the closed-loop stability and/or disturbances. The loop shaping in carried out in the frequency domain also, using a Bode diagram instead of a Nichols plot. The prefilter aims at shifting the frequency response of the closed-loop transfer function into the specifications envelope. Thus, the tracking performances of the final system are reached.

9. *Analysis and validation.* Finally, the resulting closed-loop system is analysed to make sure that the design specifications are satisfied for frequencies other than those used for computing the bounds. Time-domain simulations should be conducted, in order to assess the tracking performances.

5.7.3 Case Study (8)

A robust controller, aiming at maximizing the energy extraction, for the PMSG-based WECS modelled in Section 3.6.2, having parameters given in Appendix A and Table A.3, was designed with the QFT method, using the QFT Toolbox for MATLAB® (Borghesani *et al.* 2003). The controller imposes the generator speed in such a way that the system operates on the ORC. The control input is the equivalent load resistance R_s. The detailed step-by-step design according to the QFT method is presented in Appendix B.3 (Cutululis *et al.* 2006b).

The controller, designed to assure maximum energy extraction over a wind speed range between 3 and 11 m/s, has the transfer function (see Appendix B.3)

$$C(s) = \frac{6.84(s+62.5)(s+10.06)(s+9.93)}{s(s+1265)(s+5.24)} \quad (5.55)$$

The prefilter transfer function is (see Appendix B.3)

$$F(s) = \frac{5478.55}{(s+343.7)(s+15.94)} \quad (5.56)$$

Numerical simulations, aiming at assessing the robust controller, were conducted in MATLAB®, using the simulation block diagram in Figure 5.47.

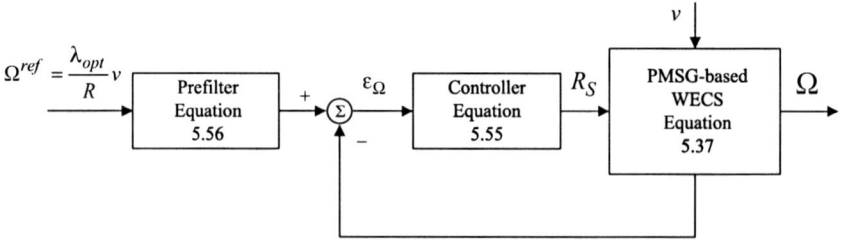

Figure 5.47. Robust control of the PMSG-based WECS – reprinted (with modifications) from Wind Energy, 9/5, Cutululis NA, Ceangă E, Hansen AD, Sørensen P, Robust multi-model control of an autonomous wind power system, 399-419, Copyright (2006), with permission from John Wiley and Sons Ltd

The first numerical simulation scheme conducted to assess the robust controller performances used as input signal a deterministic speed reference covering in steps the operating range corresponding to wind speeds from 3 to 10 m/s.

The generator speed *vs.* the reference optimal speed, Ω^{ref}, is shown in Figure 5.48, where one can see that the robust control structure manages to assure the reference tracking, but the shaft speed dynamical response is deteriorating for higher wind speeds (larger than 7 m/s). This occurs due to the large wind speed range considered – between cut-in speed and rated speed – resulting in a large parametric uncertainty range. This deterioration is more visible in the generator torque, presented in Figure 5.49.

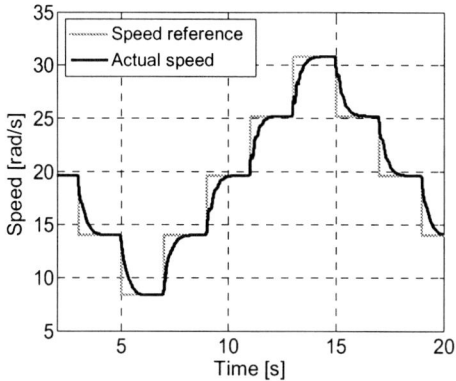

Figure 5.48. Shaft speed *vs.* reference speed – reprinted (with modifications) from Wind Energy, 9/5, Cutululis NA, Ceangă E, Hansen AD, Sørensen P, Robust multi-model control of an autonomous wind power system, 399-419, Copyright (2006), with permission from John Wiley and Sons Ltd

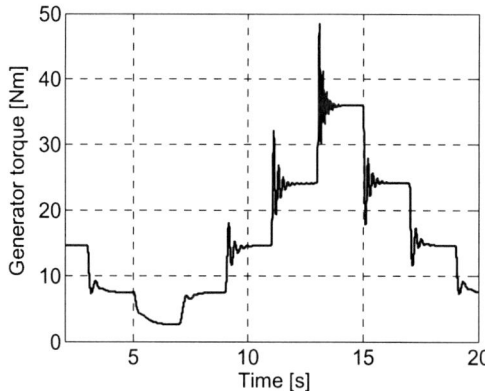

Figure 5.49. Generator torque – reprinted (with modifications) from Wind Energy, 9/5, Cutululis NA, Ceangă E, Hansen AD, Sørensen P, Robust multi-model control of an autonomous wind power system, 399-419, Copyright (2006), with permission from John Wiley and Sons Ltd

The results of a simulation with a wind sequence having the average speed of about 7 m/s and a medium turbulence intensity of $I=0.15$ (Figure 5.50), obtained using the von Karman spectrum in the IEC standard, are shown in Figure 5.51. The evolution of the power coefficient, C_p, shows that the controller manages to maximize the energy extraction in an acceptable manner.

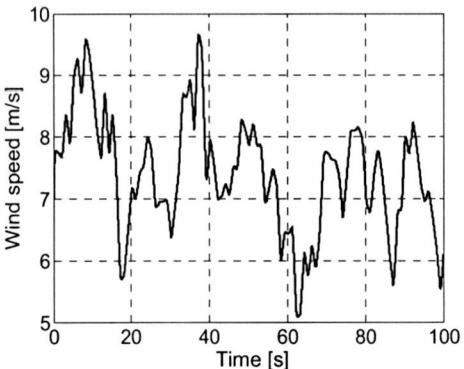

Figure 5.50. Wind profile

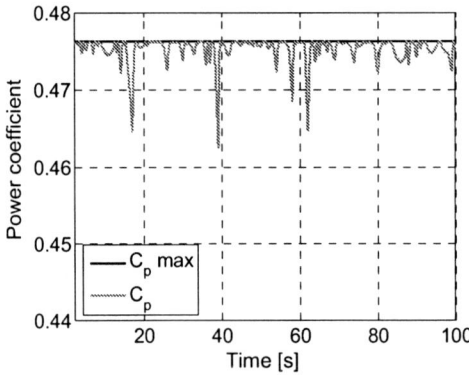

Figure 5.51. C_p vs. $C_{p_{max}}$ evolution

In order to improve the dynamical response of the shaft speed, a combination between robust QFT design method and a multi-model control structure is considered, in which every model deals with a reduced parametric uncertainty range. Therefore, two wind speed variation ranges, one between 3 and 9 m/s and one between 5 and 11 m/s, are considered. For the first wind speed variation range, the control structure with transfer functions from Equations 5.55 and 5.56 for the controller and the prefilter is used, since it has been proved to have good results, while for the second wind speed range, a new control structure (controller and prefilter) is designed using the QFT method (see Appendix B.3). The overall control law is synthesised using multi-model control theory.

5.7 QFT Robust Control

The second control structure – controller and prefilter – denoted with the subscript 2, is described by

$$C_2(s) = \frac{8332(s+61.32)(s+10.06)(s+10.04)(s+9.92)}{s(s+1265)(s+163.1)(s+73.6)(s+5.24)} \quad (5.57)$$

$$F_2(s) = \frac{252572}{(s+343.7)(s+46.16)(s+15.94)} \quad (5.58)$$

There is a wind speed range – between 5 and 9 m/s – where both control structures are active. This assures the robustness of the control law and avoids the shocks that might appear when suddenly switching from a control structure to the other (Narendra and Balakrishnan 1997). The actual operating range is determined based on the measured wind speed. The total control applied to the considered power system is computed by shaping the individual control structure command, as shown in Figure 5.52.

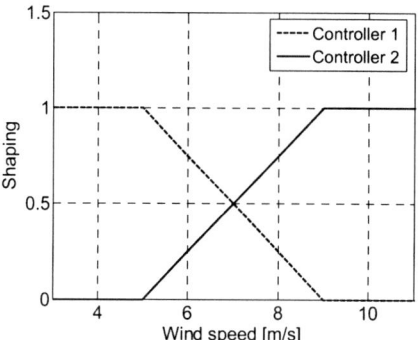

Figure 5.52. Control input shaping according to the wind speed – reprinted (with modifications) from Wind Energy, 9/5, Cutululis NA, Ceangă E, Hansen AD, Sørensen P, Robust multi-model control of an autonomous wind power system, 399-419, Copyright (2006), with permission from John Wiley and Sons Ltd

The total control input applied to the considered WECS is

$$u = \frac{w_1 \cdot u_1 + w_2 \cdot u_2}{w_1 + w_2} \quad (5.59)$$

with the shaping parameters

$$w_1 = \begin{cases} 1 & \text{for } v \leq 5 \\ 1-0.25 \cdot (v-5) & \text{for } 5 < v \leq 9 \\ 0 & \text{for } v > 9 \end{cases} \quad w_2 = \begin{cases} 1 & \text{for } v > 9 \\ 0.25 \cdot (v-5) & \text{for } 5 \leq v \leq 9 \\ 0 & \text{for } v < 5 \end{cases},$$

$$(5.60)$$

where v is the measured wind speed.

The multi-model robust control structure is presented in Figure 5.53; it was numerically simulated in MATLAB® for the same case of PMSG-based WECS (see Appendix A and Table A.3). The simulations aimed at assessing the improvement in the dynamic behaviour of the system, compared to the global robust control structure presented above.

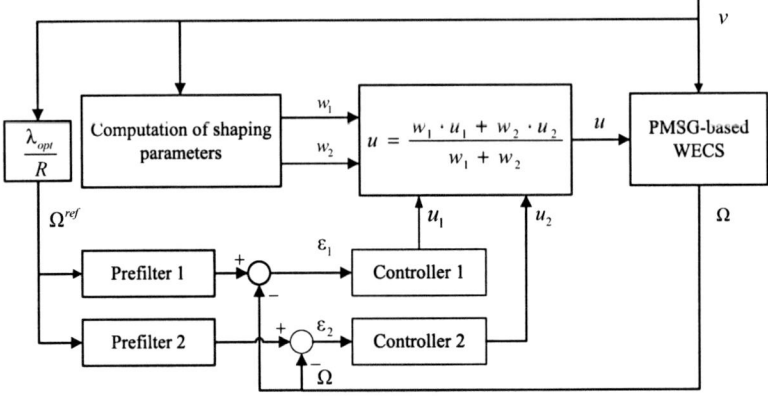

Figure 5.53. Robust multi-model control structure – reprinted (with modifications) from Wind Energy, 9/5, Cutululis NA, Ceangă E, Hansen AD, Sørensen P, Robust multi-model control of an autonomous wind power system, 399-419, Copyright (2006), with permission from John Wiley and Sons Ltd

In order to assess the improvement, the same wind profiles – first the deterministic and then the realistic one – were used. The results, compared with the global robust controller, are presented in Figures 5.54 and 5.55.

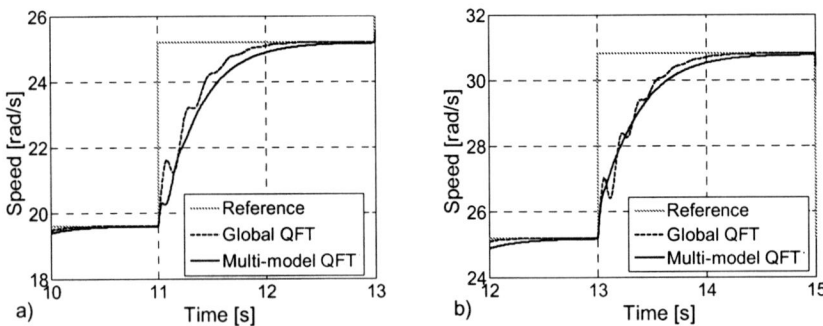

Figure 5.54. Generator speed, wind steps from 7 to 9 (**a**) and from 9 to 11 (**b**) m/s – reprinted (with modifications) from Wind Energy, 9/5, Cutululis NA, Ceangă E, Hansen AD, Sørensen P, Robust multi-model control of an autonomous wind power system, 399-419, Copyright (2006), with permission from John Wiley and Sons Ltd

The multi-model robust control structure, similar to the global robust controller, manages to assure high energy extraction, as can be seen from the power coefficient evolution in Figure 5.56.

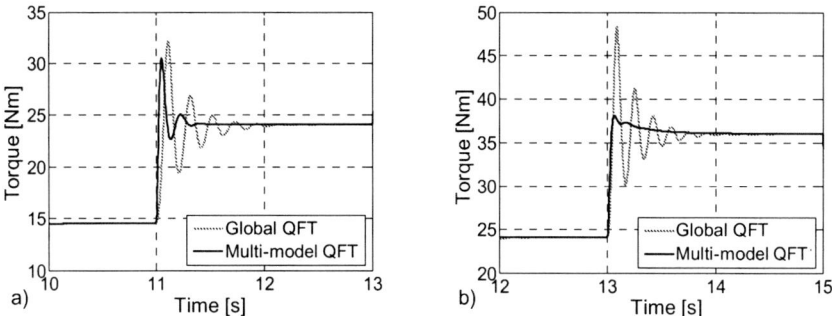

Figure 5.55. Generator torque, wind steps from 7 to 9 (**a**) and from 9 to 11 (**b**) m/s – reprinted (with modifications) from Wind Energy, 9/5, Cutululis NA, Ceangă E, Hansen AD, Sørensen P, Robust multi-model control of an autonomous wind power system, 399-419, Copyright (2006), with permission from John Wiley and Sons Ltd

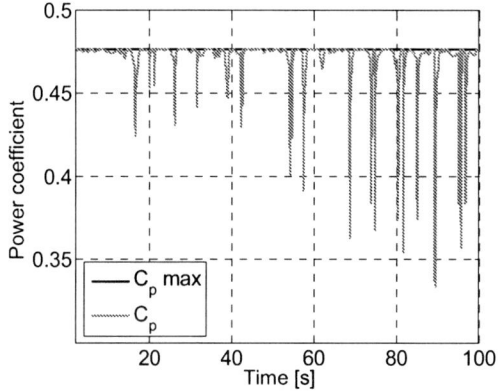

Figure 5.56. C_p vs. $C_{p\,max}$ evolution

There is a slight reduction in the energy extraction efficiency when compared to the global robust controller. This can be assessed by comparing the ratio between the captured energy, E_{cap}, and the available (optimal) energy, E_{opt}, denoted by η_{con}:

$$\eta_{con} = \frac{E_{cap}}{E_{opt}} = \frac{\int_{t_0}^{t_f} 0.5 \cdot \pi \cdot \rho \cdot R^2 \cdot v^3(t) \cdot C_p(\lambda(t)) \, dt}{\int_{t_0}^{t_f} 0.5 \cdot \pi \cdot \rho \cdot R^2 \cdot v^3(t) \cdot C_p(\lambda_{opt}) \, dt} \quad (5.61)$$

For a wind profile with similar stochastic properties like the one presented in Figure 5.50, the control efficiency – from the energy point of view – computed with Equation 5.61 over a 300-s horizon is 0.995. This is slightly lower than the global QFT control efficiency (0.997), but the main design objective of the robust multi-model control structure – improve the dynamic performance for higher wind speeds – is accomplished. The use of a multi-model control structure eliminates the oscillatory component of the system's response at high wind speeds (see Figures 5.54 and 5.55).

In order to find details about the MATLAB®/Simulink® implementation of this case study, the reader is referred to the folder `case_study_8` within the software material.

5.8 Conclusion

This chapter has aimed at presenting different control approaches dedicated to harvesting the maximum power in the partial-load regime (below the rated wind speed) using the variable-speed capability. The approaches presented employ simple and reliable wind turbine models under a reasonable set of modelling assumptions for the purpose of developing WECS controllers. To overcome the drawbacks possibly induced by a "poor modelling", some advanced control techniques have been employed, conferring robustness to the resulting controlled WECS. While the optimisation has been defined as solely harvesting the maximum power from the wind, different means of limiting the mechanical loads have been indicated in the case of each control method presented.

Two classes of approaches have been assessed here: those suitable when knowledge of the WECS parameters is rich enough and, respectively, those preferable when many parameters are unknown (or insufficiently precisely known). Following these remarks, six approaches aiming at optimizing the WECS operation from the energy viewpoint have been developed and illustrated by case studies in this chapter. An assessment of their performances and drawbacks is presented below.

First, a version of the reliable control method called *Maximum Power Point Tracking* (MPPT), requiring minimal knowledge about the system, has been proposed. Based on the extremum seeking principle, this method intends to drive the average operating point of the turbine towards the optimal one, thus ensuring optimal conversion regime, by using the turbulence component of the wind speed as search (probing) signal.

The operating point position is estimated using the average phase shift between the power coefficient and the tip speed ratio time variations. Discrete Fourier Transform has been applied to these signals in order to extract the phase of each harmonic component. Then the phase lag of components of the same frequency corresponding to the two signals is computed; their average values contain information concerning the operating point position, which is further integrated for obtaining the rotational speed reference for slowly driving the operating point towards the optimal one. Another method of obtaining the same feedback information is by convoluting the tip speed and the power coefficient signals. Since the concerned control is dealing with average operating values, the inflicted supplementary mechanical stress is negligible. An interesting remark is that this method has an improved efficiency as the average wind speed increases, because the turbulence (searching signal) level increases too, yielding better WECS energy efficiency.

MPPT-control structures generally suffer from lack of flexibility; on the other hand they are simple and robust. The MPPT version presented can gain in flexibility by designing an adaptive law of the searching speed, which must take into account the turbulence intensity. This technique of ensuring the energy maximization represents a reasonable choice in WECS control when only limited knowledge about the system is available.

In Sections 5.3 and 5.4 two intrinsically robust control methods have been presented, namely a linear one (the classical PI control) and a nonlinear one (the on-off control); both are designed based upon a linearized WECS model.

In the *PI-control* case the limitations in the control efficiency are impossible to overcome, mainly because the system is nonlinear and variant with the mean wind velocity level. The system inertia prevents the achieving of a good ORC tracking within reasonable limits of high-frequency load variations. To this end the sole solution for achieving reasonable behaviour is to employ a "weakly" tuned controller and deal with greater variation of the operating point around the ORC.

In the *on-off-control* case, the torque-controlled WECS is stabilised around the energy optimum by small-amplitude high-frequency self-oscillations. The amplitude of the on-off control input can be modified in order to alleviate the mechanical stress induced by the electromagnetic torque variations.

Sections 5.5, 5.6 and 5.7 are dedicated to advanced optimal control techniques, more versatile from the point of view of the various constraints that need to be met. Conversely, they assume quite a rich knowledge of the system being available.

Section 5.5 deals with a *sliding-mode control* law aiming at maintaining WECS operating point in a certain neighbourhood of the ORC. The system's state trajectory is restricted in this case to the sliding surface. For reasons concerning the control effort, this surface cannot be the ORC itself, but it can be designed so as to have a nonempty intersection with the ORC. In this way, the operating point has a certain freedom around the ORC, implicitly allowing the possibility of limiting the control effort. The equivalent component of the control input ensures an adjustable slope of the sliding surface, allowing it to approach the ORC more or less, thus forcing the WECS to follow as accurately as desired the optimal conversion regime, according to a trade-off coefficient. A quantitative expression of the achievable trade-off, *i.e.*, how much is the stress effectively reduced when giving up the tracking performance at a certain degree, has not yet been derived.

The on-off alternating component, adding robustness to the controlled system, has been computed based on a sigmoid hysteresis for limiting even more the undesired high-frequency torque variations (chattering). Some drawbacks of the control law have been signalled, which can however easily be overcome in practice under some realistic assumptions. Future work can envisage possible extensions of the same framework at harvested power regulation in the full-load regime.

The effectiveness of the *feedback linearization control* has been tested on a permanent-magnet-synchronous-generator (PMSG)-based WECS in isolated operation, as detailed in Section 5.6. As analytically proven, the exact linearization is not possible in this case. A partially linearized model has thereby been obtained, in order to derive the control algorithm. The case study considered has revealed good closed-loop behaviour when the wind speed ranges from the cut-in to the rated value.

Finally, for the same case, of a PMSG-based WECS in isolated operation, the *QFT robust control* technique has also been applied for purposes of conversion efficiency maximization. The linearized around the ORC model has an oscillating behaviour which strongly depends on the position of the operating point on the ORC. As suggested by numerical simulation, the closed-loop system has a good performance for the entire domain of partial-load operation, but decreases as the wind speed approaches its rated value. This remark is likely to hold irrespectively of any particular WECS configuration. Consequently, for large wind speeds, the QFT control performs better if hybridized with a multi-model control technique, in order to improve dynamics and to alleviate the electromagnetic torque variations.

This analysis could not finish without recalling that the final goal is the real-world implementation of that WECS control structure most suitable for a given situation. One can remark that this choice is not necessarily a (purely technical) optimal solution, but must represent the best trade-off between closeness to the targeted optimum on the one hand and simplicity and robustness, on the other.

Given the multiple aspects of the wind turbine global efficiency (such as its reliability, its availability, remote operation and maintenance costs, electrical generation regimes, captured power, *etc.*), control of wind turbines imply the necessity of formulating multi-purpose optimization problems.

6

WECS Optimal Control with Mixed Criteria

6.1 Introduction

The idea of conveniently sizing a trade-off between energy efficiency and increasing the service lifetime of WECS by alleviating fatigue loads is continuously being paid special attention, even when employing classical controllers like PI or PID. Inside the control loop aiming at keeping the operating point on the optimal regimes characteristic (ORC), a control requiring a drastic fulfilment of the performance criteria will induce significant stress due to important torque variations. In contrast, reasonably diminishing control performance may result in reducing the torque variations without altering the energy performance (see the results of case studies reported in Sections 5.3.3 and 5.3.4). However, this approach does not allow a rigorous control design in order to perform a fine tuning of the trade-off between the energy performance and the reliability demands.

In this chapter the optimal control using mixed criteria will be examined according to the general formulation stated in Section 4.5. In Sections 6.2–6.6 a two-term performance criterion will be considered, reflecting contradictory demands for the control input properties, namely:
– energy performance *vs.* reliability performance, when the WECS operates on the ORC;
– accuracy in ensuring the imposed values of the WECS parameters (rotational speed, torque or active power) *vs.*reliability performance, when the system works either in partial load, but with speed/torque limitation, or in full load.

In the following, operation on the ORC is considered.

Section 6.7, the last of this chapter, presents a possible extension of the multi-criterion approach, in order to embed other requirements, for example, those concerning power conditioning. This approach relies upon the frequency separation principle of the control input components associated with each of the demands represented in the global performance index.

6.2 LQ Control of WECS

6.2.1 Problem Statement

According to the basics of operating fixed-pitch variable-speed WECS, presented in Chapter 4, the WECS functioning on the ORC requires a tracking loop that can achieve one of the following functions, depending on the nature of the controlled variable:
- tracking of a variable rotational speed reference, computed as depending on the wind speed, according to Equation 4.2;
- tracking of a variable active power reference, depending on the rotational speed as to Equations 4.4 and 4.5;
- tracking of a variable generator torque reference, depending on the rotational speed by virtue of Equations 4.9 and 4.10.

In this section it is considered that the optimization is achieved by means of a rotational speed tracking loop. Generic notation for the controlled variable is $y(t)$. The control input, denoted by $u(t)$, is the torque reference. The energy performance achievement can be assessed in two ways:
- directly, through the associated tracking errors, which show the deviation of the current operating point from ORC;
- indirectly, by means of the tip speed error, $\lambda_{opt} - \lambda(t)$.

Reliability performance is expressed as reducing the variations of the control input, *i.e.*, of the generator torque, $u(t)$, which are responsible for the mechanical fatigue.

In the context described, the optimal control problem can be defined and solved by various approaches, depending on the following data.
1. plant model:
- input–output model;
- state-space model;
2. manner of assessing the energy performance:
- by means of the tracking error;
- through the deviation of the tip speed in relation to its optimal value.

6.2.2 Input–Output Approach

In the following the design of a linear quadratic optimal controller with an energy-reliability mixed criterion is presented. The procedure is applied and assessed by numerical simulation in the case of a WECS with a flexible-coupled generator, described by an input–output model. The performance criterion is defined in deterministic context. The main feature of the approach presented is that the plant model is a discrete-time one, being obtained by experimental identification. A deterministic optimal control approach of a rigid-drive-train WECS in state-space representation can be found in Section 6.4.4. In that case the energy performance is reflected by minimizing the tip speed error, $\lambda_{opt} - \lambda(t)$.

Let

$$H_p(q^{-1}) = q^{-d} \frac{B(q^{-1})}{A(q^{-1})} \qquad (6.1)$$

be the transfer function of the plant, where q^{-1} is the one-sample-time delay operator,

$$A(q^{-1}) = 1 + a_1 q^{-1} + \ldots + a_n q^{-na}, \qquad (6.2)$$

$$B(q^{-1}) = b_1 q^{-1} + \ldots + b_{nb} q^{-nb} \qquad (6.3)$$

and d is the dead lag. The corresponding input-output equation is

$$y_1(i) = -\sum_{k=1}^{na} a_i y(i-k) + \sum_{k=1}^{nb} b_j u(i-k-d), \qquad (6.4)$$

where y and u are the rotational speed and the generator torque respectively. The simplest expression of the performance criterion is

$$J = \left[y^{opt}(i+1) - y(i+1) \right]^2 + \alpha \cdot \Delta u^2(i), \qquad (6.5)$$

where α is a weighting parameter. The first term represents the deviation of the controlled variable from its reference value, y^{opt}, corresponding to operation on ORC. The second term is an expression of the mechanical fatigue loading. The optimal control input results from the condition

$$\frac{\partial J}{\partial (\Delta u)} = 0 \qquad (6.6)$$

In the criterion at Equation 6.5 the reference $y^{opt}(i+1)$ of the speed tracking loop is computed by using Equation 4.2 and depends on the predicted value of the wind speed at time $i+1$. One-step wind speed prediction can be achieved by simple linear extrapolation; the ARMA wind speed prediction is further presented in Section 6.3.3 and used in Section 6.5. Consequently, reference $y^{opt}(i+1)$ is considered as known; conversely, variable $y(i+1)$ must be replaced by a prediction, $\hat{y}(i+1)$. In order to obtain this prediction, the plant model at Equation 6.4 is rewritten (Mokhtari and Marie 1998):

$$y(i) = \left[1 - A(q^{-1}) \right] y(i) + B(q^{-1}) u(i) =$$
$$= z^{-1} A^*(q^{-1}) y(i) + q^{-1} \left[q^{-1} B^*(q^{-1}) + b_1 \right] u(i), \qquad (6.7)$$

where

$$\begin{cases} A^*\left(q^{-1}\right) = -a_1 - a_2 q^{-1} - \ldots - a_{na} q^{-na+1} \\ B^*\left(q^{-1}\right) = b_2 + b_3 q^{-1} + \ldots + b_{nb} q^{-nb+2} \end{cases} \quad (6.8)$$

Equation 6.7 can be put into the form

$$y(i) = A^*\left(q^{-1}\right) y(i-1) + B^*\left(q^{-1}\right) u(i-2) + b_1 u(i-1) \quad (6.9)$$

If replacing the discrete time i by $i+1$, one obtains the predictor equation that provides $\hat{y}(i+1)$:

$$\hat{y}(i+1) = A^*\left(q^{-1}\right) y(i) + B^*\left(q^{-1}\right) u(i-1) + b_1 u(i) \quad (6.10)$$

The performance criterion must be rearranged in order to contain the control input variation. To this end, at the right-side term of Equation 6.10 the null term is added, according to Equation 6.9:

$$y(i) - A^*\left(q^{-1}\right) y(i-1) - B^*\left(q^{-1}\right) u(i-2) - b_1 u(i-1),$$

then the predicted value at Equation 6.10 is replaced in Equation 6.5; it results in

$$J = \left\{ y^c(i+1) - \left[1 + A^*\left(q^{-1}\right)\right] y(i) + A^*\left(q^{-1}\right) y(i-1) - \right. \\ \left. - B^*\left(q^{-1}\right) \Delta u(i-1) - b_1 \Delta u(i) \right\}^2 + \alpha \cdot \Delta u^2(i) \quad (6.11)$$

By imposing the condition at Equation 6.6 the optimal control expression results as (Mokhtari and Marie 1998)

$$\Delta u(i) = \eta \left\{ y^c(i+1) - \left[1 + A^*\left(q^{-1}\right)\right] y(i) + A^*\left(q^{-1}\right) y(i-1) - B^*\left(q^{-1}\right) \Delta u(i-1) \right\}, (6.12)$$

where

$$\eta = \frac{b_1}{b_1^2 + \alpha} \quad (6.13)$$

The control input value $u(i)$ to be applied at current time is

$$u(i) = u(i-1) + \Delta u(i) \quad (6.14)$$

The optimal control law can be implemented as an R-S-T controller, illustrated in Figure 6.1, introducing the transfer functions

$$R\left(z^{-1}\right) = \eta \left[1 + a_1 + (a_2 - a_1) z^{-1} + \ldots + (a_{na} - a_{na-1}) z^{-na+1} - a_{na} z^{-na} \right] \quad (6.15)$$

$$S(z^{-1}) = \left[1 + \eta\left(b_2 z^{-1} + b_3 z^{-2} + \ldots + b_{nb} z^{-nb+1}\right)\right] \cdot \left(1 - z^{-1}\right) \quad (6.16)$$

$$T(z^{-1}) = \eta \quad (6.17)$$

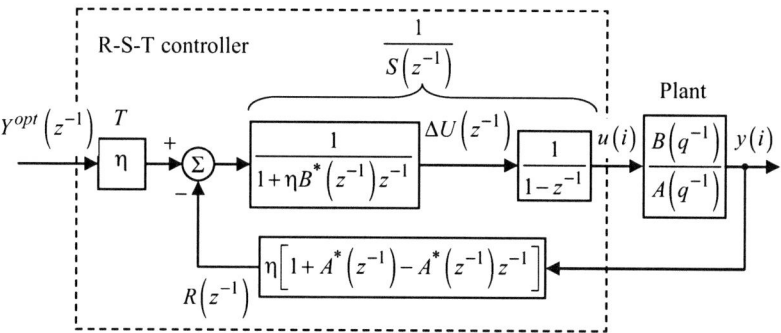

Figure 6.1. LQ optimal control structure in the form of an R-S-T controller

6.2.3 Case Study (9): LQ Control of WECS with Flexibly-coupled Generator Using R-S-T Controller

The effectiveness of the R-S-T optimal control structure has been tested on a low-power fixed-pitch SCIG-based WECS having flexible drive train – its features can be found in Appendix A and Table A.4. The input-output discrete model has resulted from a statistic identification procedure. Thus, the open-loop system has been brought to the steady-state operation corresponding to the wind speed of 8 m/s; then, a small-amplitude pseudorandom variation has been superposed on the torque reference. The variation of the rotational speed results from two actions exerted on the system: excitation through the torque loop, due to the probing signal used in identification, and excitation produced by the turbulence component of the wind speed. In Figure 6.2 one can see the evolution of the probing signal, *i.e.*, the torque reference, and the corresponding speed response of the system.

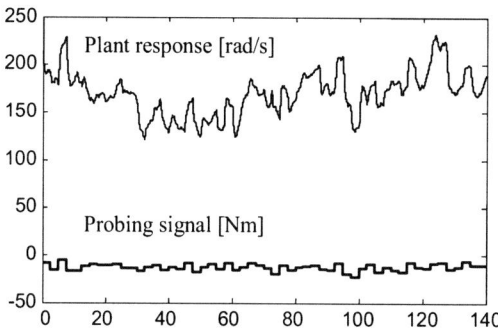

Figure 6.2. Torque reference used as probing signal and corresponding rotational speed response

One can note that the influence of the wind turbulence over the speed response is important, greater than that of the probing signal. Provided that the model from torque reference to rotational speed is sought and the output variable is affected by a non-negligible coloured noise, the instrumental variables identification method has been used. By calling the MATLAB® function iv4, one obtains the following model:

$$\begin{cases} A(q^{-1}) \cdot y(t) = B(q^{-1}) \cdot u(t) + e(t) \\ A(q^{-1}) = 1 - 1.945 \cdot q^{-1} + 1.038 \cdot q^{-2} - 0.09079 \cdot q^{-3} \\ B(q^{-1}) = 0.4224 \cdot q^{-1} - 0.7725 \cdot q^{-2} + 0.3522 \cdot q^{-3} \end{cases} \quad (6.18)$$

Transfer functions $R(z^{-1})$, $S(z^{-1})$ and $T(z^{-1})$ of the optimal controller have been computed by using Equations 6.15, 6.16 and 6.17, for two values of α, namely 10 and 2. Numeric results are given in Table 6.1.

Table 6.1. Transfer functions associated with the R-S-T optimal controller for two values of the weighting coefficient, α

α=10	$T(z^{-1}) = -0.0415$
	$S(z^{-1}) = 1 - 1.032 \cdot z^{-1} + 0.04667 \cdot z^{-2} - 0.01462 \cdot z^{-3}$
	$R(z^{-1}) = -0.1222 + 0.1238 \cdot z^{-1} - 0.04684 \cdot z^{-2} + 0.003767 \cdot z^{-3}$
α=2	$T(z^{-1}) = -0.1939$
	$S(z^{-1}) = 1 - 1.15 \cdot z^{-1} + 0.2181 \cdot z^{-2} - 0.06829 \cdot z^{-3}$
	$R(z^{-1}) = -0.571 + 0.5784 \cdot z^{-1} - 0.2189 \cdot z^{-2} + 0.0176 \cdot z^{-3}$

Figures 6.3 and 6.4 illustrate the closed-loop simulation results for α = 10 and α = 2. In Figure 6.3a,b one can see the evolution of the torque reference, which allows assessing the mechanical loads. Figure 6.3c,d offers details about the variations of the torque reference, Δu. Figure 6.4a,b presents the evolution of the tip speed, whereas Figure 6.4c,d shows the power coefficient evolution.

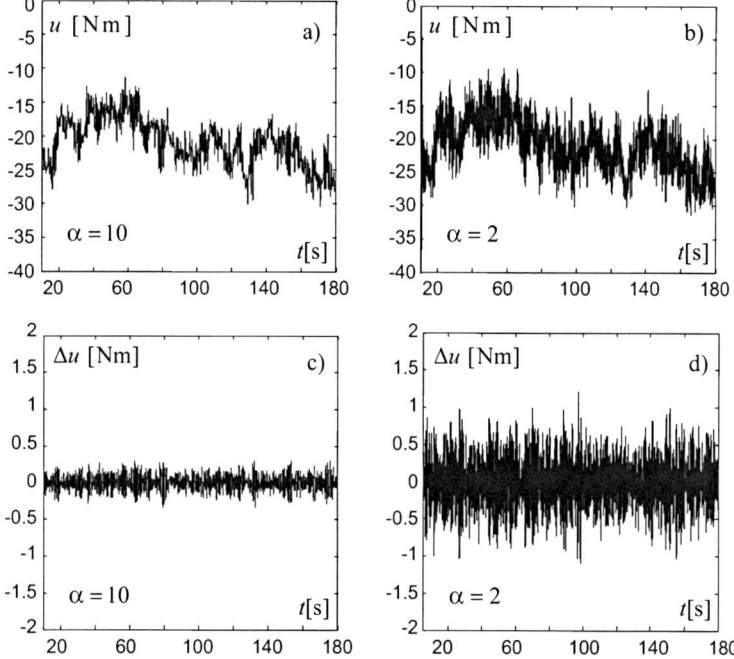

Figure 6.3. Performance of an R-S-T optimally controlled WECS: **a,b** control input evolution; **c,d** evolution of the control input's variation

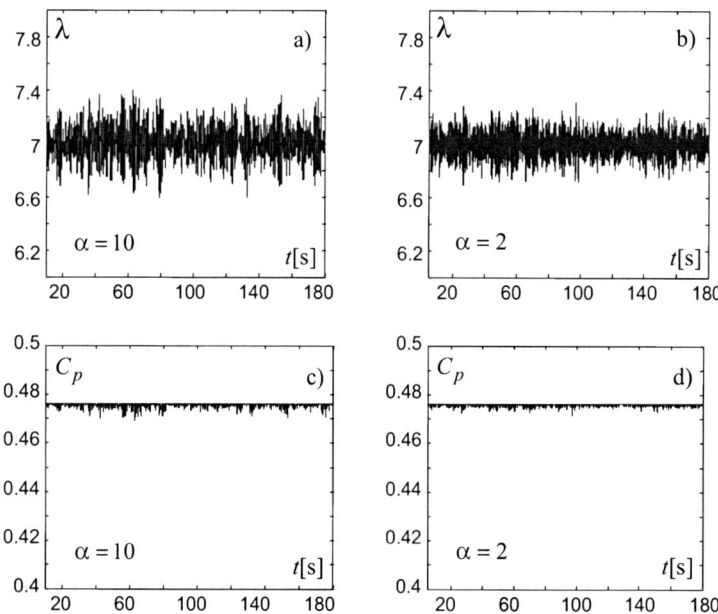

Figure 6.4. Performance of an R-S-T optimally controlled WECS: **a,b** tip speed evolution; **c,d** evolution of the power coefficient

One can note that stressing the penalization of control input variations is not a reasonable decision, because an insignificant improvement of energy performance corresponds to an important alteration of demands on mechanical loading.

For details concerning the MATLAB®/Simulink® implementation of this case study, the reader is invited to consult the folder case_study_9 from the software material.

6.3 Frequency Separation Principle in the Optimal Control of WECS

This section presents an optimal control structure for WECS, aiming at ensuring a trade-off between maximizing the energetic efficiency and minimizing the control input effort. This structure attempts to render unnecessary adaptive structures and is based upon noting that WECS have a two-time-scale dynamics, respectively corresponding to the two spectral ranges identified in the wind speed dynamics (Burton *et al.* 2001): a low-frequency dynamic due to the low-frequency component of the wind speed and a high-frequency dynamic resulted from the action of the turbulent wind speed.

6.3.1 Frequency Separation of the WECS Dynamics

The basic idea of this approach is to deal distinctly with the effects induced by the two components of the wind speed on the WECS dynamics.

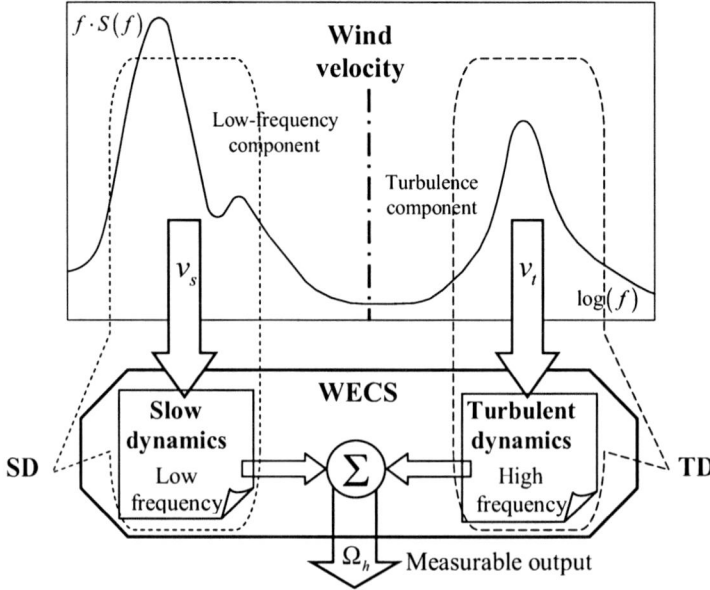

Figure 6.5. Frequency separation principle applied in the WECS dynamical behaviour analysis

The low-frequency component, v_S, determines the average position of the operating point on the wind turbine characteristic (these are slow, *low-frequency* dynamics, considering that their spectrum is entirely placed inside the turbine passband), and the turbulence component, $\Delta v(t) = v - v_S \equiv v_t$, excites the high-frequency dynamics by generating *high-frequency* variations around this point.

Therefore, the WECS global dynamical behaviour, expressed by the evolution of the measurable variables (*i.e.*, generator rotational speed, Ω_h), results from superposing some *slow dynamics* (SD), induced by the low-frequency component, and some *turbulent dynamics* (TD), due to the turbulence component, as suggested in Figure 6.5. Thus, it has led to the idea of separately compensating the two dynamics (Munteanu *et al.* 2004a, 2005), by designing a two-loop control structure, where the loops use estimated values of v_S and respectively v_t from the measured (total) wind speed, v. The system output, Ω_h, serves for estimating the necessary feedback information.

6.3.2 Optimal Control Structure and Design Procedure (2LFSP)

The WECS is torque controlled; it is considered that the electromagnetic subsystem is a first-order element providing electromagnetic torque at a time constant much smaller than that of the drive train.

The general optimization problem, stated in Sections 4.3 and 4.5, is split into two disjoint sub-problems, respectively associated with the two dynamics identified in the WECS. Because the torque variations induced by the low-frequency wind component are negligible, the optimization problem related to the slow dynamics may be reduced to a steady-state control (*i.e.*, λ_{opt} tracking). In contrast, the mechanical fatigue induced by the turbulence wind component is significant, so the optimization problem associated to the turbulent dynamics may be formulated in terms of an LQG dynamic optimization.

Briefly, the two components of the wind speed act separately within two loops of the proposed control structure: a *low-frequency loop* (referred further to as LFL), using v_S within a steady-state optimal controller (tip speed controller – TSC) to drive and maintain the system at a steady-state operating point on the ORC and a *high-frequency loop* (further referred to as HFL), using v_t within a LQG controller, for dynamic optimization, of the linearized system's behaviour around this point. In this way, both energetic optimization in relation to the low-frequency wind speed and mechanical stress alleviation due to the turbulence wind speed are possible.

According to the frequency separation principle, any variable from the system is decomposed into two components: a steady-state one, of low frequency, and a high-frequency one, representing variations around the steady-state value. These two kinds of values describe respectively the two kinds of dynamics and will be dealt with within the corresponding control loops. Let the notation adopted in Section 3.6.1 and introduced by Equation 3.74 for a generic variable x be reconsidered:

$$\overline{x} = x\big|_{\text{steady-state operating point}} \; ; \; \Delta x = x - \overline{x} \; ; \; \overline{\Delta x} = \Delta x / \overline{x}$$

Starting from WECS equations described in Chapter 3, two kinds of linearized models will be developed — one suitable for the LFL and the other one for the HFL. The first one results from a linearization around a steady-state operating point, as imposed by the low-frequency wind speed. The second one uses variables' normalized variations around the steady-state operating point fixed by the former model, such that it captures the high-frequency WECS behaviour.

The steady-state value, $\overline{x}$, belongs to the low-frequency loop and the normalized variation $\overline{\Delta x}$ is treated within the high-frequency loop. Figure 6.6 presents the block diagram of the two-loop optimal control structure based upon the frequency separation principle, referred to in the following by its acronym, 2LFSP. Each of the two loops receives as feedback $\overline{\Omega}_h$ and $\overline{\Delta\Omega}_h$ respectively and separately provides a torque reference component, $\overline{\Gamma}_G$ and $\overline{\Delta\Gamma}_G$ respectively, such that the total torque reference to be fed into the electromagnetic subsystem is the sum of them, $\Gamma_G^* = \overline{\Gamma}_G + \Delta\Gamma_G$. The slow dynamics embeds also the dynamic of the associated exogenous signal, v_s; the same holds in the case of the turbulent dynamics.

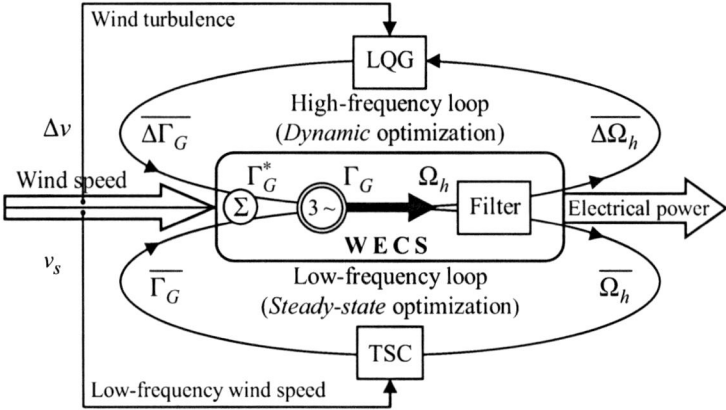

Figure 6.6. Block diagram of the two-loop optimal control structure based upon the frequency separation principle (2LFSP) – reprinted (with modifications) from Control Engineering Practice, 13, Munteanu I, Cutululis NA, Bratcu AI, Ceangă E, Optimization of variable speed wind power systems based on a LQG approach, 903-912, Copyright (2005), with permission from Elsevier

The turbulent dynamics are linearized around the steady-state operating point reached within the low-frequency loop (modelling results can be found in Ekelund 1997 and also in Section 3.6 of this work). Considering that the low-frequency loop achieves the goal of maintaining the operating point on the ORC, then the turbulent dynamics are invariant (Munteanu *et al.* 2005), because the system's most variant variable, the torque parameter, γ (Equation 3.78), is maintained at $\gamma(\lambda_{opt})$, corresponding to energy optimum. This allows solving the optimization problem within the high-frequency loop (that is, associated to the turbulent dynamics) by a LQG invariant design procedure (Munteanu *et al.* 2005), as detailed further, and eliminates the necessity of using any adaptive structures.

The initial, global optimization problem was split into two sub-problems, each of which is intended to be solved to optimality; therefore the solution obtained by aggregating the two separate solutions is in general sub-optimal. One can remark that its closeness to the ideal, optimal solution ultimately depends upon how sharply the two wind dynamics can be separated.

The low-frequency loop aims exclusively at maximizing the WECS energetic efficiency. This simply requires operating the system at variable speed, such that its (steady-state) operating point stays on the ORC. This is called steady-state optimization and tracking the low rotational speed corresponding to λ_{opt}, $\Omega_{l\ opt} = v_s \cdot \lambda_{opt}/R$, is a possibility of achieving it. A tip speed controller (TSC in Figure 6.6) is used for this purpose, whose reference is computed based upon the low-frequency component of the wind speed, v_s, as this one determines the (slow) variation of the operating point. A classical PI and an on-off controller are two possible versions of such a controller, as it has been detailed in Sections 5.3 and 5.4 respectively of this work.

Due to the high inertia of the turbine *vs.* the wind speed variation, a control exclusively based on the steady-state optimization generates large torque variations at the generator's shaft, which could damage the mechanical subsystems (the gear-box for example). It is the high-frequency loop performing the dynamic optimization of the turbulent dynamics of WECS, which can alleviate the mechanical stress, by casting the control problem into an LQ Gaussian optimization problem defined on a linearized model around the steady-state operating point. The associated performance index describes an energy-reliability trade-off in the form of Equation 4.11.

Parameters of the linearized model depend on the operating point, that is, on the low-frequency wind speed, v_s. The torque parameter, γ, depends significantly on the low-frequency wind speed through the tip speed ratio (see Equation 3.78). For example, when λ varies in the range from 5 to 8, the γ parameter changes its sign in the range from 1.5 to –1.5 (Ekelund 1997).

Therefore, the system is variant in relation to the low-frequency wind speed, which is inconvenient for the control strategies design. The gain-scheduling control solution proposed in Ekelund (1997) was supposing the exact knowledge on how parameters, especially γ, depend on the operating point. Or, as the information about $C_p(\lambda)$ is often poor or difficult to obtain, it is practically impossible to compute the value of γ.

The combined action of the two control loops inside the 2LFSP practically desensitizes the system in relation to the static operating point, by cancelling the variation of γ *vs.* the tip speed, λ, as will be illustrated in Sections 6.4 and 6.5.

In conclusion, the design procedure associated to 2LFSP has the major steps described by Algorithm 6.1.

180   6 WECS Optimal Control with Mixed Criteria

Algorithm 6.1. Design procedure of a two-loop optimal control structure based upon the frequency separation principle for WECS

---
1. Measure the wind speed, v, as being the output signal of the anemometer, and HSS rotational speed, Ω_h, from encoder.
2. Obtain the low-frequency wind speed, v_s, by low-pass filtering the signal v.
3. Obtain the turbulence component of the wind speed, $\Delta v \equiv v_t = v - v_s$, and its normalized value, $\overline{\Delta v} \equiv \Delta v / v_s$.
4. Obtain the low-frequency component of the HSS rotational speed, $\overline{\Omega_h}$, by low-pass filtering the signal Ω_h.
5. Obtain the normalized value of the HSS rotational speed high-frequency component, $\overline{\Delta \Omega_h} = \left(\Omega_h - \overline{\Omega_h}\right) / \overline{\Omega_h}$.
6. v_s and $\overline{\Omega_h}$ are fed into the low-frequency loop, whose design may follow either a linear or a nonlinear procedure. The steady-state component of the control input, $\overline{\Gamma_G}$, results as output of the TSC.
7. $\overline{\Delta v}$ and $\overline{\Delta \Omega_h}$ are fed into the high-frequency loop, whose design is based upon the classical LQG synthesis. The dynamic component of the control input, $\Delta \Gamma_G$, results.
8. The total control input (electromagnetic torque), Γ_G^*, is obtained by summing up the steady-state component and the dynamic component: $\Gamma_G^* = \overline{\Gamma_G} + \Delta \Gamma_G$.

6.3.3 Filtering and Prediction Algorithms for Wind Speed Estimation

The low-frequency component of the wind speed, v_s, can in principle be extracted from the (total) wind speed, $v(t)$, by a high-order low-pass filter, whose cut-off frequency must be at most the turbine's bandwidth. However, splitting the wind components by filtering is rather rough, providing an unrealistically delayed version of the low-frequency component and consequently a slightly deformed turbulent component. This solution cannot always be successfully used; for example, flexible drive train WECS can experience stability problems due to large phase lags, no matter the chosen control method is. It could happen that the reference torque level is significantly higher than the wind torque, thus compromising the operation of the wind turbine.

Different ways have been adopted in order to avoid this problem: increasing the cut-off frequency of the low-pass filter, then making use of the filtering properties of a classical PI controller (Munteanu *et al.* 2005), or implementing an on-off controller for zeroing the difference $\lambda - \lambda_{opt}$ (Munteanu *et al.* 2006c). A more accurate estimation of v_s is obtained by combining a reduced-order low-pass filtering (LPF) of the total wind with an ARMA-model-based prediction, as detailed in the following (Figure 6.7).

In order to ease the analysis, let us consider a typical wind sequence, like the one in Figure 6.8. Let v_s^f be the output of the LPF and v_s^p the estimation of v_s by filtering and prediction. The speed references computed using either v_s^f or v_s^p will make the system turn at two tip speed ratios, $\bar{\lambda}^f$ and $\bar{\lambda}^p$, as the turbine experiences the real low-frequency component of the wind speed, v_s (see Figure 6.8).

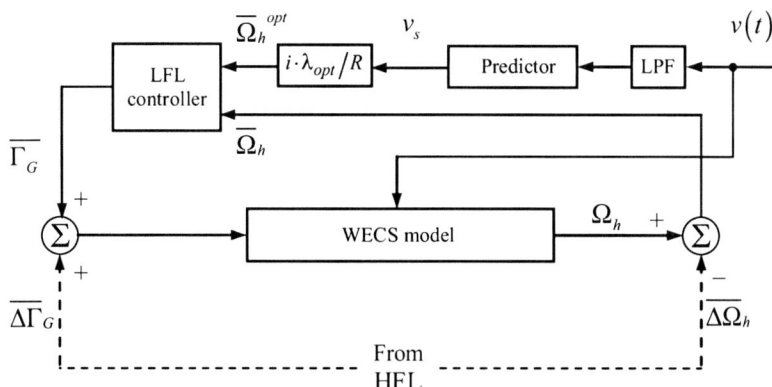

Figure 6.7. Low-frequency loop using predicted low-frequency component of wind speed

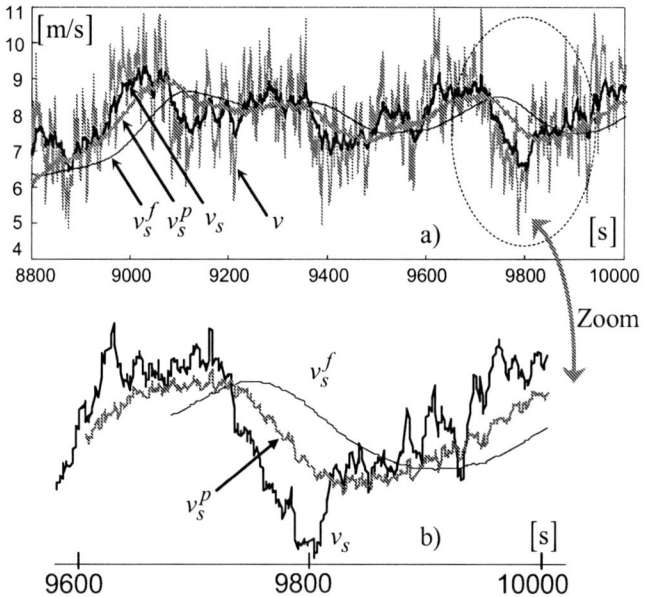

Figure 6.8. a Predicted *vs.* filtered low-frequency wind speed. **b** Zoom

Figure 6.8a shows the superiority of the presented method *vs.* simple low-pass filtering. For example, at time 9800, v_s is much less than v_s^f; a reference computed using v_s^f would be difficult to track, because the real power available is much lower (see Figure 6.8b). Meanwhile, the predicted value, v_s^p, is much closer to the real low-frequency component, v_s, ensuring a better tracking of the ORC within the LFL. One can compare the following standard deviations: $\sigma(\overline{\lambda}^f - \lambda_{opt}) = 0.72$, $\sigma(\overline{\lambda}^p - \lambda_{opt}) = 0.44$.

The ARMA model is recursively implemented for the prediction of the k-th sample of v_s^p from v_s^f:

$$v_{s_k}^p = \sum_{i=1}^{n} a_i \cdot v_{s_{k-i}}^f + \sum_{j=1}^{m} b_j \cdot \left(v_{s_{k-j}}^p - v_{s_{k-j}}^f\right) \qquad (6.19)$$

For the sake of example, one can make the following choice: the prediction horizon is of 2 min, the filter's order is $m=3$, the regressing vector dimension is $n=6$ and a_i, b_j result from a recursive least mean squares procedure with a unitary forget factor (Pike *et al.* 1996).

6.4 2LFSP applied to WECS with Rigidly-coupled Generator

This section deals with applying the 2LFSP, the control strategy detailed in the previous section, to rigid-drive-train variable-speed WECS, based upon torque controlled generators, in order to achieve a multi-purpose optimization, regarded as a balance between energy efficiency and reliability.

First, the modelling approaches for both slow and turbulent dynamics are presented. Next, the main guidelines for designing both the low-frequency and the high-frequency loop are laid on. Finally, the combined action of the two control loops within the proposed structure is largely illustrated and assessed by both numerical (MATLAB®/Simulink® – off-line) and real-time (on-line) simulation in the case of an induction-generator-based low-power WECS. The on-line experiments have been carried out on an electromechanical WECS simulator.

6.4.1 Modelling

Modelling of WECS Slow Dynamics
The slow dynamic of WECS is given by the coupling between the rotor and the rigid drive train, for which a linearized model can be deduced, assuming that the electromagnetic subsystem's dynamic is negligible *vs.* that of the mechanical subsystem. This modelling approach has been presented in Section 3.6 and resulted in a first-order model variant with the low-frequency component of the wind speed, v_s, which captures the torque-controlled WECS slow dynamics – Equation 3.75.

Modelling of WECS Turbulent Dynamics
Ekelund (1997) presented the linearized model with respect to low rotational speed and wind speed of a rigid-drive-train WECS, which has as state variables the

normalized variations of the low rotational speed and of the wind speed respectively. Another form of the linearized in variations model can be obtained by considering as state variables the normalized variation of the low rotational speed around its steady-state value, $\overline{\Delta\Omega_l}$, and the normalized variation of the wind torque around its average value (that is, around its component given by the low-frequency wind speed), $\overline{\Delta\Gamma_{wt}}$.

The state vector and the control input are respectively defined as

$$\mathbf{x}(t) = \begin{bmatrix} \overline{\Delta\Omega_l}(t) & \overline{\Delta\Gamma_{wt}}(t) \end{bmatrix}^T \quad u(t) = \overline{\Delta\Gamma_G}(t) \quad (6.20)$$

Equation 3.77, giving the analytical expression of the linearized wind torque characteristic, is time derived:

$$\dot{\overline{\Delta\Gamma_{wt}}}(t) = \gamma \cdot \dot{\overline{\Delta\Omega_l}}(t) + (2-\gamma) \cdot \dot{\overline{\Delta v}}(t) \quad (6.21)$$

In Equation 6.21 the low-speed shaft motion equation is replaced as

$$\dot{\overline{\Delta\Omega_l}}(t) = \frac{1}{J_T} \cdot \left(\overline{\Delta\Gamma_{wt}}(t) - \overline{\Delta\Gamma_G}(t)\right), \quad (6.22)$$

where

$$J_T = \frac{\overline{\Omega_l} \cdot J_l}{\overline{\Gamma_{wt}}} = \frac{\overline{\Omega_h} \cdot J_h}{\overline{\Gamma_{wt}}} \quad (6.23)$$

has dimensions of time constant, with J_h given by Equation 3.66 and J_l given by Equation 3.67. In Equation 6.21 the normalized wind speed variation, modelled as a pseudorandom process yielded by filtering a white noise, $e(t)$, according to a simplified particularisation of the procedure given in Figure 3.6 is also replaced:

$$\dot{\overline{\Delta v}}(t) = \frac{1}{T_w} \cdot \left(e(t) - \overline{\Delta v}(t)\right),$$

where

$$T_w = L_t / v_s \quad (6.24)$$

is the time constant of a first-order shaping filter (similar to Equation 3.8), with L_t being the turbulence length. Both time constants depend on the operating point, that is, on v_s. Finally, one obtains

$$\dot{\overline{\Delta\Gamma_{wt}}}(t) = \frac{\gamma}{J_T} \cdot \left(\overline{\Delta\Gamma_{wt}}(t) - \overline{\Delta\Gamma_G}(t)\right) + \frac{(2-\gamma)}{T_w} \cdot \left(e(t) - \overline{\Delta v}(t)\right) \quad (6.25)$$

As $\gamma \neq 2$ (the wind speed influences the wind torque), Equation 3.77 gives

184   6 WECS Optimal Control with Mixed Criteria

$$\overline{\Delta v}(t) = \frac{1}{(2-\gamma)} \cdot \overline{\Delta\Gamma_{wt}}(t) - \frac{\gamma}{(2-\gamma)} \cdot \overline{\Delta\Omega_l}(t), \qquad (6.26)$$

which is used in Equation 6.25. After some algebra one obtains

$$\overline{\dot{\Delta\Gamma}_{wt}}(t) = \left(\frac{\gamma}{J_T} - \frac{1}{T_w}\right)\overline{\Delta\Gamma_{wt}}(t) + \frac{\gamma}{T_w}\overline{\Delta\Omega_l}(t) - \frac{\gamma}{J_T}\overline{\Delta\Gamma_G}(t) + \frac{(2-\gamma)}{T_w}e(t) \quad (6.27)$$

Equations 6.22 and 6.27 define the linearized state model with respect to low rotational speed and wind torque:

$$\begin{cases} \overline{\dot{\Delta\Omega}_l}(t) = \frac{1}{J_T}\overline{\Delta\Gamma_{wt}}(t) - \frac{1}{J_T}\overline{\Delta\Gamma_G}(t) \\ \overline{\dot{\Delta\Gamma}_{wt}}(t) = \left(\frac{\gamma}{J_T} - \frac{1}{T_w}\right)\overline{\Delta\Gamma_{wt}}(t) + \frac{\gamma}{T_w}\overline{\Delta\Omega_l}(t) - \frac{\gamma}{J_T}\overline{\Delta\Gamma_G}(t) + \frac{(2-\gamma)}{T_w}e(t) \end{cases} \qquad (6.28)$$

Using notation at Equation 6.20 one can introduce the state matrix, **A**, the input matrix, **B** and the exogenous matrix, **L**:

$$\dot{\mathbf{x}}(t) = \underbrace{\begin{bmatrix} 0 & 1 \\ \dfrac{\gamma}{T_w} & \dfrac{\gamma}{J_T} - \dfrac{1}{T_w} \end{bmatrix}}_{\mathbf{A}} \cdot \mathbf{x}(t) + \underbrace{\begin{bmatrix} -\dfrac{1}{J_T} \\ -\dfrac{\gamma}{J_T} \end{bmatrix}}_{\mathbf{B}} \cdot u(t) + \underbrace{\begin{bmatrix} 0 \\ \dfrac{2-\gamma}{T_w} \end{bmatrix}}_{\mathbf{L}} \cdot e(t) \qquad (6.29)$$

The output variable is the normalized variation of the tip speed ratio, $z(t) \equiv \overline{\Delta\lambda}(t)$. This variable is expressed depending on the state vector

$$\overline{\Delta\lambda}(t) = \overline{\Delta\Omega_l}(t) - \frac{1}{(2-\gamma)}\overline{\Delta\Gamma_{wt}}(t) + \frac{\gamma}{(2-\gamma)}\overline{\Delta\Omega_l}(t) = \frac{2}{(2-\gamma)}\overline{\Delta\Omega_l}(t) - \frac{1}{(2-\gamma)}\overline{\Delta\Gamma_{wt}}(t),$$

introducing the output matrix, **C**:

$$z(t) \equiv \overline{\Delta\lambda}(t) = \underbrace{\begin{bmatrix} \dfrac{2}{(2-\gamma)} & -\dfrac{1}{(2-\gamma)} \end{bmatrix}}_{\mathbf{C}} \cdot \underbrace{\begin{bmatrix} \overline{\Delta\Omega_l}(t) \\ \overline{\Delta\Gamma_{wt}}(t) \end{bmatrix}}_{\mathbf{x}(t)} \qquad (6.30)$$

This model is described by the block diagram in Figure 6.9.

The state model has the same form as the one in Ekelund (1997), except for the form of the state vector and of matrices involved. It can be shown that the matrix pair (**A**,**B**) is controllable.

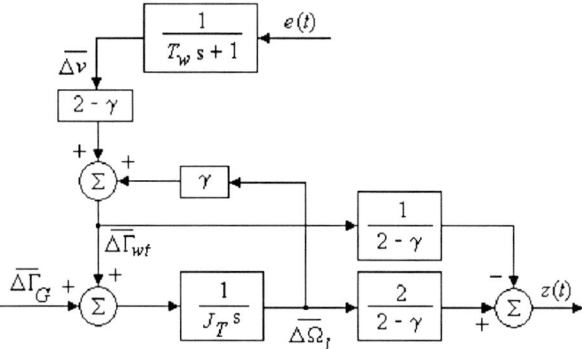

Figure 6.9. Linearized model of a rigid-drive-train WECS with respect to low rotational speed and wind torque

6.4.2 Steady-state Optimization Within the Low-frequency Loop

The steady-state optimization aims at maintaining the optimal tip speed, λ_{opt}, either by tracking the low speed shaft's rotational speed that corresponds to the low-frequency wind speed, v_s, $\overline{\Omega_l}^{ref} = \lambda_{opt}/R \cdot v_s$, or by tracking the maximum wind power. Thus, the low-frequency loop can be based either on a speed controller – of PI type, as detailed in Section 5.3.2, or of on-off type, as shown in Section 5.4 – or on a power controller, as presented in Section 5.3.2.

The simplified block diagram of LFL can be seen in Figure 6.7. The low-frequency component, v_s, is usually obtained by low-pass filtering the signal $v(t)$ provided by the anemometer. A fourth-order Butterworth low-pass filter having the cut-off frequency at most equal with the WECS passband may be used for this goal (Munteanu *et al.* 2005). Estimation of the low-frequency wind speed may be improved by prediction, as briefed in Section 6.3.3.

The low-frequency (steady-state) component of the low rotational speed, $\overline{\Omega_l}$, is obtained also by filtering and its normalized variation results as $\overline{\Delta\Omega_l} = (\Omega_l - \overline{\Omega_l})/\overline{\Omega_l}$ and is applied to the high-frequency loop (HFL) (see the design procedure in the form of Algorithm 6.1 in Section 6.3.2).

The empirical Ziegler–Nichols procedure (Ziegler and Nichols, 1942; Hautier and Caron, 1997) is another possible way for computing the PI controller's parameters. In any case, the choice of these parameters is not critical, because the low-frequency wind speed varies much more slowly comparatively to the WECS dynamics.

6.4.3 LQG Dynamic Optimization Within the High-frequency Loop

Optimization Criterion
The dynamic optimization within the HFL is expressed by a performance index in the Equation 4.11 associated to the linearized state model of Equation 6.29, with

$\mathbf{x}(t) = \begin{bmatrix} \overline{\Delta\Omega_l}(t) & \overline{\Delta\Gamma_{wt}}(t) \end{bmatrix}^T$ as state vector and $u(t) = \overline{\Delta\Gamma_G}(t)$ as control input, thus reflecting a trade-off between minimizing both the output variations and those of the control input variable (Munteanu *et al.* 2004a, 2005). This is a linear quadratic Gaussian optimization (LQG) problem (Athans and Falb 1966). The first component results as a quadratic form of the state variable

$$I_1 = E\left\{\alpha \cdot \overline{\Delta\lambda}^2(t)\right\} \to \min \Leftrightarrow I_1 = E\left\{\mathbf{x}^T(t) \cdot \mathbf{C}_\alpha^T \mathbf{C}_\alpha \cdot \mathbf{x}(t)\right\} \to \min,$$

where matrix **C** is given by Expression 6.30 and $\mathbf{C}_\alpha = \sqrt{\alpha} \cdot \mathbf{C}$.

The reliability constraint is solved by minimizing the control input variations (Maximization of Energy with minimization of control input – MEmci) (Munteanu *et al.* 2004a). The second component is defined as

$$I_2 = E\left\{\overline{\Delta\Gamma_G}^2(t)\right\} = E\left\{\mathbf{u}^T(t) \cdot \mathbf{N} \cdot \mathbf{u}(t)\right\} \to \min,$$

where **N**=1 (the vector notation for *u*(*t*) is used for formal conformity). Therefore, the global index is

$$I = E\left\{\mathbf{x}^T(t) \cdot \left(\mathbf{C}_\alpha^T \cdot \mathbf{C}_\alpha\right) \cdot \mathbf{x}(t) + \mathbf{u}^T(t) \cdot \mathbf{N} \cdot \mathbf{u}(t)\right\} \to \min \quad (6.31)$$

LQG Controller Design
Provided that the LFL is working properly, the parameters of the turbulent dynamic are time invariant; therefore the stated LQG dynamic optimization problem is also time invariant.

Existence and uniqueness of solution are guaranteed if the open-loop system meets a well-known set of structural properties (Lublin and Athans 1996).

The solution of this problem is the torque control input normalized variation obtained as full-state feedback:

$$u(t) \equiv \overline{\Delta\Gamma_G} = -\mathbf{K} \cdot \mathbf{x}(t), \quad (6.32)$$

(see Figure 6.10), with **K** being the feedback matrix gain:

$$\mathbf{K} = \mathbf{R}^{-1} \cdot \mathbf{B}^T \cdot \mathbf{S},$$

where **S** is the unique, symmetric and positive semi-definite matrix satisfying the Riccati algebraic matrix equation:

$$\mathbf{S} \cdot \mathbf{A} + \mathbf{A}^T \cdot \mathbf{S} + \mathbf{C}_\alpha^T \cdot \mathbf{C}_\alpha - \mathbf{S} \cdot \mathbf{B} \cdot \mathbf{R}^{-1} \cdot \mathbf{B}^T \cdot \mathbf{S} = 0$$

The asymptotic stability of the closed-loop system from Figure 6.10, described by the equation $\dot{\mathbf{x}}(t) = (\mathbf{A} - \mathbf{B} \cdot \mathbf{K}) \cdot \mathbf{x}(t)$ (whose elements are previously defined), is guaranteed (Lublin and Athans 1996).

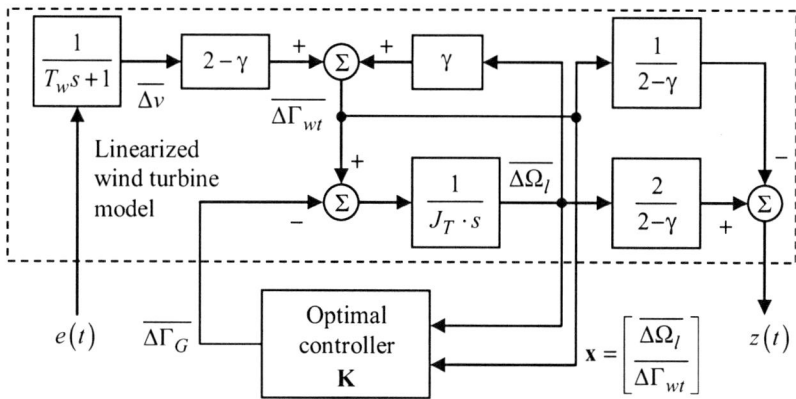

Figure 6.10. High-frequency loop block diagram in the case of a rigid-drive-train WECS – reprinted (with modifications) from Control Engineering Practice, 13, Munteanu I, Cutululis NA, Bratcu AI, Ceangă E, Optimization of variable speed wind power systems based on a LQG approach, 903-912, Copyright (2005), with permission from Elsevier

6.4.4 LQ Dynamic Optimization Within the High-frequency Loop

In Cutululis *et al.* (2006a) the LQ optimization is approached in a *deterministic* context – this can be another way of achieving the dynamic optimization within the HFL. Based upon Equations 6.22 and 3.77, the state model of the system can be written as

$$\dot{\mathbf{x}} = \mathbf{A}\mathbf{x} + \mathbf{B}u + v, \qquad (6.33)$$

where

$$\begin{cases} \mathbf{x} = \overline{\Delta\Omega_l}(t) \quad u = \overline{\Delta\Gamma_G}(t) \\ \mathbf{A} = \dfrac{\gamma}{J_T} \quad \mathbf{B} = -\dfrac{1}{J_T} \quad v = \dfrac{2-\gamma}{J_T}\overline{\Delta v} \end{cases} \qquad (6.34)$$

The optimization criterion is defined here as

$$I = \int_0^\infty \left[\alpha \cdot \overline{\Delta\lambda}^2(t) + \overline{\Delta\Gamma_G}^2(t)\right] dt = \int_0^\infty \left[\alpha \cdot \overline{\Delta\lambda}^2(t) + u^2(t)\right] dt,$$

with $\overline{\Delta\lambda}(t) = \Delta\lambda(t)/\lambda_{opt}$. The role of the positive weighting coefficient, α, is for adjusting the trade-off between energy performance and power quality.

Recall that, by linearizing the tip speed, $\lambda(t) = R \cdot \Omega_l(t)/v(t)$, one obtains $\overline{\Delta\lambda}(t) = \overline{\Delta\Omega_l}(t) - \overline{\Delta v}(t)$; therefore, the above criterion can be rewritten as

$$I = \int_0^\infty \left\{\alpha\left[\overline{\Delta\Omega_l}^2(t) - 2\overline{\Delta\Omega_l}(t)\overline{\Delta v}(t) + \overline{\Delta v}^2(t)\right] + u^2(t)\right\} dt$$

Because $\overline{\Delta v}^2$ is an independent variable, which does not depend on the state, $\mathbf{x} \equiv \overline{\Delta \Omega_l}$, nor on the control input, $u(t)$, it can be eliminated from the criterion, resulting in

$$I_1 = \int_0^\infty \left[\alpha\left(\overline{\Delta \Omega_l}^2 - 2\overline{\Delta \Omega_l} \cdot \overline{\Delta v}\right) + u^2\right] dt = \int_0^\infty \left[\alpha\left(x^2 + 2qx\right) + u^2\right] dt,$$

where

$$q(t) = -\alpha \cdot \frac{J_T}{2-\gamma} \cdot v(t) \tag{6.35}$$

In the general case of a system $(\mathbf{A},\mathbf{B})$, for which an optimal control problem is defined as minimizing a linear quadratic criterion depending on weighting matrices $\mathbf{Q}$ and $\mathbf{R}$, the general solution is

$$u(t) = \hat{u}(t) - \mathbf{L}_1 \cdot x(t),$$

where

$$\hat{u} = -\mathbf{R}^{-1}\mathbf{B}^T p(t) \quad \mathbf{L}_1 = \mathbf{R}^{-1}\mathbf{B}^T \mathbf{P},$$

with $\mathbf{P}$ being the solution of the algebraic Riccati equation and $p(t)$ resulting from the associated algebraic equation:

$$\left(\mathbf{A}^T - \mathbf{PBR}^{-1}\mathbf{B}\right)p(t) + \mathbf{P}v(t) + q(t) = 0$$

Taking account of Equations 6.33 and 6.34 of parameters $\mathbf{A}$ and $\mathbf{B}$ and identifying $\mathbf{Q} = \alpha$ and $\mathbf{R} = 1$, one easy solves the Riccati equation (Cutululis et al. 2006a):

$$\mathbf{P} = J_T\left(\gamma + \sqrt{\gamma^2 + \alpha}\right) \tag{6.36}$$

The associated equation becomes

$$\left(\mathbf{A} - \mathbf{P} \cdot \mathbf{B}^2\right)p(t) + \mathbf{P}v(t) + q(t) = 0 \tag{6.37}$$

or, by replacing $\mathbf{P}$ with Expression 6.36 and $q(t)$ with Expression 6.35:

$$\left[\mathbf{A} - \left(\gamma + \sqrt{\gamma^2 + \alpha}\right) \cdot \frac{1}{J_T}\right] \cdot p(t) +$$

$$+ J_T\left(\gamma + \sqrt{\gamma^2 + \alpha}\right) \cdot v(t) - \alpha \frac{J_T}{2-\gamma} \cdot v(t) = 0$$

6.4 2LFSP applied to WECS with Rigidly-coupled Generator

Because $\mathbf{A} = \dfrac{\gamma}{J_T}$ and $v = \dfrac{2-\gamma}{J_T} \cdot \overline{\Delta v}$,

$$p(t) = \dfrac{\alpha - \left(\gamma \pm \sqrt{\gamma^2 + \alpha}\right)(2-\gamma)}{\sqrt{\gamma^2 + \alpha}} J_T \cdot \overline{\Delta v}(t)$$

Finally, the solution for the state feedback is obtained:

$$\mathbf{L}_1 = -\left(\gamma + \sqrt{\gamma^2 + \alpha}\right) \qquad (6.38)$$

and the open-loop correction is

$$\hat{u}(t) = -\mathbf{R}^{-1}\mathbf{B}^T \cdot p(t) = \mathbf{L}_2 \cdot \overline{\Delta v}(t), \qquad (6.39)$$

where

$$\mathbf{L}_2 = \dfrac{\left(\gamma + \sqrt{\gamma^2 + \alpha}\right)(2-\gamma) - \alpha}{\sqrt{\gamma^2 + \alpha}} \qquad (6.40)$$

The optimal control scheme from Figure 6.11 synthesizes the result of this approach; $\mathbf{L}_1$ and $\mathbf{L}_2$ here are scalars, but vector-matrix notation has been preserved for formal conformity.

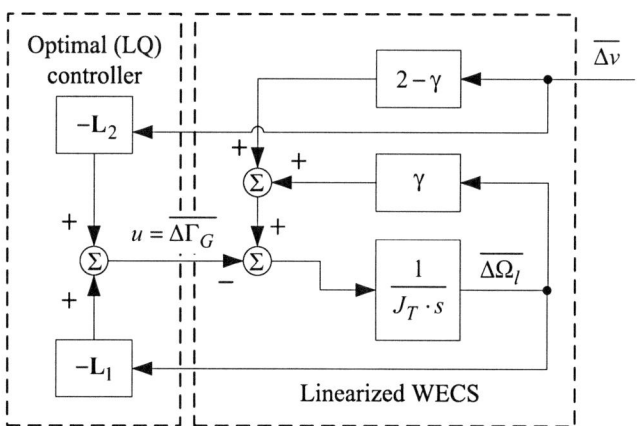

Figure 6.11. LQ optimal control scheme – deterministic approach

For the sake of example, the variations of the tip speed and of the induction generator torque, for two values of the trade-off coefficient, $\alpha = 0.1$ and $\alpha = 10$, are shown in Figure 6.12 in the case of a low-power variable-speed fixed-pitch WECS (Cutululis *et al.* 2006a).

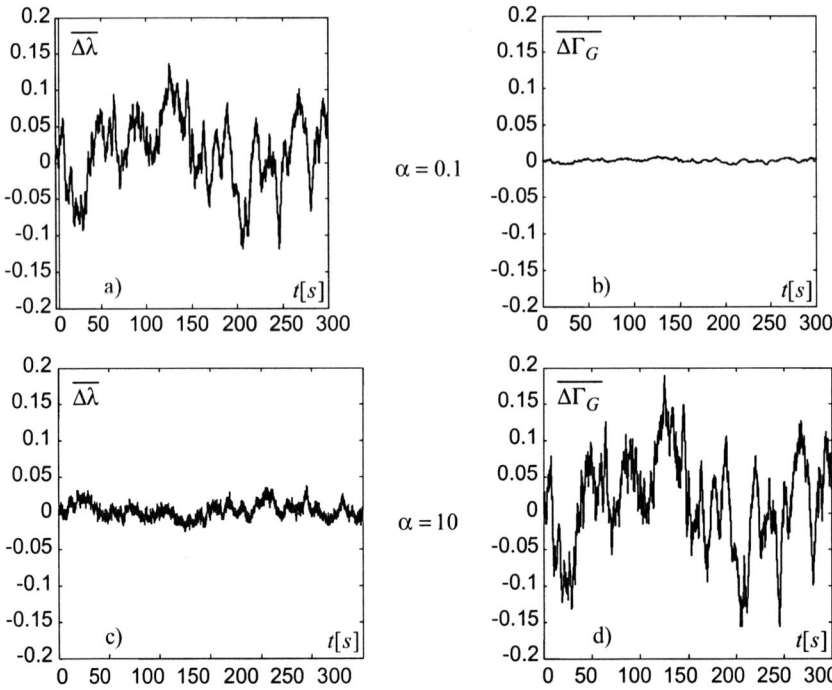

Figure 6.12. Tip speed (**a,c**) and torque (**b,d**) variations for $\alpha = 0.1$ and $\alpha = 10$

6.4.5 Case Study (10)

The inter-conditioned operation of the LFL and HFL into the same optimal control structure has been validated on a low-power (6 kW) variable-speed fixed-pitch rigid-drive-train induction-generator-based WECS by numerical (off-line) simulation. Appendix A and Table A.2 provide features of the used WECS.

The control structures presented above for respectively low- and high-frequency WECS dynamics structures have been combined to obtain an instance of the general Figure 6.6, shown in Figure C.1 from Appendix C.

Below are reported MATLAB®/Simulink® simulations performed on this block diagram within the following setup:
- the total wind speed is obtained by summing up the outputs of two first-order shaping filters fed with the same white noise (similar to the more general procedure suggested in Figure 3.6);
- the low-frequency wind speed results as the output of a fourth-order Butterworth low-pass filter, with 0.1 Hz as cut-off frequency;
- the steady-state electromagnetic torque, $\overline{\Gamma_G}$, is generated by a PI controller, whereas the dynamic electromagnetic torque, $\Delta\Gamma_G$, is provided based on the LQG controller's output;
- filtering blocks have been used for splitting the involved wind and rotational speed components;

- the closed-loop system has been simulated for a 10-min time horizon; for sake of consistency, in all simulation cases, the WECS was excited by the same wind sequence.

Time evolutions of some relevant variables are presented, which show the closeness of the operating point to the optimal one and also the mechanical load intensity induced by the torque variations for constant PI parameters and various values of the trade-off parameter, α.

Figure 6.13 puts into evidence how the trade-off between minimizing the tip speed normalized variations, $\overline{\Delta\lambda(t)}$, and minimizing the electromagnetic torque normalized variations, $\overline{\Delta\Gamma_G(t)}$, is managed, by using different values of α.

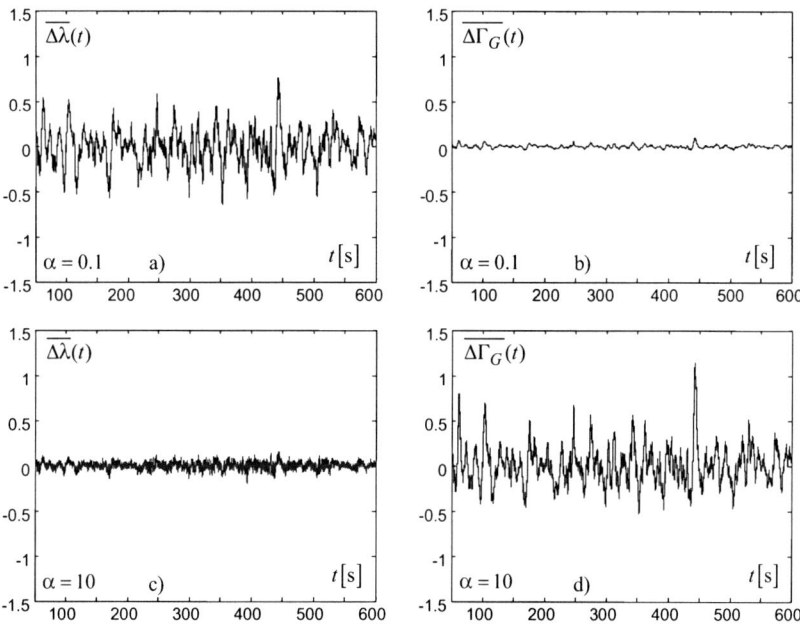

Figure 6.13. Tip speed (**a,c**) and respectively electromagnetic torque (control input – **b,d**) normalized variations for different values of the weighting coefficient, α

Figure 6.14 presents the tip speed and electromagnetic torque evolutions for a small (Figure 6.14a,b) and respectively a large (Figure 6.14c,d) value of α; the conversion regime is as far from optimality as the weighting coefficient is small. The other way round, using a small α ensures reduced torque variations, as the corresponding electromagnetic torque evolutions for the same two values of α show (Figure 6.15).

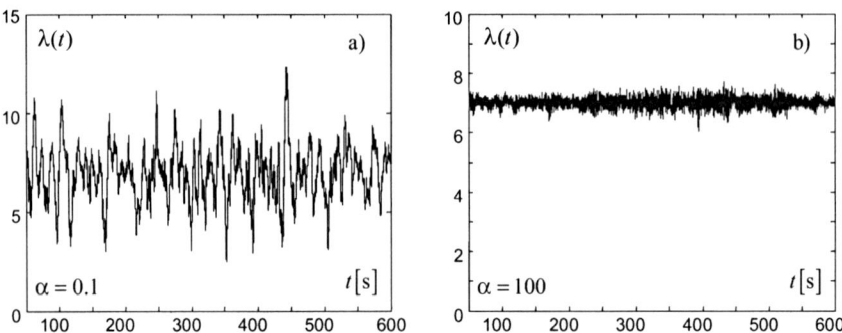

Figure 6.14. Influence of the weighting coefficient, α, on the tip speed evolution for the 2LFSP global operation: small (**a**) and large (**b**) α

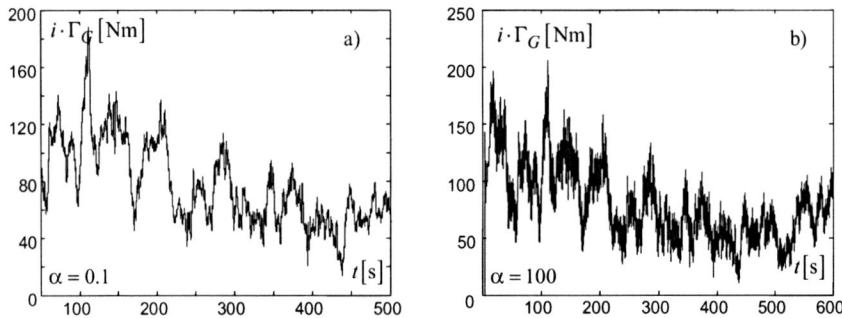

Figure 6.15. Influence of the weighting coefficient, α, on the electromagnetic torque evolution for the 2LFSP global operation: small (**a**) and large (**b**) α

The standard deviations, denoted by $\sigma_\Delta(\cdot)$, of the normalized variations $\overline{\Delta\lambda}(t)$ and $\overline{\Delta\Gamma_G}(t)$ for various values of α are given in Table 6.2 and suggest that a saturation limit is finally reached in increasing α for obtaining a better conversion efficiency.

Table 6.2. Standard deviations of the normalized variations $\overline{\Delta\lambda}(t)$ and $\overline{\Delta\Gamma_G}(t)$ vs. α

σ_Δ/α	α=0.1	α=1	α=2	α=5	α=10	α=20	α=100	α=200
$\sigma_\Delta(\overline{\Delta\lambda})$	0.2	0.12	0.087	0.052	0.037	0.029	0.02	0.019
$\sigma_\Delta(\overline{\Delta\Gamma_G})$	0.019	0.10	0.14	0.19	0.21	0.22	0.24	0.25

Figure 6.16 presents how the operating point moves around the ORC in the plane $\Omega_l - P_{wt}$ for different values of α. Its deviation around the ORC becomes non-significant for hundred-sized values of α, denoting very good aerodynamic efficiency.

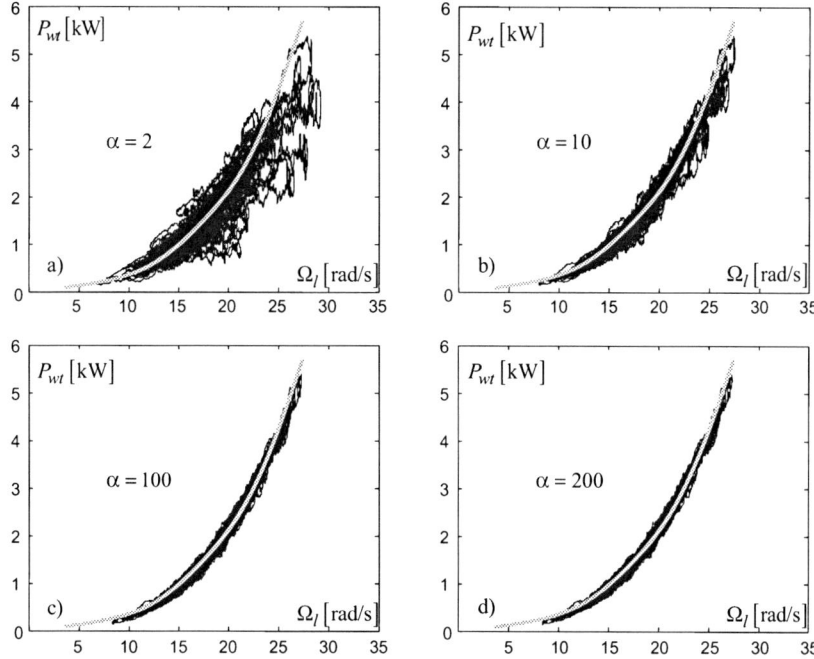

Figure 6.16. Operating point's evolution under the 2LFSP global operation: tracking the ORC

One must note that the weighting coefficient variation range must be limited for the following reasons. For small α (Figure 6.16a,b) the WECS operating point largely varies around the ORC, the conversion regime is far from optimality. Moreover, the system might operate incorrectly, that is, the weak control input variations might induce motoring regimes of the generator. On the other hand, excessively large values of α (Figure 6.16c,d) do not greatly increase the harvested power amount, but can induce significant supplementary mechanical stress.

The reader can find details about the MATLAB®/Simulink® implementation of this case study by consulting the folder case_study_10 in the software material.

6.4.6 Global Real-time Simulation Results

The control strategy based on the frequency separation principle is also validated by real-time simulation using a small-scale WECS electromechanical simulator, whose block diagram can be viewed in Figure C.2 from Appendix C. In Figure C.3 one can see a photo of it and identification of the main components. This simulator is conceived based upon the hardware-in-the-loop simulation concept (detailed in Chapter 7 of this book), having the generator rotational speed as driving variable (Diop *et al.* 1999; Cutululis 2005). Both structures – namely 2LFSP with PI-controller-based low-frequency loop and with on-off-controller-based LFL – have been tested.

194 6 WECS Optimal Control with Mixed Criteria

Global On-line Simulation Results for PI-controller-based LFL
Real-time simulations have been performed using the same parameter values as in the off-line simulation; the control block diagram used here is essentially the same as the one in Figure C.1 (except for some supplementary elements required for interfacing the physical subsystem).

The low-frequency loop is excited by the low-frequency component of the wind speed (Figure 6.17a), taking values in the usual range, from 4 to 10 m/s. Its role – of maintaining the operating point on the ORC, corresponding to the tip speed of λ_{opt} – is illustrated in Figure 6.17b, where the variable γ, which is strongly dependent of λ (see Equation 3.78), exhibits reasonably small variations around the value $\gamma(\lambda_{opt}) = -1$. The amplitude of these variations depends on the low-frequency component of the wind, v_s, which is the output of the separating filter.

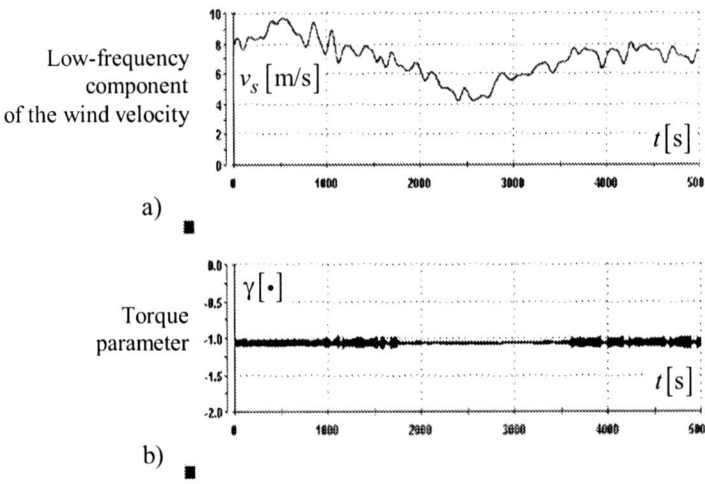

a)

b)

Figure 6.17. Evolution of the PI-controller-based LFL – reprinted (with modifications) from Control Engineering Practice, 13, Munteanu I, Cutululis NA, Bratcu AI, Ceangă E, Optimization of variable speed wind power systems based on a LQG approach, 903-912, Copyright (2005), with permission from Elsevier

Concerning the high-frequency loop, the variables of interest are normalized variations around a static operating point – characterizing the smooth, steady-state regime – i.e., that of the tip speed ratio, $z(t) = \overline{\Delta\lambda(t)}$, and that of the generator torque, which is the control input, $u(t) = \overline{\Delta\Gamma_G}$. Maximum variations of the operating point can be reasonably considered within ±20% around the optimal values.

Simulations have been performed on the test rig described above for several values of the weighting coefficient, α. For each value of α, the state feedback, **K**, has been computed based upon the model's parameters obtained for v_s in the middle of its variation range, e.g., **K**=[–0.11 –0.02] for α=0.2 and **K**=[–6.49 2.34] for α=100.

Figure 6.18 shows how the normalized variations of the tip speed ratio and of the electromagnetic torque depend on α. The suggested qualitative interpretation is that, as was expected, the amplitude of the tip speed ratio normalized variation, $\overline{\Delta\lambda}$, decreases with the value of α, while that of the electromagnetic torque, $\overline{\Delta\Gamma_G}$, increases. These variations are both placed within the band of ±20% around zero, meaning that the unnormalized values are oscillating around the operating point with the same amplitude.

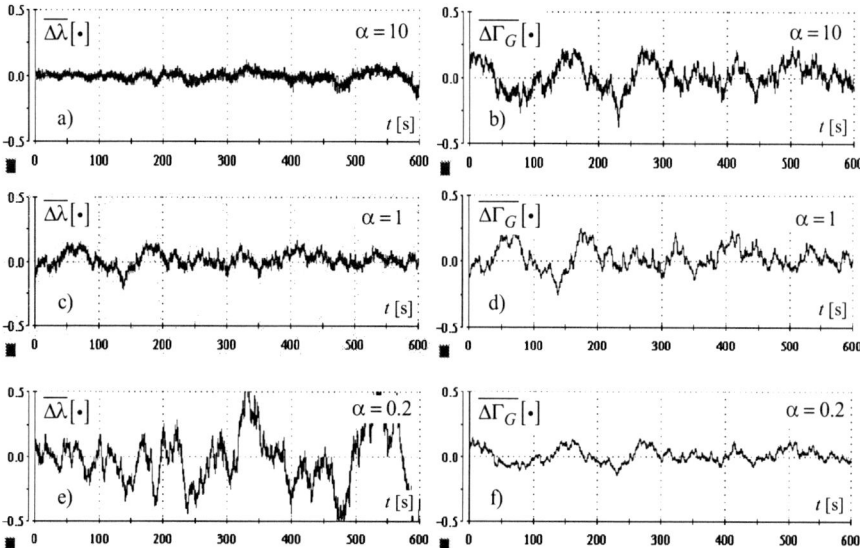

Figure 6.18. Evolution of the tip speed ratio (**a,c,e**) and of the generator torque (**b,d,f**) normalized variations for three values of the weighting coefficient, α, inside the 2LFSP – reprinted (with modifications) from Control Engineering Practice, 13, Munteanu I, Cutululis NA, Bratcu AI, Ceangă E, Optimization of variable speed wind power systems based on a LQG approach, 903-912, Copyright (2005), with permission from Elsevier

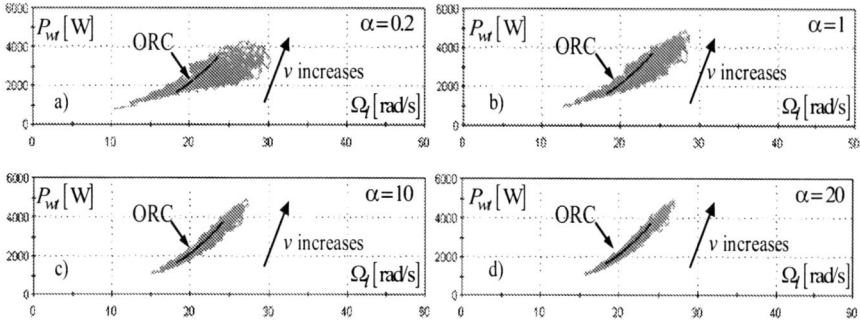

Figure 6.19. Evolution of the operating point: tracking the power optimum within 2LFSP – case of PI-controller-based LFL

196   6 WECS Optimal Control with Mixed Criteria

Figure 6.19 shows the combined functioning of the two loops, namely the variations of the operating point around the ORC for four values of α. As expected, for small α (Figure 6.19a,c), these variations are significantly larger than those for large α (Figure 6.19b,d) and increase as the wind speed increases.

Global On-line Simulation Results for On-off-controller-based LFL
For real-time validation of the 2LFSP having the low-frequency loop built upon an on-off controller, the simulation diagram from Figure C.1 has been used, where the LFL is replaced with the scheme from Figure 5.25. The on-off controller parameters are the same as in the case study reported in Section 5.4.2. Generally speaking, the on-line simulation results in this case are similar with those obtained when using PI-controller-based LFL under the same conditions.

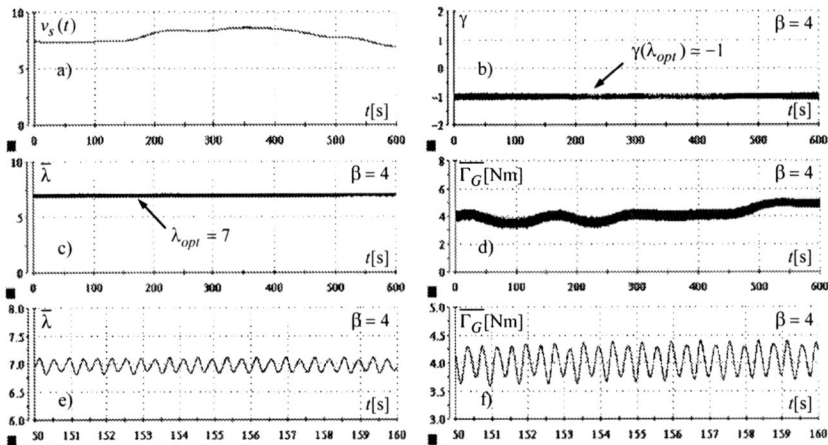

Figure 6.20. Evolutions of relevant variables within the on-off-controller-based LFL

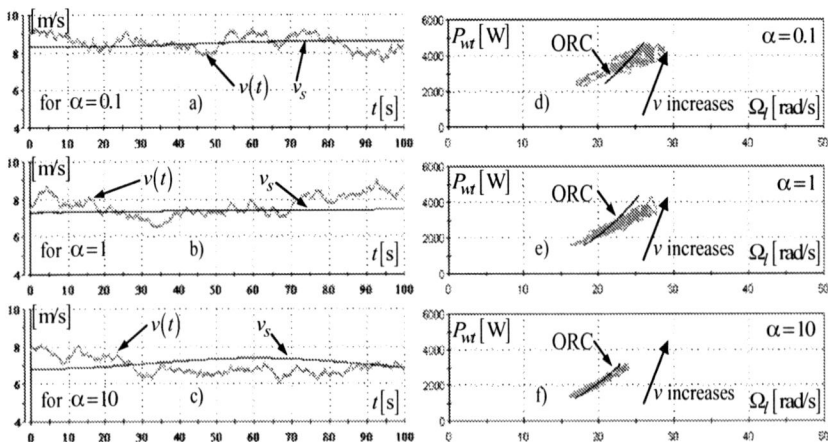

Figure 6.21. Evolution of the operating point: tracking the ORC within 2LFSP – case of on-off-controller-based LFL

The switched control amplitude is fixed at $\beta = 4$; the modified (robustified) controller has been used in simulation tests (see Section 5.4.1). The low-frequency wind speed evolution and that of the torque parameter, γ, can be seen in the first row of Figure 6.20. One can note electromagnetic torque oscillations of sufficiently reduced amplitude, as well as tip speed oscillations under 2% (these are sensibly larger than those noted in off-line simulation conditions).

Figure 6.21 depicts the excursion of the operating point around the ORC (Figure 6.21d–f) for different wind sequences (Figure 6.21a–c) and for three different values of the weighting coefficient ($\alpha \in \{0.1; 1; 10\}$).

6.5 2LFSP Applied to WECS with Flexibly-coupled Generator

In this section it is shown how the 2LFSP, the control strategy described in Section 6.3 of this work, can be applied to flexible-drive-train variable-speed torque-controlled-generator-based WECS, for trading off the energetic requirements and those of reliability. The mechanical stress and electrical power harmonics induced by the wind turbulences are usually alleviated if the turbine rotor interacts with the electromagnetic subsystem through an elastic-coupling-based drive train. The main differences of this approach from the application to rigid-drive-train WECS, exposed in the previous section, consist of adopting a different performance index for the LQG dynamic optimization in the HFL, in using an observer-based state feedback and finally in using a prediction method for estimating the low-frequency wind speed (in LFL), instead of simply low-pass filtering it from the total wind, as presented in Section 6.3.3 (Figure 6.22).

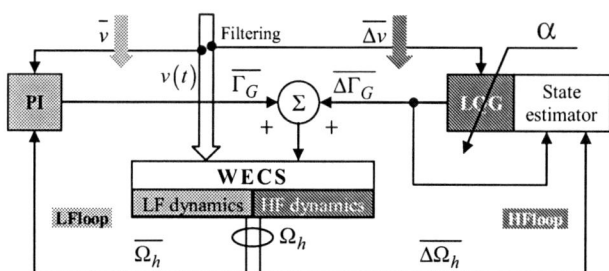

Figure 6.22. General 2LFSP control diagram for flexible-drive-train WECS

In order to assess the overall control efficiency, MATLAB®/Simulink® simulation results are discussed for a low-power flexible-drive-train induction-generator-based WECS case study.

6.5.1 Modelling

Modelling of WECS Low-frequency Dynamics
The drive train exhibits high-frequency dynamics in the sense adopted by the frequency separation principle in the WECS behaviour. Therefore, the drive train

dynamics are to be considered within the HFL and, consequently, the low-frequency dynamics of WECS will be modelled using the same approach as in the rigid-drive-train case (see Section 3.6), resulting in the transfer function at Relation 3.75. Consequently, the associated steady-state optimization control can be treated in the same manner, *e.g.*, by using a PI controller.

Modelling of WECS High-frequency Dynamics
The modelling of the high-frequency (turbulent) dynamic of a flexible drive train WECS uses here a linearized model in normalized variations around the static operating point established by the LFL (Munteanu *et al.* 2006a).

The model of a flexible drive train has been obtained in Section 3.4.2; it is given by Equation 3.68. The nonlinear wind torque characteristic of a fixed-pitch turbine (Equation 3.23) is linearized around a steady-state operating point, yielding Equation 3.77, where the linearized model of turbulence is the same as in the rigid-drive-train case:

$$\dot{\overline{\Delta v}}(t) = -1/T_w \cdot \overline{\Delta v}(t) + 1/T_w \cdot e(t),$$

with T_w being the time constant of the first-order shaping filter (Equation 6.24).

Choosing $\mathbf{x}(t) = \begin{bmatrix} \dot{\overline{\Delta \Omega_l}} & \dot{\overline{\Delta \Omega_h}} & \dot{\overline{\Delta \Gamma}} & \dot{\overline{\Delta \Gamma}}_{wt} \end{bmatrix}^T$ as state vector, $u(t) = \overline{\Delta \Gamma}_G$ as control input and the normalized variation of the tip speed ratio, $z(t) = \overline{\Delta \lambda}(t)$, as output (measure) variable, by coupling Equations 3.77 and 3.68, one obtains the following state space matrix equation:

$$\begin{cases} \dot{\mathbf{x}}(t) = \mathbf{A} \cdot \mathbf{x}(t) + \mathbf{B} \cdot u(t) + \mathbf{L} \cdot e(t) \\ z(t) = \mathbf{C} \cdot \mathbf{x}(t) \end{cases}, \qquad (6.41)$$

with

$$\begin{cases} \mathbf{A} = \begin{bmatrix} 0 & 0 & -1/J_T & 1/J_T \\ 0 & 0 & 1/J_G & 0 \\ K_A & -K_A & -B_S\left(1/J_B + 1/J_g\right) & B_S/J_B \\ \gamma/T_w & 0 & -\gamma/J_T & \gamma/J_T - 1/T_w \end{bmatrix} \\ \mathbf{B} = \begin{bmatrix} 0 & -1/J_G & B_S/J_g & 0 \end{bmatrix}^T \\ \mathbf{L} = \begin{bmatrix} 0 & 0 & 0 & (2-\gamma)/T_w \end{bmatrix}^T \\ \mathbf{C} = \begin{bmatrix} 2/(2-\gamma) & 0 & 0 & -1/(2-\gamma) \end{bmatrix} \end{cases}, \qquad (6.42)$$

where γ is the torque parameter (Equation 3.78), $J_T = \overline{\Omega_l} \cdot J_{wt} / \overline{\Gamma}_{wt}$ and $J_G = \overline{\Omega_l} \cdot i^2 \cdot J_g / \overline{\Gamma}_{wt}$ are time constants, whereas $K_A = \overline{\Omega_l} \cdot K_s \cdot i^2 / \overline{\Gamma}_{wt}$ is the inverse of a time constant. J_T, J_G and K_A depend on the static operating point, $\left(\overline{\Omega_l}, \overline{\Gamma_{wt}}\right)$. Equations 6.41 and 6.42 represent the linearized model of a fixed-pitch flexible-drive-train WECS, describing the high-frequency dynamics around the static operating point, impressed by the turbulence component of the wind speed, $\Delta v(t)$. The parameters of the model depend on the steady-state operating point (established in LFL), so ultimately on the wind speed.

6.5.2 Steady-state Optimization Within the Low-frequency Loop

The control problem associated with the LFL concerns the steady-state optimization, which consists of operating a wind turbine at variable speed such that its static operating point remains on the ORC. As in the rigid-drive-train case (Section 6.4.2), this goal is achieved by tracking the rotor speed corresponding to λ_{opt}, $\overline{\Omega_l}^{ref} = \lambda_{opt}/R \cdot v_s$, by means of a PI controller, to compensate the (weak) nonlinearities in the neighbourhood of an usual operating point. The control practice shows that the PI choice is not critical, even more in the LFL, where the low-frequency wind speed varies sufficiently slowly in relation to the wind turbine dynamics. As was previously shown, a general property of the LFL is that the mostly variable parameter of the linearized system, γ, is maintained constant with respect to the low-frequency wind speed at $\gamma = -1$, and so the system at Equation 6.41 is invariant in relation to this parameter (Munteanu *et al.* 2005).

The LFL provides the static component of the electromagnetic torque, $\overline{\Gamma_G}$ (applied to the high-speed shaft), necessary to drive the operating point of the low-speed shaft on the ORC (Figure 6.7). One can note that an estimation of the low-frequency component of the wind speed, v_s, by combining the usual low-pass filtering (LPF) of the total wind with an ARMA-model-based prediction (as detailed in Section 6.3.3), is more suitable in this case, in order to avoid stability problems due to time delays.

6.5.3 Dynamic Optimization Within the High-frequency Loop

Optimization Criterion
The most stressed mechanical part of a WECS is the drive train, supporting both the wind torque variations, $\Delta\Gamma_{wt}$, at the input shaft, and those of the electromagnetic torque, $\Delta\Gamma_G$, at the output shaft. Here the energy-reliability trade-off will be expressed differently from the rigid-drive-train case, taking account of the energy accumulation in the flexible drive train, namely between the minimization of the tip speed variations around λ_{opt} and the minimization of the total load excitations, that is, $\overline{\Delta\Gamma_{wt}}(t) + \overline{\Delta\Gamma_G}(t)$. This means to track the wind torque variations (Maximization of Energy with wind torque tracking – MEwtt) (Munteanu *et al.* 2004a) and the associated performance index becomes

$$I = E\left\{\alpha\left(\overline{\Delta\lambda}\right)^2\right\} + E\left\{\left(\overline{\Delta\Gamma_{wt}}(t) + \overline{\Delta\Gamma_G}(t)\right)^2\right\} \to \min, \quad (6.43)$$

where the weighting coefficient, α, plays the well-known role, of adjusting the energy-reliability trade-off. The first component results as a quadratic form of the state variable

$$I_1 = E\left\{\alpha \cdot \overline{\Delta\lambda}^2(t)\right\} \to \min \Leftrightarrow I_1 = E\left\{\mathbf{x}^T(t) \cdot \mathbf{C}_\alpha^T \mathbf{C}_\alpha \cdot \mathbf{x}(t)\right\} \to \min,$$

where $\mathbf{C}_\alpha = \sqrt{\alpha} \cdot \mathbf{C}$, with matrix $\mathbf{C}$ given in Equation 6.42. The global performance criterion may be then put into the following form:

$$I = E\left\{\mathbf{x}^T(t) \cdot \underbrace{\left[\left(\mathbf{M}^T \cdot \mathbf{M}\right) + \left(\mathbf{C}_\alpha^T \cdot \mathbf{C}_\alpha\right)\right]}_{\mathbf{R}_{xx}} \cdot \mathbf{x}(t) + \\
+ 2 \cdot \mathbf{x}^T(t) \cdot \underbrace{\mathbf{M}^T}_{\mathbf{R}_{xu}} \cdot u(t) + u^T(t) \cdot \mathbf{R}_{uu} \cdot u(t) \right\} \to \min \quad (6.44)$$

where $\mathbf{M} = [0\ 0\ 0\ 1]$ and $\mathbf{R}_{uu} = 1$.

LQG Controller Design
Supposing that the LFL is working, form at Equation 6.44 corresponds here also, to a linear quadratic invariant Gaussian optimization problem. The existence and the uniqueness of the solution are guaranteed if and only if a well-known set of conditions concerning the structural properties of the controlled system is verified (Lublin and Athans 1996). The unique optimal control input minimizing index at Equation 6.44 for the dynamic system given by Equations 6.41 and 6.42 is the full-state feedback law, $u(t) = -\mathbf{K} \cdot \mathbf{x}(t)$, with $\mathbf{K} = \mathbf{R}_{uu}^{-1} \cdot \left(\mathbf{R}_{xu}^T + \mathbf{B}^T \cdot \mathbf{S}\right)$, where $\mathbf{S}$ is the unique, symmetric and positive semi-definite matrix satisfying the Riccati matrix equation

$$\mathbf{S} \cdot \mathbf{A}_r + \mathbf{A}_r^T \cdot \mathbf{S} + \left(\mathbf{R}_{xx} - \mathbf{R}_{xu} \cdot \mathbf{R}_{uu}^{-1} \cdot \mathbf{R}_{xu}^T\right) - \mathbf{S} \cdot \mathbf{B} \cdot \mathbf{R}_{uu}^{-1} \cdot \mathbf{B}^T \cdot \mathbf{S} = 0,$$

where $\mathbf{A}_r = \mathbf{A} - \mathbf{B} \cdot \mathbf{R}_{uu}^{-1} \cdot \mathbf{R}_{xu}^T$. The asymptotic stability of the closed loop, described by $\dot{\mathbf{x}}(t) = (\mathbf{A} - \mathbf{B} \cdot \mathbf{K}) \cdot \mathbf{x}(t)$, is guaranteed.

This approach assumes that all the states are available for measure, which is not the case here. In fact, only $\overline{\Delta\Omega_h}$ from the state vector is measurable with reasonable costs. Therefore, an observer is used to compute state variables that are not accessible from the plant. Similar to the above design of the LQ controller, the observer design consists of computing a constant vector, $\mathbf{L_o}$, so that its transient response is to be faster than the response of the controlled loop (HFL), in order to yield a rapidly updated estimation of the state vector.

The $\mathbf{L_o}$ vector can be obtained using a pole-placement procedure and the observer canonical form (see, for example, the design methodology given in Nise (2000). The observer embeds the model at Equation 6.41 of the high-frequency WECS behaviour and has the expression below:

$$\begin{cases} \dot{\hat{\mathbf{x}}}(t) = \mathbf{A} \cdot \hat{\mathbf{x}}(t) + \mathbf{B} \cdot u(t) + \mathbf{L_o} \cdot \left(y(t) - \hat{y}(t) \right) \\ \hat{y}(t) = \mathbf{C_o} \cdot \hat{\mathbf{x}}(t) \end{cases}$$

In this case the transient response of the observer is required to be much faster than that of the uncontrolled plant, e.g., 100 times for the normalized variation of the wind torque, $\overline{\Delta\Gamma_{wt}}$, and 15 times for the other state variables.

6.5.4 Case Study (11)

MATLAB®/Simulink® simulations have been carried out on the low-power flexible-drive-train fixed-pitch SCIG-based WECS featured in Appendix A and Table A.4, in order to test the 2LFSP performance. The general block diagram is essentially that depicted in Figure 6.22. Both LFL and HFL are built around the corresponding controllers and designed according to suitable specifications, namely λ_{opt} and the weighting coefficient, α, respectively.

First of all, simulations showing the observer efficiency are presented in Figure 6.23. A very good performance of the internal torque estimation, $\widehat{\Delta\Gamma}$, can be noted in Figure 6.23a; the same result applies also for $\overline{\Delta\Omega_l}$ and $\overline{\Delta\Omega_h}$ variables from the state vector. For the wind torque variation, $\overline{\Delta\Gamma_{wt}}$, the estimation performance is satisfactory, as suggested by Figure 6.23b.

The observer model has been embedded in the global simulation diagram; in the following the most representative simulation results picked up from the two loops are presented.

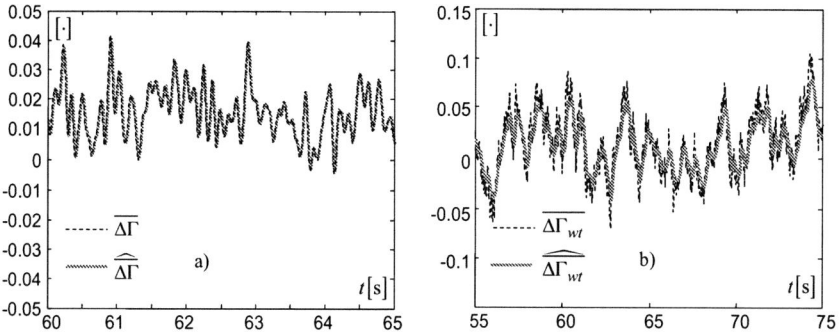

Figure 6.23. Simulation results showing the observer performance (flexible-drive-train case)

202   6 WECS Optimal Control with Mixed Criteria

The wind speed sequence exciting the turbine rotor can be noted in Figure 6.24a. The LFL performance can be assessed from Figure 6.24b: the steady-state value of the tip speed is maintained around its optimal value ($\lambda_{opt}=7$) with satisfactory small deviations (under 5%).

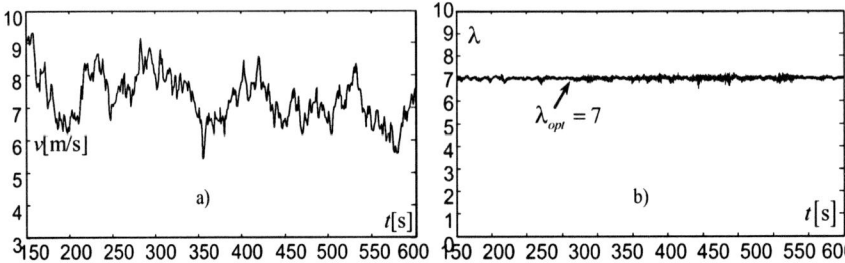

Figure 6.24. 2LFSP for flexible-drive-train WECS: steady-state tip speed ratio kept optimal (LFL)

Figure 6.25 shows how the normalized variations of the tip speed ratio, $\overline{\Delta\lambda}$ (Figure 6.25a,b), and of the sum of torque variations, $\overline{\Delta\Gamma_{wt}}(t)+\overline{\Delta\Gamma_G}(t)$ (Figure 6.25c,d), depend on α for the same wind sequence (Figure 6.24a).

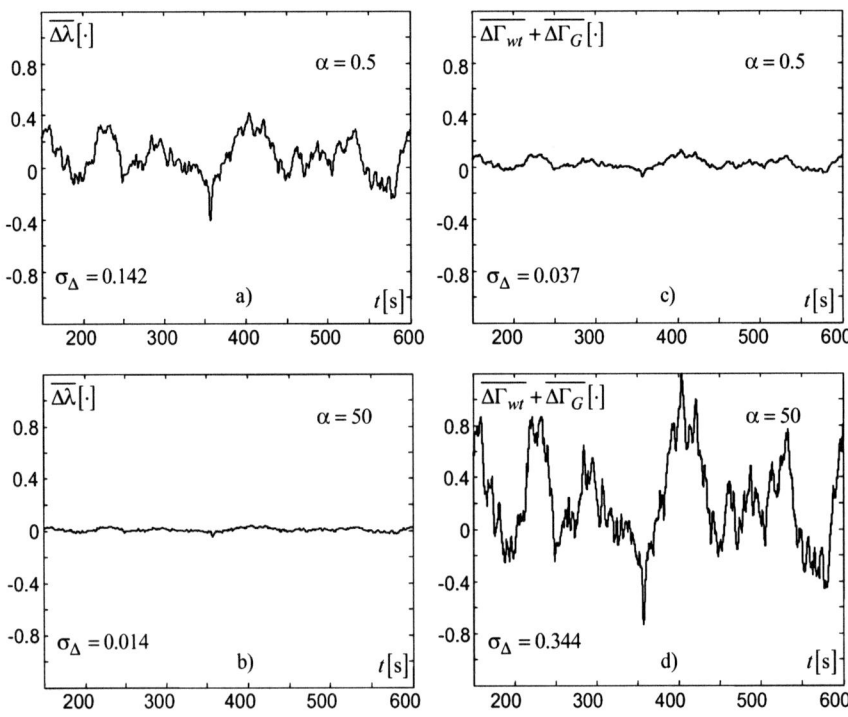

Figure 6.25. HFL operation for flexible-drive-train WECS: **a,b** evolution of the speed ratio variations; **c,d** evolution of the sum of torque variations

As was expected, the standard deviation of the tip speed normalized variation (expressing the energetic efficiency) decreases with α, while that of the sum of torques standard deviation increases (and so does the mechanical stress). One can note that, when α increases from 0.5 to 50, the standard deviation of $\overline{\Delta\lambda}$ decreases from 0.142 to 0.014, while the standard deviation of $\overline{\Delta\Gamma_{wt}(t) + \Delta\Gamma_G(t)}$ increases from 0.037 to 0.344.

Figure 6.26 shows the aerodynamic efficiency evolutions (Figure 6.26a,b) and the excursion of the operating point in relation to the ORC (Figure 6.26c,d) for the wind speed sequence in Figure 6.24a and for the same values of α as in Figure 6.25. Small α (Figure 6.26a,c) leads to large deviations from optimal operating regime, so poor energetic efficiency; in contrast, good efficiency is obtained for large values of α (Figure 6.26b,d), when the deviations from $C_{p\,max}$ become insignificant. Also, one can note that the control law performance depends on the wind speed, since the operating point trajectory in the (Ω_l, P_{wt}) plane is not "centred" on the ORC; the controller becomes unreliable for large wind speed values.

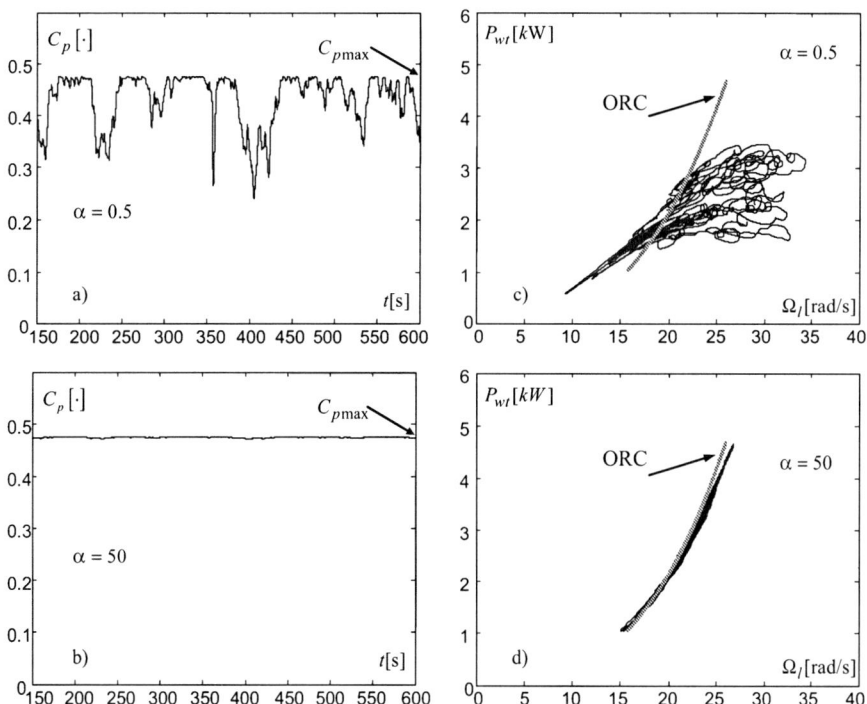

Figure 6.26. 2LFSP operation: tracking the optimal operating regime (flexible-drive-train case)

The reader can find the MATLAB®/Simulink® implementation files of this case study in the folder case_study_11 from the software material.

6.6 Concluding Remarks on the Effectiveness of 2LFSP

The frequency separation principle control approach developed in the previous sections supposes the separate processing of the two components identified in the wind speed spectrum and employs a two-loop control configuration (2LFSP). This structure is intended to be a more successful alternative to the previously developed adaptive optimal WECS control using LQG controllers.

Using the low-frequency wind speed component – obtained by low-pass filtering – one makes a steady-state optimization by maintaining the operating point on the ORC; the induced mechanical stress is insignificant as there are no high-frequency torque variations involved. This is done by employing either a PI controller or an on-off controller for maintaining the average tip speed ratio at its optimal value; this results in fact in a rotational speed tracking structure. This low-frequency loop has a double role: beside the above stated one, it maintains constant the most variable parameter of the linearized (in variations) model describing the high-frequency WECS dynamics, namely the torque parameter. The turbulent, high-frequency wind component is treated within a high-frequency loop built around a LQG controller. This controller ensures optimal dynamic behaviour of the WECS around the steady-state operating point established by the low-frequency loop. Its parameters are computed by solving a LQG optimization problem associated to the linearized model describing the high-frequency WECS behaviour, whose performance index expresses a trade-off between maximisation of the captured wind energy and minimization of the drive train mechanical fatigue (given by the control or/and wind variations).

Defining the quadratic performance index depends upon the drive train type, which can be either rigid or flexible. In the first case, the goal is to limit the generator torque variations, while optimizing the energy regime. The associated LQG problem has been developed on a suitable linearized model of the rigid-drive-train WECS. In the second case the optimization index expresses the scope of limiting the cumulated variations of the electrical generator and wind torques – which constitutes the total mechanical fatigue experienced by the drive train – while optimizing the energy regime. The associated LQG problem has been developed on the linearized model of the WECS with flexible drive train.

The 2LFSP approach confers flexibility in WECS operation, as the parameters ensuring the energy-reliability trade-off can be changed by the operator when encountering variable wind conditions (higher turbulent winds will require a controller based on a smaller trade-off coefficient). For example, in the 2LFSP case, this can be done by switching the LQG controller – its parameters being off-line *a priori* computed – such that torque variations remain within some admissible limits provided by the turbine manufacturer.

The 2LFSP control structure is quite sensitive to parametric or modelling uncertainties, but provides an easier to handle control solution. Also, the separation of the two wind components represents an inconvenient because of the delays induced by filtering.

The frequency separation approach exhibits some points to improve as follows. For guaranteeing a certain service lifetime without failure, the value of the trade-off coefficient must be related to admissible values of the high-frequency torque variations provided by the drive train manufacturer (or wind turbine integrator) for a specific WECS. Considering the electrical generator efficiency captured into the mathematical expression of the energy sub-criterion could also be a point of interest for future works. These remarks hold for both rigid- and flexible-drive-train WECS.

It appears that the frequency separation application to the flexible-drive-train WECS results in quite important deviations of the operating point from the operating regimes characteristic for high winds. This can happen due to some non-negligible prediction errors, to PI controller loss of robustness or may be a combined dysfunction of the two loops. If parametric variations of the high-frequency WECS model are the main cause of the above-mentioned effect, a gain-scheduling approach may be envisaged for correction.

6.7 Towards a Multi-purpose Global Control Approach

6.7.1 Control Objectives in Large Wind Power Plants

Unlike in the case of small- and medium-power WECS, the control objectives of megawatt-power WECS are more diversified. At this level of rated power the tower dynamics need to be taken into account, because they may decisively influence the global dynamic properties. The modes induced by the blade dynamics and by the drive train must also be included in the global mathematical model. The tower, the turbine, the drive train and the electrical generator compose an interaction chain exhibiting complex dynamic behaviour, both in relation with its internal variables (rotational speed, power factor, mechanical loads, *etc.*) and with the grid. Under these circumstances, the control demands must be reconsidered.

Some of the WECS control functions have been previously analyzed, among them:
- basic functions: active power limitation at the rated value in full load, along with shaft rotational speed and sometimes torque limitation;
- energy conversion optimization in partial load;
- mixed-criterion optimization, expressing a trade-off between ORC tracking and alleviation of electromagnetic torque variations.

In general, the requirements listed above do not cover all the technical and economical demands imposed by the exploitation practice, especially in the case of high-power WECS. A "good control" (Leithead *et al.* 1991) must comply with a larger number of demands, either generic or depending on each particular system. Taking account of the control structure depicted in Figure 6.27, one can ask in addition to the harvested power maximization that (De La Salle *et al.* 1990; Leithead *et al.* 1991):
- the global (multi-megawatt) system stability be ensured, provided that the turbine and tower dynamics are considered;
- the shaft mechanical loads be minimized;
- the mechanical loads at tower base also be minimized;
- the gearbox loads be minimized;

– the generator loads be minimized;
– the power fluctuations, especially the flicker, be minimized.

The above-mentioned mechanical loads can be either fatigue or extreme loads.

In the literature various requirements from those above are analyzed, which are either distinctly or in concurrent pairs dealt with. Thus, Veers and Butterfield (2001) highlight the importance of the extreme loads for the multi-megawatt WECS operating in the turbulence field; some solutions for alleviating the mechanical loads at the shaft and gearbox are presented in Leithead *et al.* (1991) and Burton *et al.* (2001). Sørensen *et al.* (2002, 2006) and Cutululis *et al.* (2007) analyze the problem of assessing power fluctuations of large wind farms, whereas the flicker is treated in Saad-Saoud and Jenkins (1999) and Camblong *et al.* (2002a,b).

In Sections 6.4 and 6.5 of this chapter the LQG optimal control problem has been approached, in a version based upon the *frequency separation principle*, in order to solve contradictory demands within the pair composed of the tip speed ratio variation, $\Delta\lambda(t)$, and the electromagnetic torque variation, $\Delta\Gamma_G(t)$. As a matter of fact, most of the above control objectives can be reached by frequency separation control design. The general framework of this approach, as well as the control solutions for some specific goals, is presented in the following.

6.7.2 Global Optimization vs. Frequency Separation Principle for a Multi-objective Control

In the general case, a simplified control block diagram illustrating the control objectives associated to high-power WECS can look like Figure 6.27. This diagram shows the main parts of the controlled system, as well as two control channels: the pitch-control and the speed-/torque-control channel respectively (Burton *et al.* 2001).

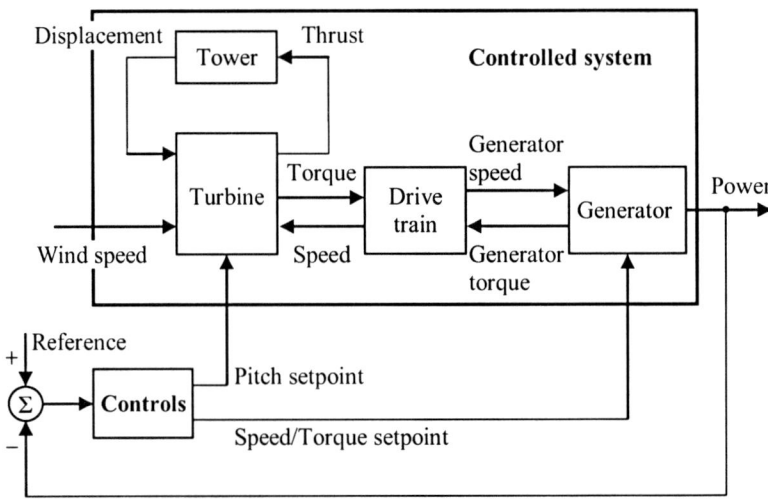

Figure 6.27. Main control subsystems in the case of a high-power WECS

In addition to having accurate information about the low-frequency wind speed component, correct knowledge of the statistical properties of the wind speed fluctuations due to wind-turbine interaction is of crucial importance when aiming at meeting the afore stated requirements.

Under variable-speed operation the system can be either torque- or speed-controlled in partial load in order to track the ORC. In high winds, the WECS is pitch-controlled in order to limit the active power at its rated value. Under fixed-speed operation the WECS is also pitch-controlled. Obviously, the two control channels, which are visible in Figure 6.27, possess different spectral properties, due to the different physical nature of the control variables involved.

The WECS optimal control problem can be stated as the extremization of a *global performance index* including all the control objectives. Such an approach, mentioned by Leithead *et al.* (1991), involves an integral index being used in the form

$$I = \int_{-\infty}^{\infty} \left\{ \left[\sum_i a_i |H_i(j\omega)|^2 \right] \cdot S_E(\omega) + a_c \cdot |H_c(j\omega)|^2 \cdot S_E(\omega) \right\} d\omega, \quad (6.45)$$

where
- $S_E(\omega)$ is the power spectral density of the effective wind speed, the one experienced by the system;
- $H_i(s)$ are the transfer functions from the equivalent wind speed input to output variables (mechanical loads, gearbox torque variation, fluctuation of power delivered to the grid, *etc.*), which make the object of performance demands mentioned in the previous section;
- $H_c(s)$ is the transfer function from the equivalent wind speed input to control variable;
- a_i are weighting factors expressing the relative importance of terms within the global index; parameter a_c weights the control input effort, in order for it to range between feasible limits.

An alternative to this approach is the frequency separation principle, which proposes the "modular" design of the control law: each term of the total control law responds to one or to a group of performance demands (the latter when the demands are grouped as contradictory pairs). The argument behind this solution is the fact that one can establish disjoint regions in the passband of the frequency model, each of which correspond to specific dynamics, envisaged by the control goals. In other words, it is assumed that the control problems can be frequency decoupled. Thus, if a specific goal addresses properties within a well-delimited frequency range and the system is excited by one or many of the effective wind speed components in this range, then the control law can be designed independently. Under these conditions, the global control law is obtained by summing up the components corresponding to the various specific goals contained in the global index.

Application of the frequency separation principle requires knowledge of the frequency models of the system and of the exogenous signals. To this end, it is

important to know:
- the system modes;
- how the major spectral components of the effective wind speed are placed in relation with the system modes;
- the bandwidth of the two control channels illustrated in Figure 6.27.

These problems are basic for designing control laws by means of this approach; they are overviewed below.

6.7.3 Frequency-domain Models of WECS

Apart from the low-frequency model (*i.e.*, the turbine rotor motion equation), the WECS dynamics include some other components like, for example:
- dynamics of the tower movement. These correspond to two degrees of freedom: movement in the wind speed direction ("tower fore-aft") and in the rotation plane of the turbine rotor ("tower side-to-side") respectively. The first dynamic has a great impact on the global dynamic behaviour of WECS;
- dynamics induced by the blade moving. The blade movement in the rotation plane of the rotor is termed "blade edge", whereas the movement in the direction perpendicular to this plane is termed "blade flap". In general, the blade movement is modelled as a distributed-parameter dynamic system, which is usually described by a finite umber of dominant modes (Dixit and Suryanarayanan 2005);
- dynamics of the drive train;
- dynamics of the electromagnetic subsystem of the generator, having the electromagnetic torque as output variable. In general, this is a fourth-order model, but it is habitually approximated by a second-order one (Iov 2003). The approximation is justified by the fact that the stator modes are placed at high frequencies, well beyond the open-loop passband.

Essential information concerning the oscillating modes induced by the dynamic components mentioned above is provided by the natural frequencies and the damping factors. These dynamics are often described individually, ignoring their interaction with each other. Thus, for the tower fore-aft one can use the simple equation

$$M \ddot{y} + D \dot{y} + K y = F, \qquad (6.46)$$

where y is the tower displacement, F is the exerted force, and M, D and K are the tower modal mass, the damping factor and the modal stiffness respectively. Such a model captures the resonant mode characterizing the tower oscillation, but it is far from completely describing the influence of the tower dynamic on the whole system properties. Various couplings between the presented dynamic components are used, aiming at obtaining improved frequency models. Thus, the tower fore-aft mode is strongly coupled with the blade flap mode. Dixit and Suryanarayanan (2005) consider the coupling between the blade edge motion and the drive train dynamic, while Iov (2003) analyses the coupling between the fourth- and second-order model of the electromagnetic subsystem and the drive train dynamics.

A unitary and general modelling approach relies on an equation of the form

$$[M]\ddot{x}+[D]\dot{x}+[K]x = f, \quad (6.47)$$

where $\ddot{x}$, $\dot{x}$ and x are respectively the vectors of accelerations, speeds and displacements and f is the vector of forces/torques exerted on the system. Vector f can be composed of:
- disturbing forces/torques due to the effective wind speed;
- forces/torques generated by the control inputs respectively acting on the two control channels depicted in Figure 6.27.

Dominguez and Leithead (2006) use a model of the form at Equation 6.47, showing two modes for the tower dynamics, two modes for the blade dynamics and two modes for the drive train.

Starting from the general model at Equation 6.46 one can deduce the linearized model, which allows the system modes to be identified. The goal is to analyze whether the control input components are frequency separable, along with identifying those frequency ranges reflecting interesting properties from the control objective viewpoint.

6.7.4 Spectral Characteristics of the Wind Speed Fluctuations

Let us consider a point on the turbine blade, placed at distance r from the rotation axis. As presented in Section 3.2.3, the wind speed fluctuations in the point considered have a deterministic component and a stochastic one, which will be analyzed in the following.

Deterministic Fluctuations
The deterministic components of the effective wind speed fluctuations are induced by the wind shear and by the tower shadow respectively.

The *wind shear* is due to the wind speed variation at the given point in relation to the height from the ground. It is determined by the friction forces exerted by the air masses moving near the Earth. The wind shear is usually characterized through the following model:

$$v_s(z) = v_s\left(z^{ref}\right) \cdot \frac{\ln(z/z_0)}{\ln\left(z^{ref}/z_0\right)}, \quad (6.48)$$

where z is the height from the ground where the wind speed is computed, z^{ref} represents the reference height (typical value: $z^{ref} = 10$ m) and z_0 is the roughness length. When a blade is turning on at Ω_l, the height to the ground, $z(t)$, has a periodic variation in the range $[h-r, h+r]$, modelled as $z(t) = h + r\cos(\Omega_l t + \varphi_0)$, where h is the hub height to the ground and φ_0 defines the initial position of the blade. By replacing this expression of $z(t)$ in Equation 6.48, one sees that the wind speed, $v_s(z(t))$, contains a periodic component.

The *tower shadow* effect results in a decreasing fluctuation of the wind torque, when the blade passes in front of the tower. The wind speed at the point considered point, *i.e.*, at distance r from the rotation axis, decreases considerably when the blade is in front of the tower. Sørensen *et al.* (2002) use the following tower shadow model for an up-wind turbine:

$$v(t) = v_s a^2 \cdot \frac{x^2(t) - y^2(t)}{\left[x^2(t) + y^2(t)\right]^2}, \quad (6.49)$$

where a is the radius of the tower cylindrical section, with $x(t)$ and $y(t)$ being the current values of the components of the distance from the considered point to the tower centre in the lateral and the longitudinal directions, respectively. Another simple model of the tower shadow is proposed by Bianchi *et al.* (2006), whereas Dolan and Lehn (2005) give the normalized wind torque model, taking account of both the tower shadow and of the wind shear effect.

Stochastic Fluctuations

The analytical determination of the rotationally-sampled power spectrum of longitudinal wind speed fluctuation relies upon the von Karman's fixed-point turbulence spectrum. Let us consider the model at Equation 3.3 of the longitudinal component of the fixed-point turbulence. The rotationally-sampled power spectrum can be determined by a three-step procedure (Burton *et al.* 2001), as follows.

1. Determine the autocorrelation function of the fixed-point turbulence in a point placed in front of the turbine. This can be obtained from Equation 3.3 by applying the Inverse Fourier Transform to the power spectral density:

$$K_{v_t}(\tau) = \frac{2\sigma^2}{\Gamma(1/3)} \cdot \left(\frac{\tau/2}{1.34 L_t v_s}\right)^{1/3} \cdot K_{1/3}\left(\frac{\tau}{1.34 L_t v_s}\right), \quad (6.50)$$

where σ is the standard deviation, L_t is the turbulence length, $\Gamma(\cdot)$ is the gamma function and $K_{1/3}(\cdot)$ is the modified Bessel function of second kind and order 1/3.

2. Deduce the autocorrelation function of the wind speed fluctuation in the given point on the blade, *i.e.*, at distance r from the rotation axis. It is assumed that, after a time interval τ, the point in front of the turbine covers longitudinally the distance s and reaches at the considered point on the blade. By analyzing the cross-correlation function between the wind turbulent fluctuations at the mentioned points, one deduces the autocorrelation function at the point on the blade (Burton *et al.* 2001):

$$K_{v_t}(r,\tau) = \frac{2\sigma^2}{\Gamma(1/3)} \left(\frac{s/2}{1.34 L_t}\right)^{1/3} \left[K_{1/3}\left(\frac{s}{1.34 L_t}\right) + \frac{s}{2(1.34 L_t)} K_{2/3}\left(\frac{s}{1.34 L_t}\right) \cdot \left(\frac{2r \sin(\Omega_l \tau/2)}{s}\right)^2 \right], \quad (6.51)$$

where

$$s = \left[v_s^2 \tau^2 + 4r^2 \sin^2\left(\frac{\Omega_l \tau}{2}\right) \right]^{1/2} \quad (6.52)$$

3. Determine the power spectral density based upon the autocorrelation function (Equation 6.51).

Figure 6.28 presents the rotationally-sampled power spectrum for the next parameters: $L_t = 70$ m, $\Omega_l = 30$ rpm, $v_s = 9$ m/s. Two points have been considered, namely at distances $r_1 = 15$ m and $r_2 = 30$ m from the rotation axis. One can see that the effect of power concentration at frequencies $i \cdot \Omega_l/(2\pi)$ is as important as distance r is larger. The power spectral density of the fixed-point turbulence is obtained for $r = 0$ and is represented by dotted line in Figure 6.28.

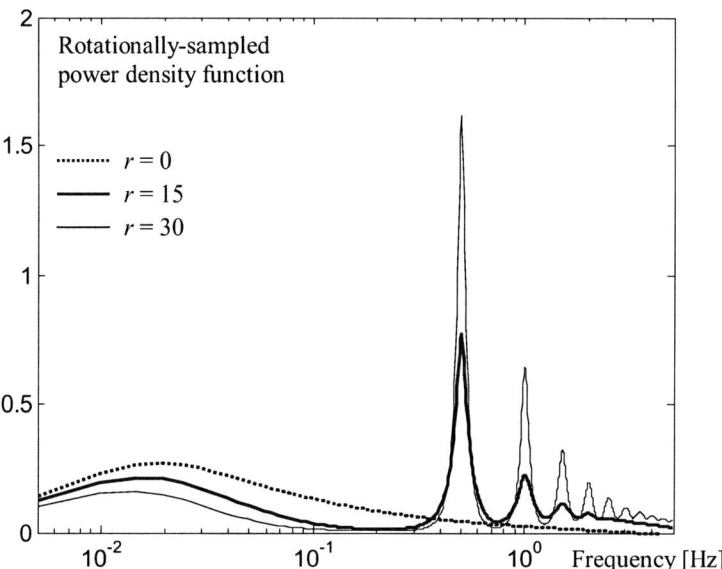

Figure 6.28. Rotationally-sampled power spectrum for different values of the distance to the rotation axis

6.7.5 Open-loop Bandwidth Limitations of WECS Control Systems

In this section the interest is focused on assessing the way in which the two control inputs shown in Figure 6.27 can implement various components of a control law designed through the frequency separation principle. To this end, the closed-loop dynamic properties will be analyzed both for the pitch-controlled, as well as for the speed-/torque-controlled system.

WECS Control by Pitch Setpoint

Pitch control is achieved through the pitch actuator. Limitations imposed on the pitch rate sensibly diminish the passband of the system. The basic function of this control channel is to limit the power at the rated, whereas the power loop operates mainly in the low-frequency range. The dynamic properties of this latter loop, synthetically expressed through the open-loop passband, are more or less adequate to limit power variations due to wind gusts and to medium-frequency wind turbulence components. A supplementary function likely to be realized based upon the frequency separation principle is the damping of the tower vibration (Burton *et al.* 2001). In the case of multi-megawatt turbines, the capacity for implementing an effective control significantly reduces when applied beyond the low-frequency range, due to the combined effect of the turbine modes and tower modes on the WECS global dynamics. These changes render the control system stability the most important issue. This problem is detailed below.

In the case of high-power WECS, the interaction between the turbine and the tower has a great influence on the control system properties. One of the consequences is the non-minimum phase global behaviour, due to right-half-plane zeros being present in the transfer functions of the linearized model. Dominguez and Leithead (2006) have analyzed the dynamic processes inducing right-half-plane zero pairs. One of these pairs appears on the channel from the pitch angle to the generator rotational speed; another one is present in the transfer function from the rotor torque to the hub torque.

The natural frequency of the first complex zero pair is practically constant, whereas its very small damping factor depends on wind speed. For certain wind speed values, the right-half-plane zeros may migrate to the left-hand-side of the complex plane, as a consequence of the pitch angle adjustment influencing the flap and edge mode. This zero pair induces a strong limiting effect on the power loop stability margin. Physically, losing stability can be explained in the following way: when the wind speed increases, the controller increases the pitch angle, in order to limit the torque and implicitly the power (limitation by pitch-to-feather). But when the pitch angle increases, the thrust decreases, and so does the tower deflection; therefore, the turbine attempts to move forward. Thus, the instantaneous wind speed relative to the rotor increases and the controller will attempt to increase further the pitch angle.

The second zero pair has both the damping factor and the natural frequency strongly dependent on the structural characteristics of the turbine and on the wind speed. Unlike in the case of the zeros induced by the tower dynamics, in this case the damping is sensibly larger.

Dominguez and Leithead (2006) presented a method for determining the maximum value of the open-loop cross-over frequency that ensures a certain stability margin of the system having right-half-plane zeros. To this end, the power open-loop transfer function is written as

$$H(s) = H_{mp}(s) \cdot H_{nmp}(s), \qquad (6.53)$$

where $H_{mp}(s)$ and $H_{nmp}(s)$ describe the minimum, respectively non-minimum phase subsystems. The non-minimum phase subsystem embeds all-pass filters, having transfer functions of the form

$$H_{apf}(s) = \frac{s^2 - 2\varsigma\omega_n s + \omega_n^2}{s^2 + 2\varsigma\omega_n s + \omega_n^2} \quad (6.54)$$

and phase characteristics

$$\varphi_{apf}(\omega) = -\mathrm{atan}\left(\frac{2\varsigma\omega\omega_n}{\omega_n^2 - \omega^2}\right) \quad (6.55)$$

One can consider that, in the neighbourhood of the cross-over frequency of system $H_{mp}(s)$, the Bode plots can be approximated through ideal curves: the magnitude plot is linear, with slope $-20 \cdot k$ dB/dec, and the phase plot is practically constant, at $-k \cdot \pi/2$. To system $H_{mp}(s)$ one consecutively adds the all-pass filters embedded in $H_{nmp}(s)$. Let us assume that the filter at Equation 6.54 is added first and impose that the system have the phase margin γ:

$$-k \cdot \pi/2 + 2\varphi_{apf}(\omega_c) = -\pi + \gamma, \quad (6.56)$$

where ω_c is the open-loop cross-over frequency. By replacing ω by ω_c in Equation 6.55, one obtains

$$\left(\frac{\omega_c}{\omega_n}\right)^2 - \frac{2\varsigma}{\tan(\varphi_{apf}(\omega_c))} \cdot \left(\frac{\omega_c}{\omega_n}\right) - 1 = 0, \quad (6.57)$$

where $\varphi_{apf}(\omega_c)$ is deduced from Equation 6.56:

$$\varphi_{apf}(\omega_c) = -0.5\pi(1 - k/2) + \gamma \quad (6.58)$$

By solving Equation 6.57, one obtains a generic solution, $(\omega_c/\omega_n) = S(k,\gamma)$, which must have physical meaning. Let ω_π be the phase cross-over frequency. Similar to Equation 6.57, from the definition of ω_π one gets

$$\left(\frac{\omega_\pi}{\omega_n}\right)^2 - \frac{2\varsigma}{\tan(0.5\pi(1 - k/2))} \cdot \left(\frac{\omega_\pi}{\omega_n}\right) - 1 = 0, \quad (6.59)$$

whose generic solution is $(\omega_\pi/\omega_n) = R(k)$. By virtue of the assumption of linear magnitude plot with slope $-k \cdot 20$ dB/dec, one obtains that the amplitude margin, m_{dB}, verifies the relation $m_{dB} = 20k \cdot (\log(\omega_\pi) - \log(\omega_c))$, from which it results that

$$\left(\frac{\omega_\pi}{\omega_c}\right)^k = \left(\frac{\omega_\pi/\omega_n}{\omega_c/\omega_n}\right)^k = \frac{R(k)}{S(k,\gamma)} = 10^{(m_{dB}/20)} \quad (6.60)$$

By imposing γ and m_{dB}, one can solve Equation 6.60 for k, then deduce the maximum cross-over frequency, ω_c, from $(\omega_c/\omega_n) = S(k,\gamma)$. This value corresponds to the situation when the right-half-plane zero pair associated with the all-pass filter (Equation 6.54) has been added in the system model. The above procedure continues for all the all-pass filters from the system (Dominguez and Leithead 2006).

WECS Control by Generator Speed/Torque Setpoint
This type of control takes place by means of static power converters, being effective over a very large frequency range. In general, this kind of control can embed all the components that are possibly designed based upon the frequency separation principle. A special problem is the power (and eventually speed) limitation function, which is usually achieved by controlling the pitch angle.

As shown in Chapter 4, a variable-speed wind turbine can be stall-controlled by decreasing the rotational speed reference and moving the operating point in the wind torque-speed plane to the left. In high winds, the power controller imposes the speed reference decreasing; thus, the turbine will operate at a smaller than the optimal incidence angle. The aerodynamic efficiency decreases, such that the power limitation effect is obtained. This power control solution is mentioned by Burton *et al.* (2001), but it is poorly illustrated in the literature. At present, in the case of high-power variable-speed WECS, the power regulation by pitch control is considered a usual solution, being the most widely real-world implemented. Power control is more frequently performed by pitch-to-feather than by pitch-to-stall, despite the fact that the pitch actuator is more intensively solicited. A problem related to pitch-to-stall is the evolution of the lift and drag coefficients: when the incidence angle increases, the lift coefficient decreases and so does the aerodynamic damping in the blade bending modes. In this way, unstable blade movement is possible (Burton *et al.* 2001). In spite of all this the passive-stall regulation often represents a widely-accepted solution. Thus, the stall control achieved by generator speed control at underrated values deserves a more detailed analysis. Obviously, one must take account of all the system variables: the lift and drag forces, the axial thrust force and the rotational torque, and also the modes resulting from coupling the turbine dynamics with all the other dynamics in the system. This approach is motivated by the advantage of the generator speed/torque control: covering all the spectral ranges involved in the control objectives.

6.7.6 Frequency Separation Control of WECS

The LQG optimal control problem approached in Sections 6.4 and 6.5 has been defined in the spectral range of the wind turbulence and "notched-off" the global control problem by frequency separation. It has been shown that the high-power WECS have a richer set of control requirements, which are generated by two main types of factors:
– intrinsic *factors*: existence of some local properties of the frequency-domain mathematical model allowing specific control solutions for improving the dynamic behaviour in the concerned frequency range;

- extrinsic *factors*: presence of periodic or near-periodic components in the wind speed spectrum, as has been shown in Section 6.7.3; the control requirements aim at alleviating the influence of these components on power variations and flicker.

The first category of factors includes the dynamic modes of the mechanical structure, including the tower. Considering the tower vibration and the drive train torsional resonance is of great importance. Damping the tower oscillations is also important, not only from the point of view of mechanical loads, but also for ensuring the control system stability within a sufficiently large frequency range (see the previous section).

Let us consider the simplified mathematical model of the tower in the form of the second-order differential equation (Burton *et al.* 2001):

$$\ddot{y} + 2\varsigma\omega_n \dot{y} + \omega_n^2 y = k_T (F + \Delta F), \qquad (6.61)$$

with

$$\omega_n = \sqrt{K/M} \quad \varsigma = 0.5D\sqrt{M/K} \quad k_T = 1/M, \qquad (6.62)$$

where y is the tower displacement, F is the exerted force, M, D and K are the tower modal mass, the damping factor and the modal stiffness, and ΔF is the thrust variation due to controlling the pitch angle. A solution for increasing the damping is to apply a control input which is proportional to $-\dot{y}$. Let $-k_D \dot{y}$ be the thrust variation produced by this component. In this case, the tower damping will be

$$\varsigma_D = 0.5 \cdot (D + k_D) \cdot \sqrt{M/K} \qquad (6.63)$$

Using a control input proportional with $-\dot{y}$ involves either a state estimator being employed (*e.g.*, a Kalman filter), or the displacement variation speed being measured by means of some reliable devices (accelerometer followed by integrator – Burton *et al.* 2001). Frequency separation of this control component is both useful and necessary too, in order to eliminate the noisy components from the speed transducer output signal.

The same control principle can be adapted to the drive train torsional vibration control. Let

$$J\ddot{\theta} + D\dot{\theta} + K\theta = \Gamma_{wt} - \frac{i}{\eta}\Gamma_G \qquad (6.64)$$

be the turbine motion equation, where J, D and K have already been introduced. In this case it is necessary that the generator torque contains a term proportional to the generator speed: $k_D(i/\eta)\dot{\theta}$. According to Equation 6.64, this term produces an increase of the damping factor; its weight in the system dynamics is established through parameter k_D.

To implement this idea, in Bossanyi (2000) and Burton *et al.* (2001) it is proposed that the torque reference is modified by a term acting in a certain frequency range, namely aiming at drive train damping. This term is obtained from processing the encoder signal by using the filter

$$H_{DT}(s) = \frac{k_D s(\tau s + 1)}{s^2 + 2\varsigma_1 \omega_{nt} s + \omega_{nt}^2}, \quad (6.65)$$

where the ω_{nt} frequency is equal to the resonance frequency of the torsional oscillations. The Bode plots of this filter can look like Figure 6.29; its main features are:
- high resonance gain;
- low gain at low frequency; in this way, the filter output signal is separated from the low-frequency dynamics of the system;
- zero-phase at resonance, ω_{nt}; this requirement is achieved by the derivative component inducing a phase lag close to 90° for frequencies smaller than ω_{nt}.

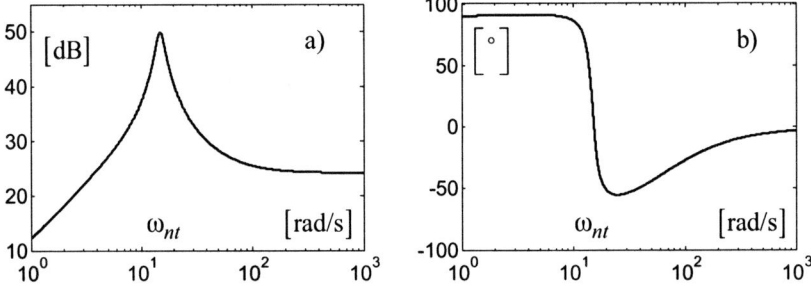

Figure 6.29. Possible Bode plots of a filter used in order to damp drive train oscillations

Dixit and Suryanarayanan (2005) have adopted a gain-scheduling-type solution in order to damp oscillations, provided that the natural frequency, ω_{nt}, changes with the pitch angle, β. To this end, two selectable filters are used, both having the transfer function of the form at Equation 6.65, with parameters computed for two average values of β. These values correspond to two different operating conditions, determined by *a priori* established mean wind velocities.

A similar solution is employed by van Engelen *et al.* (2003) in order to damp two oscillatory modes of a multi-megawatt turbine. For this purpose, the control input added to the torque reference is designed by means of pole-placement technique and involves a Kalman filter being used for state estimation.

Another approach for controlling torsional vibrations consists of using a notch filter tuned on the resonance frequency of the drive train (Burton *et al.* 2001). Such a filter can be realized through a simple transfer function:

$$H_{NF}(s) = \frac{s^2 + 2\varsigma_1 \omega_{nt} s + \omega_{nt}^2}{s^2 + 2\varsigma_2 \omega_{nt} s + \omega_{nt}^2}, \quad (6.66)$$

where the ω_{nt} frequency is equal to the resonance frequency and $\varsigma_1 < \varsigma_2$ (for $\varsigma_1 = 0$, the transfer through filter is zeroed). Figure 6.30 presents the Bode plots of a notch filter defined by means of a discrete transfer function. When taking into account the variation of ω_{nt} vs. the pitch angle, one can use either a gain-scheduling technique or adopt a series of elementary filters of the type at Equation 6.66, having natural frequencies close to each other. In this case, the Bode plot of the notch filter offers a larger frequency range in order to compensate the open-loop frequency characteristic at resonance.

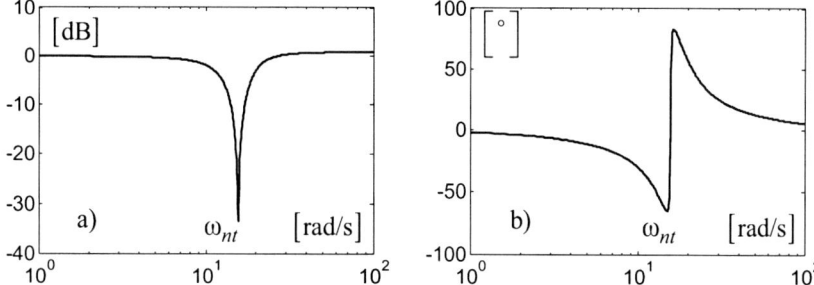

Figure 6.30. Bode plots of a notch filter employed for torsional vibration control

Another type of control demand concerns the system behaviour in relation to extrinsic factors, *e.g.*, periodic disturbances at the blade-passing frequency or disturbances due to power concentration at $i \cdot \Omega_l/(2\pi)$ frequencies from the rotationally-sampled power spectrum (see Figure 6.28). In this case, the performance requirements refer to minimizing the active power fluctuations and the flicker. Given that the disturbance input of the system has a spectrum with maxima at discrete frequencies depending on the turbine rotational speed, the frequency-domain control approach appears as normal. In Camblong *et al.* (2002a,b) an RST-controller is designed by shaping the frequency characteristic of the sensibility function in relation to the disturbance. Thus, minimal sensibility values are imposed at frequencies where disturbances must be rejected. The control problem design can be approached also by classical resonant control techniques (Lenwari *et al.* 2006; Zmood *et al.* 2001), similar to those used in active filter control design.

The global dynamics of the turbine-tower interaction is generally described by a six degree of freedom model. Therefore, the couplings between the various dynamics (tower fore-aft motion, blade edge motion, blade flap motion, *etc.*), determine the multimodal character of the model (*e.g.*, the drive-train torsional motion is strongly coupled to the blade edge-wise motion). On the other hand, when operated at variable-speed, the system is excited by the spectral wind speed components at the blade-passing frequency and at frequencies from the rotationally-sampled power spectrum, $N_b \cdot i \cdot \Omega_l/(2\pi)$. These spectral components possess variable frequencies that may excite the resonant modes of the system.

In conclusion, ensuring "good control" of the multi-megawatt WECS obviously requires advanced control techniques being involved. Often, this is an argument for giving up and adopting simple design solutions, like the constant-speed operation using softstarter and capacitor bank ("Danish concept" – Hansen and Hansen 2007). Nevertheless, the advantages of variable-speed operation cannot be neglected, so that going beyond the above-mentioned difficulties can remain a challenge for WECS control engineering.

7

Development Systems for Experimental Investigation of WECS Control Structures

7.1 Introduction

The main motivation of using physical simulators for wind energy applications is very simple. A laboratory setup (or test rig) can provide something that does not exist in real-world applications: controllable wind velocity. A wind turbine physical simulator (WTPS) offers a "wind-turbine-powered-like shaft" (Nichita *et al.* 2002) which allows one to obtain the static and dynamic characteristics of a mathematically modelled wind turbine.

Historically speaking, the wind turbine simulators belong to one of the following classes, depending on the approach used for their building:
– WTPS based on controlling a DC motor (Enslin and van Wyk 1992; Nuñes *et al.* 1993; Battaioto *et al.* 1996); here it is generally considered that the electrical variables (current and voltage) are electric images of the mechanical variables (torque and respectively rotational speed);
– WTPS having a general structure, which can be implemented based upon any type of servomotor; this structure comprises two subsystems (Nichita *et al.* 1994, 1998a; Nichita 1995; Diop *et al.* 1999; Cutululis *et al.* 2002; Teodorescu *et al.* 2003): the first is a real-time simulator software implementing the wind turbine mathematic model, including the wind speed generator, and the second is a electromechanical tracking subsystem, which interacts with the real-time software simulator;
– WTPS conceived using the *hardware-in-the-loop* architecture (Nichita *et al.* 1998a; Steurer *et al.* 2004; Bouscayrol *et al.* 2005). This approach is currently widely-used and makes the object of the analysis developed in the following.

In the literature one can find two types of papers dealing with small-scale WECS simulators for different generation configurations. The first category focuses on test rig building aspects (Leithead *et al.* 1994; Nichita 1995; Battaioto *et al.* 1996; Rodriguez-Amenedo *et al.* 1998; Diop *et al.* 1999a; Akhmatov *et al.* 2000; Cardenas *et al.* 2001; Rabelo and Hofmann 2002; Teodorescu *et al.* 2003; Kojabadi *et al.* 2004; Steurer *et al.* 2004; Bouscayrol *et al.* 2005). The second category underlines the role of test rigs for preliminary validation of WECS control laws (Enslin and van Wyk 1992; Cárdenas et al. 1996, Peña et al. 1996; Kana et al. 2001; Munteanu et al. 2005; El Mokadem et al. 2005; Camblong et al. 2006).

7.2 Electromechanical Simulators for WECS

In the last few years, the preliminary design and testing of modern industrial control systems have been based on the use of the hardware-in-the-loop-simulation (HILS) techniques. A hardware-in-the-loop (HIL) simulator essentially consists of closed-loop connection of physical and software parts, aiming at reproducing the dynamical behaviour of an industrial process in a controllable environment. HILS concept was originally introduced for developing and testing control structures for mechanical equipments (Hanselmann 1993) and has proved its efficiency in low-cost and rapid prototyping.

The use of HIL simulators to control law preliminary validation is justified in certain situations, *e.g.*, when the nature of the controlled process does not allow deterministic tests in its natural environment, or experiments involve important risks for the global operation, or the controlled plant is not available and its manufacturing cost is very high, *etc.* In this context, the HILS concept may be useful for a wide range of applications requiring research and development effort, such as spacecraft missions, aeronautic industry, *etc.* For example, in Hanselmann (1996) and Kiffmeier (1996) HILS applications in the automotive industry are reported, using the MATLAB®/Simulink® software on a dSPACE development kit. The power systems are another field where HILS may be used, from which renewable energy conversion is currently a rich investigation area, particularly suitable to using this kind of simulators. Their extensive use in wind power systems research makes that this topic finds its place within this book.

Because of the fast development of software and real-time information processing devices, the technological base of HILS systems has grown more than its conceptualisation. In the following some guidelines of a systematic design of HIL simulators will be provided, with application to real-time physical simulation of WECS. The approach presented in this chapter relies upon the concepts, terminology and methodological aspects introduced by Nichita et al. (1998b) and partially used in Diop (1999) and Nichita et al. (2006).

7.2.1 Principles of Hardware-in-the-loop (HIL) Systems

Let a *basic physical system* (BPS) be considered, for which a generic control problem is formulated. The associated mathematical model can be written as

$$\begin{cases} \dot{x} = F(x, u) \\ y = G(x, u) \end{cases},$$

where **u**, **x** and **y** are respectively the input, state and output vectors, **F** and **G** being generally nonlinear vector fields.

A HIL simulator is required to allow generally a cost-effective and safe reproduction of the BPS dynamic behaviour, for as realistic as possible closed loop deterministic experiments under laboratory conditions.

The basic idea used in HILS structures generally supposes that the BPS can be naturally divided into two subsystems which interact one with the other, namely a first subsystem – *e.g.*, a primary mover – and a second subsystem, undertaking the control action. The first subsystem is such that the closed loop experiments are very expensive and deterministic experiments are almost impossible. Therefore, it will be this subsystem whose behaviour must be replaced by a physical simulator and it will be called an *emulated physical system* (EPS). The second subsystem will exist in the HIL simulator exactly as it is in the BPS, thus allowing laboratory experiments under realistic conditions. Being the object of research, it will be called an *investigated physical system* (IPS).

The interaction between EPS and IPS, corresponding to a power transfer between them, is characterised by a pair of so-called *interaction variables*, further denoted as z_1 and z_2. Let us assume that in the interaction EPS-IPS the only "active" participant – *i.e.*, able of providing energy – is the EPS, whereas the IPS absorbs this energy. Having made this assumption, the interaction from the EPS point of view is depicted in Figure 7.1. The physical nature of the interaction variables depends on BPS in a biunique manner. z_1 is the cause variable, whose variation initiates the energy imbalance, and z_2 is the response variable, common to the EPS and IPS, by virtue of their coupling; their product has always power dimension.

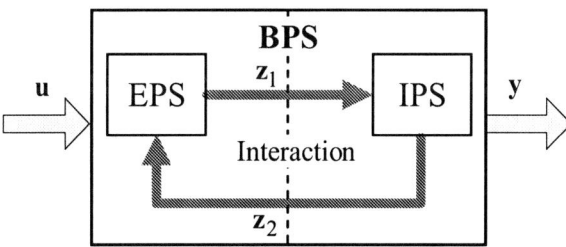

Figure 7.1. Basic physical system (BPS) structure illustrating the interaction between the investigated physical system (IPS) and the emulated physical system (EPS)

The aim is to replace the EPS by a so-called *real-time physical simulator* (RTPS). The IPS remains the same as in the BPS, as its study represents one of the main purposes of the HIL simulator building. The RTPS must offer the "natural" environment for IPS and must reproduce the models of the EPS and of the interaction EPS-IPS, such that the resulted HIL simulator will approximate the BPS dynamics.

In short, the RTPS must physically provide one of the interaction variables based on the measure of the other one and on the EPS model. This is achieved by means of a tracking loop at the output of the RTPS, which is called the *effector* (EFT); the controlled variable is called *driving variable* and the other one – *response variable*. The EFT reference is established by the so-called *real-time software simulator* (RTSS), which embeds a model of the EPS, further called EPSM. Figure 7.2 shows the RTPS structure for the simplified case when the interaction variables are scalars.

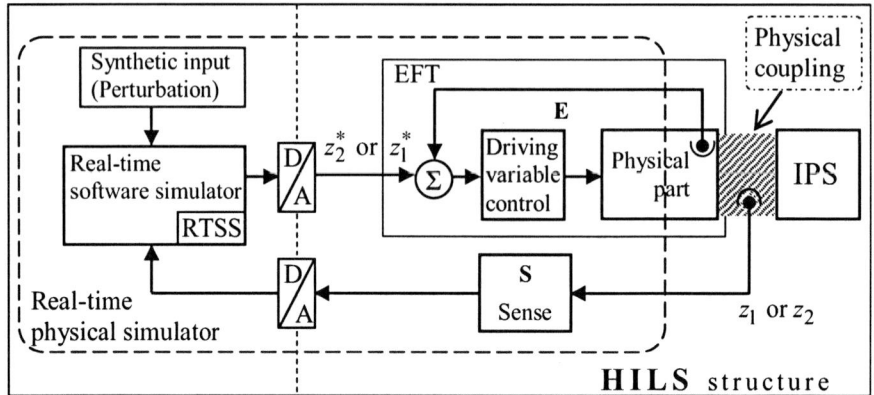

Figure 7.2. General structure of the real-time physical simulator (RTPS)

When choosing the generic driving and response variables, two situations may happen, as follows.

Case 1: the variable chosen as driving variable is an output/state of the BPS (z_2); therefore an *effect* variable, the model implemented in the RTSS is strictly causal and is obtained directly from the EPS model.

Case 2: the variable chosen as driving variable is a *cause* variable (z_1); the model implemented in the RTSS is non-causal and is fed by a measure of the effect, z_2.

Both cases have disadvantages. Thus, in the first case the EFT dynamic is quite slow, whereas getting the second case into practice is difficult because temporal derivatives must be computed, increasing the measurement noise.

This approach raises some questions. First, the IPS is generally "active" inducing in BPS variations of potential energy. Its influence on the RTPS should be considered as further detailed. Also, in Figure 7.2 one can note that the response variable is affected by the transducer dynamic (described by vector field **S**) and the driving variable by the effector dynamic (described by vector field **E**). Therefore, these variables have slightly modified instantaneous values, affecting the accuracy with which the HILS emulates the BPS.

The RTPS successfully replaces EPS if some conditions – called *basic reproducibility conditions* – are met:
- computation inside the RTSS must be sufficiently faster than the dynamic of the EPS (real-time condition);
- the EFT dynamic must be faster than that of the EPS (the tracking loop is sufficiently fast);
- the passband of the transducers must be sufficiently large.

7.2.2 Systematic Procedure of Designing HIL Systems

This section aims at concluding the previous discourse by providing an operational tool for designing HIL structures. Thus, considering that the interaction variables are scalars, a general design methodology may consist of several steps, as follows:

1. *Design of the modelling structure of the basic physical system* (BPS), *according to the stated control problem.*

At this step one must deduce the BPS mathematical model, by identifying its subsystems and the input, state and output variables.

2. *Delimitation of the* IPS *and of the* EPS *in the structural diagram of the* BPS *and choice of the interaction variables.*

This step is guided by the requirements imposed on the HIL structures, concerning the investigated physical system (IPS), which is taken as it is from the BPS. It is this system that represents the starting point, since its real environment must be "transposed" in laboratory conditions. The EPS results by taking the IPS out from the BPS and the interaction nature of these two subsystems completely determines the choice of the interaction variables.

3. *Configuration of the* RTPS.

There are two operations to be performed at this step, namely: to establish the physical part of the EFT, which must be directly coupled to IPS and, respectively, to choose the driving and response variables.

4. *Development of the detailed model of the* EPS (EPSM), *which will be implemented in the* RTSS.

This step depends on the chosen driving variable and on how the RTSS is realized. The EPSM can be implemented either in direct form, if the possibility or the wish exists that the driving variable is naturally defined as output variable (effect), or in inverse form, if taking a cause as driving variable.

5. *Putting into practice the* RTSS.

This is achieved by using the EPSM, deduced in step 4. If this model describes a fast dynamic and if, in addition, it is non-causal, then the problem of simultaneously ensuring the real-time operation of the RTSS and its immunity to disturbances becomes critical.

6. *Design of the tracking system contained in the physical part of the* RTPS (EFT).

The difficulty of designing a control law for this loop is in general due to the nonlinearity of the process and to the fast variations of its dynamic parameters. Therefore, one must guarantee that the EFT settling time is much smaller than the EPS one.

7. *Building of the* IPS, which is strongly context dependent.

7.2.3 Building of Physical Simulators for WECS

There are multiple possibilities for building HIL simulators for WECS. Usually, one considers that the IPS and EPS interact at the rotating high speed shaft; hence, the RTPS physical part is based on a rotating electrical machine (servomotor). DC motors are often employed (Nuñes *et al.* 1993; Battaioto *et al.* 1996), although AC machines can be used with similar performances (Steurer *et al.* 2004; El Mokadem *et al.* 2005; Munteanu 2006). The IPS depends on the BPS electromechanical system, is typically based on synchronous or induction machine and may include power electronics converters and control systems if operating at variable speed.

The interaction variables in this case are the rotational speed, $\Omega_h \equiv z_2$ and the mechanical "effective" torque, $\Gamma_{ef} \equiv z_1$ of the high speed shaft. Therefore the RTPS output is the high speed shaft (HSS) dynamical characteristic. This means that the algorithm within RTSS will implement models of the aerodynamics and drive train. The wind velocity feeding the turbine model is also part of this algorithm as a stochastic sequence with statistical parameters corresponding to a certain wind site. Deterministic test signals can also be employed for testing the simulated WECS behaviour. The EFT role is to convert the output of the RTSS into real mechanical variables.

The RTPS can be configured either for the *speed-driven* case (choosing an effect as driving variable) – see Figure 7.3 or for the *torque-driven* case (a cause variable is used for driving the EFT) – see Figure 7.4.

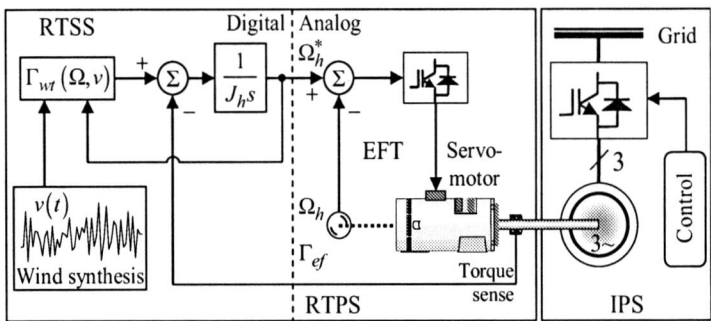

Figure 7.3. Block diagram of the electromechanical HIL simulator – speed-driven case

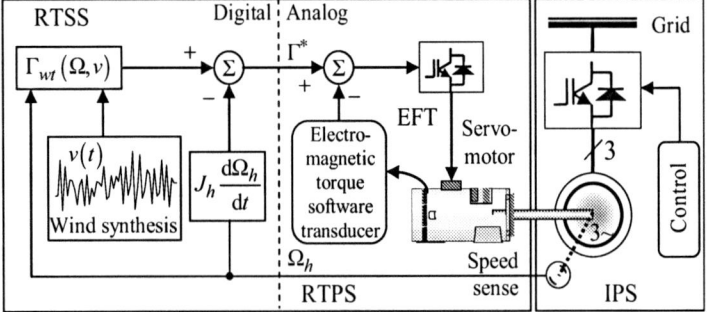

Figure 7.4. Block diagram of the electromechanical HIL simulator – torque-driven case

These figures show a generic architecture of the two simulators, as the RTSS is depicted in its simplest form and the general structure of the electromechanical tracking loop representing the EFT is presented. However, the RTSS algorithm may contain a very detailed model of the EPS, for example variable pitch aerodynamics, three-mass transmission model (flexible drive train), structural dynamics, *etc.*

The RTSS is usually built on digital signal processors having a comfortable number of I/O ports for controlling the RTPS physical part and sufficiently high computing power for EPSM real-time execution. It outputs the signal reference for the electromechanical tracking loop, *i.e.*, the rotational speed for case 1 (speed-driven simulator) and the HSS torque for case 2 (torque-driven simulator).

The RTPS based on speed drives use high-dynamics machines – *i.e.*, the electrical and mechanical time constants are comparable – for reducing the EFT response time. The use of the effective mechanical torque (Γ_{ef}) transducers or estimators is mandatory in this case. The other type of RTPS is based on the torque drives and must use motors that accurately provide high values of electromagnetic torque. In this case the effective mechanical torque transmitted to IPS is obtained by subtracting the dynamic (acceleration) torque. Therefore the EPSM embeds an inverse model of the HSS motion equation, and the gradient computation of the rotational speed (Ω_h) is necessary. In order to limit the induced noise, filtering of Ω_h is required, especially if an optical encoder is used as speed transducer. This potentially reduces the bandwidth of the EFT, with implications on the HILS emulation accuracy.

A supplementary difficulty is generated because IPS actively interacts with EPS/RTPS, thereby modifying the steady-state BPS operating point and the control loop within EFT is changing its dynamic proprieties. This suggests the use of adaptive control structures (*i.e.*, gain scheduling) for the electromechanical tracking loop, such that the tracking error remains within acceptable limits.

7.2.4 Error Assessment in WECS HIL Simulators

The electromechanical wind turbines simulator is a laboratory tool very useful for applicative research envisaging control subsystem design or grid interfacing. Thus, the simulator performance computation and the analysis of the simulation error are mandatory phases prior to its employing in real-time experiments (Diop *et al.* 2000). Figure 7.5 presents the errors describing the main differences between real and simulated behaviour of a physical system (Nichita 1996).

The simpler the model, the larger the modelling errors. An ideal physical simulator implements only the dynamic behaviour described by the adopted model. Therefore, the physical system modelling is of crucial importance, because it must be suited to the goal of HIL-based experimental investigation. As regards the simulation errors, they are practically negligible if the real-time computing system is properly configured. Unlike these two kinds of errors, the tracking errors due to real-world implementation depend on the way of choosing the driving variable. In the following the tracking errors in WECS HIL simulators are analysed in the frequency domain, based on the linearized model of wind turbine (see Section 3.6).

The high-speed shaft motion equation is

$$J_h \cdot \frac{d\Omega_h}{dt} = \Gamma_t(\Omega_h, v) - \Gamma_l(\Omega_h, c), \qquad (7.1)$$

where Γ_l is the load torque, depending on the so-called control variable, c, and Γ_t is the torque fed into the HSS by the wind turbine (it embeds the aerodynamic and drive train dynamics). This model is linearized around a steady-state operating point (see also Figure 3.24).

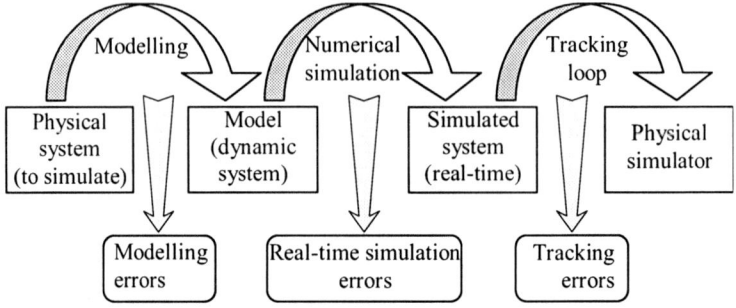

Figure 7.5. Nature of errors in real-time physical simulators

In order to describe the model suitably, the following notations are employed for the slopes:

$$K_{\Gamma\Omega} = \frac{\overline{\partial \Gamma_t}}{\partial \Omega_h}, \quad K_{\Gamma v} = \frac{\overline{\partial \Gamma_t}}{\partial v} \quad (7.2)$$

and for the transfer functions describing the influence of the control variable and of the rotational speed Ω_h respectively, on the load torque:

$$H_c(s) = \frac{\Delta \Gamma_l(s)}{\Delta c(s)}, \quad H_l(s) = \frac{\Delta \Gamma_l(s)}{\Delta \Omega_h(s)} \quad (7.3)$$

The general linearized model of the turbine-generator (load) coupling is given in Figure 7.6a.

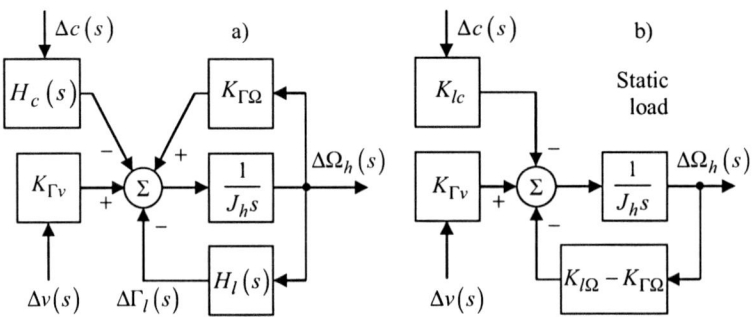

Figure 7.6. Linearized model of the WECS: **a** general case; **b** static load

In the case of a static load characteristic, both dynamics from Relation 7.3 come down to purely algebraic formulae:

$$H_c(s) = K_{lc} = \overline{\frac{\partial \Gamma_l}{\partial c}}, \quad H_l(s) = K_{l\Omega} = \overline{\frac{\partial \Gamma_l}{\partial \Omega_h}}, \quad (7.4)$$

the corresponding linear model being presented in Figure 7.6b. Interest is focused here on the transfer from the wind velocity to the rotational speed, $\Delta v \to \Delta \Omega_h$, therefore $\Delta c \equiv 0$. This transfer is described by the relation

$$H_{ws}(s) = \frac{\Delta \Omega_h(s)}{\Delta v(s)} = \frac{K_t}{T_t s + 1}, \quad (7.5)$$

where $K_t = K_{\Gamma v}/(K_{l\Omega} - K_{\Gamma\Omega})$ and $T_t = J_h/(K_{l\Omega} - K_{\Gamma\Omega})$. If considering the load dynamic with respect to the rotational speed, a more general relation holds:

$$H_{ws}(s) = \frac{K_{\Gamma v}}{J_h \cdot s + H_l(s) - K_{\Gamma\Omega}} \quad (7.6)$$

Thus, the transfer function of the linearized WECS is deduced. In the following the linearized mathematical model of the HIL simulator will be obtained, for both speed-driven and torque-driven cases, $H^\Omega(s)$ and $H^\Gamma(s)$. The purpose is to compare the simulator models $H^{\Omega,\Gamma}(s)$ with that in Equation 7.6, by means of a frequency-domain error analysis. The linearized models of the electromechanical simulator derive from Figures 7.3 and 7.4 and are depicted in Figure 7.7a,b.

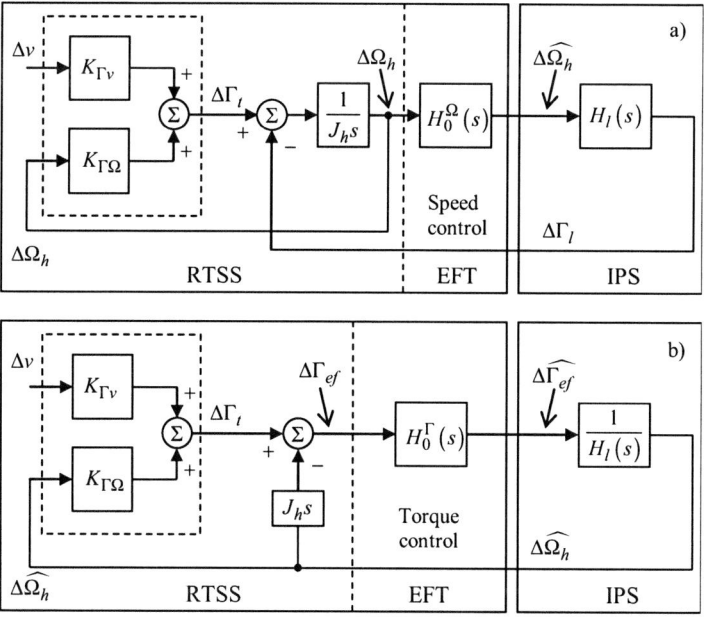

Figure 7.7. Linearized models of the electromechanical simulator: **a** speed-driven case; **b** torque-driven case

In the speed-driven case the EFT reference is the output of the model in Figure 7.6a (recall that the influence of Δc is zeroed), whereas in the torque-driven case the reference is given by the effective mechanical torque:

$$\Gamma_{ef} = \Gamma_t - J_h \cdot \frac{d\Omega_h}{dt} \qquad (7.7)$$

Due to the non-ideal realisation of the electromechanical tracking systems (EFT), their outputs differ from the corresponding references and therefore will be denoted by $\widehat{\Delta \cdot}$ (Figure 7.7). The EFT transfer functions are denoted by $H_0^\Omega(s)$ and $H_0^\Gamma(s)$; generally, they are non-unitary and their main time constants can sensibly differ one from the other, both depending on the servomotor features.

After calculations, the simulator transfer function in the speed-driven case is

$$H^\Omega(s) = \frac{\widehat{\Delta \Omega_h}(s)}{\Delta v(s)} = K_{\Gamma v} \cdot \frac{H_0^\Omega(s)}{J_h s + H_0^\Omega(s) \cdot H_l(s) - K_{\Gamma\Omega}} \qquad (7.8)$$

Similarly, the simulator transfer function in the torque-driven case becomes

$$H^\Gamma(s) = \frac{\widehat{\Delta \Omega_h}(s)}{\Delta v(s)} = K_{\Gamma v} \cdot \frac{H_0^\Gamma(s)}{H_l(s) + H_0^\Gamma(s) \cdot (J_h s - K_{\Gamma\Omega})} \qquad (7.9)$$

If the EFT is chosen with a very fast dynamic in relation to those of WECS, its transfer function can be considered unitary, $H_0^\Omega(s) = H_0^\Gamma(s) = 1$, and the transfer function of the simulator in speed- and torque-driven cases will be the same as in Expression 7.6; thus

$$H^\Omega(s) = H^\Gamma(s) = H_{ws}(s) \qquad (7.10)$$

An unwise choice of the EFT dynamic, namely too close to the main dynamic of WECS, can determine large simulation errors and even unstable behaviour of the HIL simulator. Figure 7.8a shows the correct relative position of the (linearized) WECS amplitude characteristic and that of the HILS. The frequency range in which an HIL simulator can emulate a WECS is $[0; \omega_{EFT}]$ (Diop et al. 2000). The simulation error is described by the difference of the rotational speeds between the WECS and the HIL simulator under the same wind velocity sequence. Thus, the errors can be frequency-domain characterised by the following relation:

$$H_\varepsilon(s) = H_{ws}(s) - H^{\Omega,\Gamma}(s) \qquad (7.11)$$

Further calculations give interpretable results only for particular transfer functions of the load and EFT, respectively. In general, the error $|H_\varepsilon(j\omega)|$ increases with the frequency in the EFT bandwidth (Figure 7.8b). However, for higher frequencies, the error is continuously decreasing due to the strictly causal nature of the system $H_\varepsilon(s)$ (Diop et al. 2000). Furthermore, the error in speed-driven simulators grows faster as the frequency increases than in the torque-driven simulator case. But the use of high dynamic servomotors can significantly reduce this difference, thus allowing similar dynamic performances being obtained.

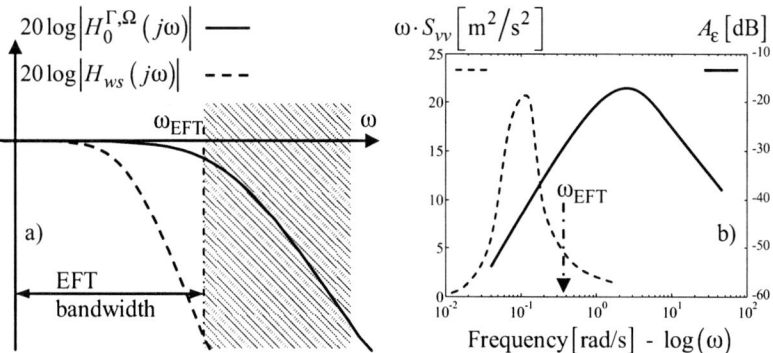

Figure 7.8. Performance assessment of the simulation accuracy: **a** bandwidth of the EFT; **b** frequency characterization of the model errors

A general approach of minimizing the simulation dynamic errors is to introduce an integral criterion defined on a large time horizon. In this case it is justified to minimize the error power spectrum, $S_{\varepsilon\varepsilon}(\omega)$. As the main exogenous is the wind velocity, its power spectrum $S_{vv}(\omega)$, heavily influences the error. Figure 7.8b shows the relative position of this spectrum to the error frequency characteristic. The EFT bandwidth must be chosen such that it includes the wind velocity spectrum and the WECS passband, ensuring small enough value of the error $A_{\varepsilon}(\omega_{EFT})$. Consequently, in the $S_{\varepsilon\varepsilon}(\omega)$ calculation, the error $|\varepsilon(j\omega)|$ must be weighted by the wind velocity power spectrum:

$$S_{\varepsilon\varepsilon}(\omega) = \left| H_{ws}(j\omega) - H^{\Omega,\Gamma}(s) \right| \cdot S_{vv}(\omega) \qquad (7.12)$$

In conclusion, an optimal physical simulator for WECS, guaranteeing minimal tracking errors due to real-world implementation, could be obtained by minimizing the following integral criterion (Diop et al. 2000):

$$I = \frac{1}{\pi} \cdot \int_0^\infty \left| H_{ws}(j\omega) - H^{\Omega,\Gamma}(s) \right| \cdot S_{vv}(\omega) d\omega \qquad (7.13)$$

7.3 Case Study (12): Building of a HIL Simulator for a DFIG-based WECS

Building an experimental test rig, in a certain hardware structure, is sensible to the particular configuration of the generation scheme. The WECS considered here as case study has the main features listed in Appendix A and Table A.5 and corresponds to the block diagram of Figure 3.1b, where the generation subsystem contains a doubly-fed induction generator (DFIG) and the rotor has fixed pitch. In this case, the variable-speed regime of the wind turbine is allowed by using back-to-back power electronics (PWM) converters in the rotor circuit (Lubosny 2003). Also, a rigid single-ratio speed multiplier is used as mechanical transmission.

Two kinds of control systems are used for the converted power transfer to the electrical grid: the torque/speed control of the induction generator and the control of the parameters of the electric power fed to the grid. An upper level of control for global WECS management should also be considered.

From the point of view of the HILS methodology, the system described above is the BPS, for which the maximisation of the captured power from the wind is defined as a control problem.

7.3.1 Requirements Imposed to the WECS Simulator

A 1:1 (same power) simulator for the chosen WECS, allowing reproducible experiments, must be built. It must reproduce the behaviour of different wind turbines belonging to the same WECS class when conditions about the wind sites and the wind/weather vary, in different operating regimes and under various control inputs.

As the physical system is fixed, the simulator must have the ability of changing the scale factors in order to allow different wind turbines being simulated. The parameters of the wind turbines and of the wind site must also be easily changed to obtain a flexible test rig. As its final scope is to experiment various control laws in variable-speed regimes, the test rig must allow these control laws being interchanged and the manipulation of the parameters and of the experimental results through a user-friendly interface. The WECS simulator must render the experiments insensitive to weather conditions, allowing to easily changing the features of the wind site.

The main methodological steps listed in Section 7.2.2 are referred to later, in order to build a HIL simulator that complies with the requirements stated above for the considered variable-speed WECS.

7.3.2 Building of the Real-time Physical Simulator (RTPS)

This section aims at illustrating the use of the main methodological steps to the construction of a physical simulator for a variable speed WECS having the main characteristics previously stated.

Modelling of the BPS and Identification of the Interaction Variables
It is proposed that the BPS from Figure 3.1b be structured from the HILS point of view as follows.

The energy conversion process is dealt with by means of the generator control, thus the IPS will contain the electrical part of the WECS, *i.e.*, the electromagnetic subsystem, taken as it is in the original system.

The use of synthetic wind velocity is envisaged; therefore a wind generator will be included in the simulator. For flexibility and economical reasons, the aerodynamic parts and transmission behaviour must also be emulated, and so included in the RTPS. Thus, the frontier between EPS and IPS is established at high-speed shaft, the interaction variables have a mechanical nature, namely the mechanical torque and the rotational speed, both corresponding to the high-speed shaft.

The modelling of the BPS is made according to Chapter 3, where the WECS components are identified in the particular case study.

The wind velocity generator can output standard test waveforms (*e.g.*, ramp, square) and pseudorandom sequences with stochastic parameters corresponding to a certain wind site. For the latter case a nonstationary wind velocity generation structure is employed (Figure 3.6). By using the approximation at Equation 3.12, the turbulence has the Von Karman's spectrum, while its intensity is computed using the IEC standard. The low-frequency wind speed component is provided by a time series generator.

For the aerodynamic subsystem, the average model employing the rotor aerodynamic performance is used, yielding the wind (aerodynamic) torque provided by the turbine rotor. As the rotor has fixed pitch, Relations 3.24 and 3.25 are the basis of its modelling.

For the rigid drive train one employs the single-mass model represented by the high-speed shaft motion equation (see Relations 3.64 and 3.66).

All these elements are included within EPS and should be emulated in the RTPS. The following components are enclosed in the IPS and are present in the HILS as they are. Thus, their modelling is not critical for the simulator building process.

The generator is modelled by Equations 3.31, 3.34 and 3.35, and the machine side inverter is considered without dynamics. The grid-side inverter behaviour is represented by Equation 3.71. Certain control inputs are envisaged, namely for the generator variable speed operation and for the grid-side inverter running.

Configuration of the Physical Part of the RTPS and Determination of the EPSM
The choice of the driving (and consequently the response) variable(s) considerably affects both the hardware and the software parts of the test rig. Therefore, one must clearly establish the driving variable before starting building the simulator. In this case study the driving variable will be a cause, namely the high-speed shaft mechanical torque. The response variable will then be the high speed shaft rotational speed, Ω_h (measured by means of an encoder); therefore, EFT is a torque control loop.

Thus, the associated block diagram is the one in Figure 7.4. The physical part of the EFT is implemented here by a torque (current)-controlled permanent-magnet DC motor (whose parameters are given in Appendix C), rigidly coupled to the DFIG shaft in the IPS by means of a speed reduction gear. This motor must reproduce the mechanical characteristics of the high speed shaft of the wind power system.

Consequently to choosing the driving variable as being a cause variable, the resulted EPSM (to be implemented by the RTSS) is non-causal. Employing some algebra in Equation 7.7, the value to impose as reference to the tracking loop inside the EFT results as

$$\Gamma_{ef}^* = \frac{\eta \cdot \Gamma_{wt}}{i} - \left(\frac{J_{wt} \cdot \eta}{i^2} - J_{DCM}\right) \cdot \dot{\Omega}_h, \qquad (7.14)$$

where η and i are the drive train efficiency and ratio respectively, J_{DCM} is the DCM inertia, and Γ_{wt} results from the EPS aerodynamics.

In the general case of a 1:n simulator, with $n = P_n/P_s$, P_n being the rated power of the real system and P_s being the rated power of the simulator, the simulator inertia results from the motion equation as

$$J_s = \frac{J}{n} \cdot \left(\frac{\Omega_n}{\Omega_s}\right)^2 \qquad (7.15)$$

with J, Ω_n and Ω_s being the system inertia and rated speed and the simulator rated speed, respectively. Therefore, when building a scaled down simulator, the physical system inertia must be amended with the factor from Equation 7.15, to obtain the correct dynamic torque.

Building of the RTSS and Design of the Tracking System within the Effector
Remember that the role of RTSS (software part of the RTPS) is to compute the imposed value of the driving variable, fed as a reference to the EFT. Models of the wind velocity, EPS aerodynamics and Equation 7.14 form the EPSM. Appendix C contains diagrams briefly showing the RTPS hardware and software components.

The RTSS is a component of a software application running on a floating point digital signal processor (DSP) that is a PowerPC 750 of the DS1005 board (dSPACE 2003). The other component of the program is dedicated to implementing the PWM control algorithms of the elements of IPS, which will be detailed in Section 7.3.3. The program on DS 1005 runs two tasks (Figure C.5a):
- a *principal (interruption) task*, which has the highest priority, in charge with real-time operations, related to RTSS: reading the analogical inputs from transducers, running the EPSM and providing the output reference of EFT;
- a *background task*, which updates the DSP memory used for the modified variables by means of a user interface.

The simulation diagrams corresponding to the two tasks have been built in MATLAB®/Simulink® and implemented by means of the Real Time Interface Toolbox (dSPACE RTI for target processor DS1005). The bidirectional information flux between the physical and the software part is supported by the data acquisition interface of the DSP. The user supervising interface has been realized through a ControlDesk® panel for monitoring the interest variables.

The physical realization of the driving variable, Γ_{ef}^*, is carried out using the DC machine; based upon its proportionality with the induced current (Leonhard 2001), a current tracking loop is employed. It is composed of a current transducer, a numerical PI regulator supported by a fixed point digital signal processor, TMS320F240, and an actuator composed of a PWM IGBT-chopper (four quadrants) connected with a diode rectifier (see Figure C.5b).

Experimental Results to Validate the RTPS Operation
This section will present the performance of the effector in different conditions.

In Figure 7.9a one can see the step response of the current loop within the EFT, when the direct-current machine current reference step covers the entire domain. The rising time of the response is very short, as expected, around 4 ms. As the time constant of the EPS is generally large (for low-power wind turbines it can be around 0.1 s), the speed of the torque tracking loop is high enough for ensuring the second basic reproducibility condition (see Section 7.2.1).

In Figure 7.9b one can see the dynamical behaviour of the turbine rotor for a pseudo-random wind speed sequence having an average value of 7 m/s and medium turbulence intensity of $I=0.17$, obtained by using the von Karman spectrum and the IEC standard.

The upper part of Figure 7.9b shows that the rotational speed of the low-speed shaft is maintained almost constant by the IPS. The induction generator speed evolves within a restraint domain in oversynchronous regime and presents some variations due to a sufficiently high slip.

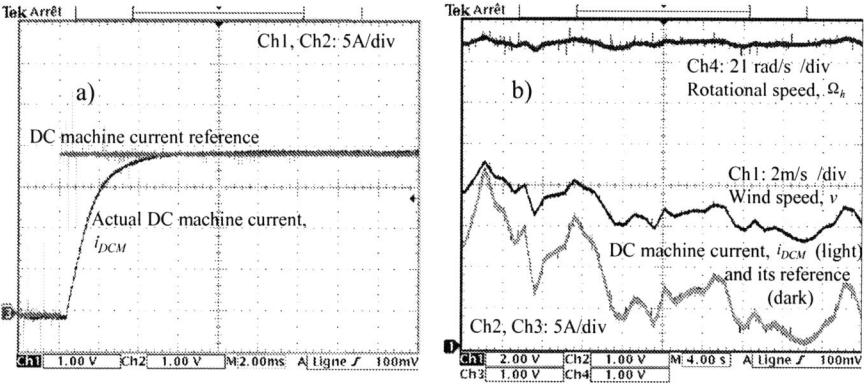

Figure 7.9. Oscilloscope captures illustrating the RTPS operation within the HIL simulator

The current evolution corresponding to high-speed shaft mechanical torque is given in the lower part of Figure 7.9b. One can note the perfect tracking of the high-speed shaft torque (current) reference, obtained using the computer-synthesized wind speed and the measured rotational speed and the EPSM.

7.3.3 Building of the Investigated Physical System (IPS) and Electrical Generator Control

The IPS in the HIL simulator is the same as in WECS. Here, its construction is oriented to the control loops realizing the transfer of the DFIG power to the electrical grid. As this transfer is performed in two stages, two control subsystems are to be implemented, one for controlling the generator and the other for transferring the DC-link power to the grid. Each structure acts on the correspondent voltage source (PWM) inverter (see Figure 3.1b). A multitude of transducers (encoder, current and voltage sensors) is employed for obtaining information from EMS. This information is fed to the software environment responsible for the WECS control. The same DS 1005 board is used for handling this task, the controllers being implemented in the same MATLAB®/Simulink® block diagram as the RTSS.

Control of the Wind Turbine

The variable speed operation of WECS having rigidly-coupled drive train can be generally achieved by controlling either the speed or the torque of the high-speed shaft. In order to obtain a smoother behaviour of the drive train and therefore less mechanical fatigue, the DFIG is controlled by means of a vector control structure. It is obtained using a stator field oriented (d,q) model for the wound rotor induction machine under some assumptions (Bose 2001). The control structure is based on the separate control of the electromagnetic torque (motion control) and rotor flux (flux alignment onto d axis) (Lubosny 2003; Cárdenas and Peña 2004). PID controllers are used for this purpose, *e.g.*, employing the procedures presented in Section 4.7.1 (Åström and Hägglund 1995).

The speed control adds another outer loop governed by a third PID controller, designed by using the symmetrical optimum criterion (Bose 2001; Leonhard 2001).

Figure 7.10 (ControlDesk®/MATLAB® captures) shows the behaviour of the torque-controlled WECS over a 20-s time horizon when the wind speed is constant at 7 m/s, for step changes in the electromagnetic torque reference. One can see an instantaneous (millisecond-sized) torque response of the DFIG (Figure 7.10a). The evolution of the rotational speed (high-speed shaft) is presented in Figure 7.10b.

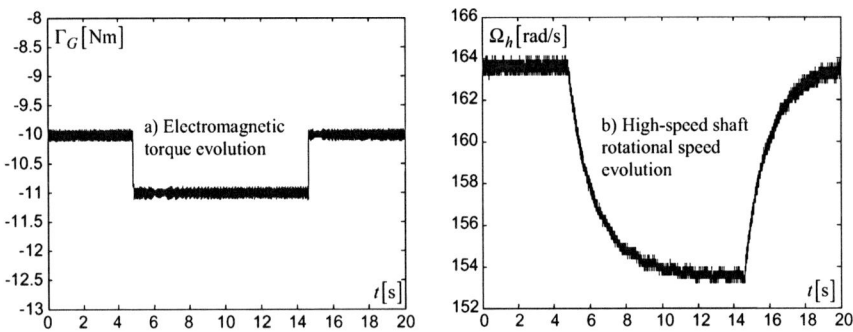

Figure 7.10. Torque control of the WECS within the HIL simulator (v=7 m/s)

The dynamic response of the system (rotational speed) is very slow because of the turbine rotor inertia and of the fact that the load characteristic has zero slope in the (Ω_h, Γ_G) plane. However, higher dynamics are expected if the generator mechanical characteristic has an accentuated slope. The noise induced by the encoder is also visible. This makes mandatory low-pass filtering in order to obtain reliable data for the Ω_h gradient computation.

In Figure 7.11 one can see the WECS response to a step change of the rotational speed reference, under constant wind speed, of 8 m/s. Figure 7.11a shows differences between tracking the rising edge and respectively the falling

7.3 Case Study (12): Building of a HIL Simulator for a DFIG-based WECS

edge of the reference, mainly due to the torque limitations, as visible in Figure 7.11b in the evolution of i_{2q} (rotor current, proportional with the electromagnetic torque). Current i_{2d} is zeroed; hence, the rotor flux has a constant value depending on i_{1d}.

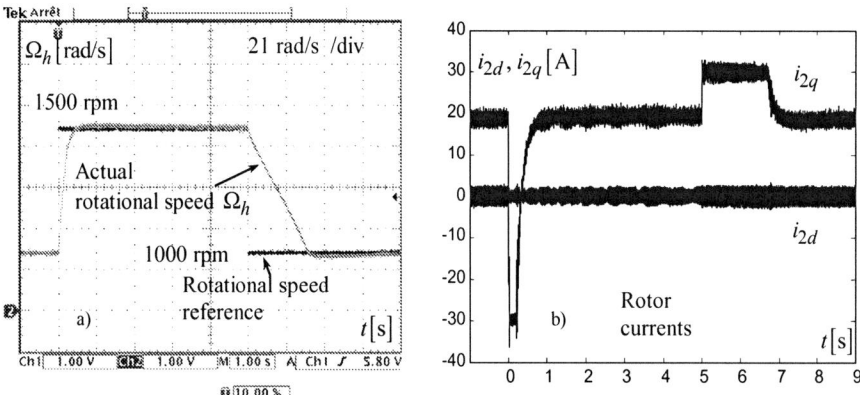

Figure 7.11. Speed control of the WECS within the HIL simulator ($v=8$ m/s): **a** rotational speed step response; **b** corresponding evolution of the rotor currents

Figure 7.11 also shows that the DFIG motoring regimes are allowed, which is useful for starting the turbine when the wind speed is not sufficiently high. In conclusion, a fraction of the RTPS output power is converted – depending upon the turbine driving regime – at the DFIG rotor level, being further retrieved within the DC-link (as also suggested in Figure 3.1b).

Transfer of the Generator Rotor Power to the Grid
A second stage for transferring the DFIG rotor's power to the grid is realized using another inverter (grid-side) acting as an interface, as shown in Figure 3.1b. This is performed by feeding tri-phased currents into the mains, while controlling the DC-link voltage at a constant value, V_{bus}. The control structure is synthesized using the (d,q) model of the tri-phased electrical system and the DC-link modelling, having an outer voltage control loop and two current control loops for q and d current components (Peña *et al.* 1996; Pöller 2003).

These loops employ anti-windup PI controllers, tuned by using modulus optimum criterion. The inner loops output the grid voltage references in the (d,q) frame, further transformed into a tri-phased system and then applied to the inverter. A phase-locked loop is used for synchronising the rotor voltage system angle with the grid one (Kaura and Blasko 1997; Rabelo and Hofmann 2002). The reference of the voltage controller is constant at 350 V, and its output feeds the i_q current loop. The i_d reference is zero, as the reactive power control is not of interest here.

In Figure 7.12a, one can see the evolution of the power that can be maximally harvested, $P_{wt_{max}}$, corresponding to a wind speed sequence of an average 7 m/s and medium turbulence intensity I=0.17 (obtained by using the von Karman spectrum in the IEC standard). The power fed to the grid, P_{grid}, is also plotted, exhibiting a total efficiency of about 40%. Figure 7.12b shows the DC-link voltage control efficiency and the variations of the DC-link current, i_{bus}, feeding the grid-side inverter, under the same wind conditions.

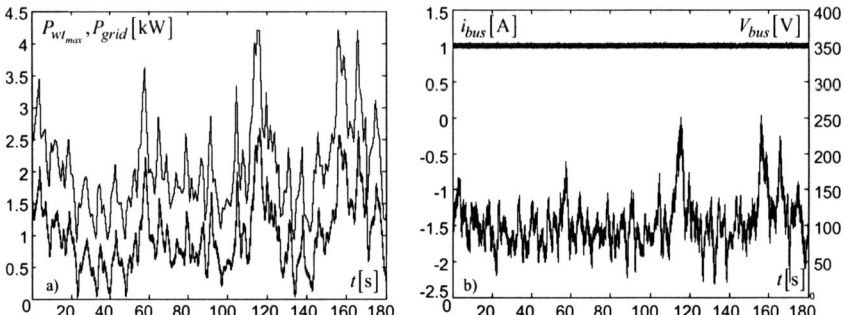

Figure 7.12. Grid-side inverter control within the HIL simulator: **a** wind power and power delivered to the grid; **b** DC-link current and voltage

7.3.4 Global Operation of the Simulated WECS

As a corollary of the developed approach, the complete functional diagram of the WECS real-time simulator, designed according to the HILS methodology can be found in Figure C.6; a photo of it is given in Figure C.7. One can note that the conversion chain, shown in Figure 3.1b, is fully replicated by the HILS structure.

Simulation of Fixed-speed Operating Regime
Figure 7.13 presents simulation results concerning the replication of the fixed-speed operation of WECS (Figure 7.13d) under a random wind speed sequence (Figure 7.13a). One can note the non-optimal evolution of the tip speed, λ, (Figure 7.13b), of the generator torque, Γ_G (Figure 7.13c), of the power coefficient, C_p, expressing the aerodynamic efficiency high deviations from the optimal value (Figure 7.13e), and of the electrical power delivered to the grid, P_{grid} (Figure 7.13f).

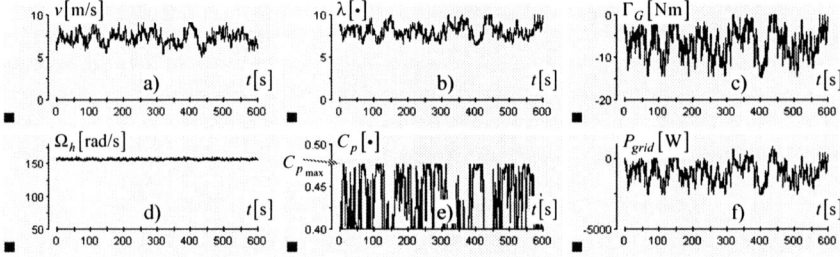

Figure 7.13. Global fixed-speed operation of the HIL simulator – ControlDesk® captures

Variable-speed Operation

Figure 7.14 illustrates the variable speed operation of the considered WECS. In this case, the IPS is torque controlled (Figure 7.14c), in order to track the energetic optimum (Figure 7.14b), the maximum power coefficient (Figures 7.14e,h) and to harvesting wind power around the ORC (Figure 7.14i). The random wind speed sequence used here (Figure 7.14a) has the same spectral properties as that in Figure 7.13a. The high speed shaft's rotational speed varies in a large domain (Figure 7.14d).

One can note the evolution of the power captured from wind (Figure 7.14g) and of the power provided to the grid, P_{grid} (Figure 7.14f); the difference between the two results from a non-unitary global efficiency of the system.

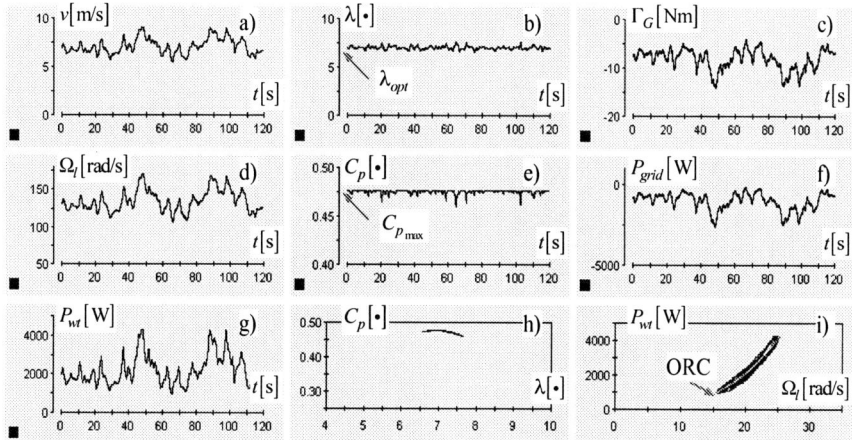

Figure 7.14. Global variable-speed operation of the HIL simulator – ControlDesk® captures

In the fixed-speed case the tip speed presents significantly larger deviations from its optimal value, λ_{opt}, than in the variable-speed case (Figures 7.13b and 7.14b). The same remark holds for the power coefficient (see Figures 7.13e and 7.14e). The IPS is controlled in sliding mode, as it has been extensively presented in Section 5.5 and also in Munteanu *et al.* (2006b).

7.4 Conclusion

The approach discussed in this chapter concerns the building methodology and practical aspects dedicated to the real-time physical simulation of WECS. Its construction is based upon using the HILS principle, which generally consists of closed loop connecting hardware and software components, in order to reproduce the dynamic behaviour of an industrial process in a laboratory controlled environment. The insertion of the physical elements contributes to a better reproduction of the real phenomena and to the reduction of the information processing time within the simulator. The HILS structure consists of a real-time physical simulator (RTPS) which interacts with an investigated physical subsystem (IPS). The interaction between the two is controlled by the software part of the simulator, which incorporates a model of the simulated physical subsystem.

As it uses synthetic wind speed, such a simulator can be used for wind turbine performance analysis and sitting, for wind turbine testing in various functioning (or failure) regimes, for validating various control strategies of the turbine generator and for studying the transfer of the provided electric power to the grid.

8

General Conclusion

At present, the horizontal-axis WECS are characterized by a great variety, due to the different solutions adopted as regards their structure, components and operating regimes. These solutions depend largely on the WECS rated power. Thus, low-power (one to tens of kW) and medium-power WECS are integrated either into distributed power production systems or into hybrid renewable energy conversion systems operating in isolated grids. Grid-connected WECS rated powers have moved from the range 600–1000 kW to values of 3–6 MW in the last decade. The multi-megawatt WECS are nowadays representative for the technological state-of-the-art. On the other hand, the impact of the small-scale distributed power production systems, operating close to the consumption location, is getting higher and higher. Consequently, the whole range of WECS rated powers is of interest, their configurations and operation modes being very diverse. For all the mentioned WECS the control problem is important, both for power limiting in high winds and also for energy optimization below rated winds.

The variable-speed concept applied to wind turbines is considered to be a significant advance in WECS technology. Applying this concept continues to have a positive impact on the efficiency of wind turbines for economic reasons which are easy to understand. Important factors of WECS technology evolution were the control solutions, allowing crucial functions to be implemented. This book has aimed primarily at designing control structures dedicated to WECS energy conversion optimization. This is an optimal control problem stated for the partial-load regime, which can be dealt with distinctly only for low-power WECS. Still, even in this case one must take care to alleviate the control input effort in order to avoid excessive mechanical loads that shorten the WECS service lifetime. As a consequence, minimization of mechanical loads is mandatory when formulating control goals.

Studies indicate and integration experience confirms that increasing the energy capture without affecting the service lifetime is not a trivial task, given the erratic nature of wind. From a more technical point of view, a wind turbine is a nonlinear time variant system excited by stochastic inputs which significantly affect its reliability. Under these circumstances, a suitable partial load control objective is the optimization of a performance index containing two contradictory demands:

energy efficiency and mechanical reliability. A similar control problem can be formulated for the full-load operation; this involves a criterion that trades off the stabilisation of the state variables at their rated values and the fatigue loads alleviation.

Methodologically speaking, the control problem can be unitarily solved, *i.e.*, over the entire wind speed operating range, by gain-scheduling techniques. Unlike this approach, the point of view expressed in this book has emphasized the possibility of separately designing each component of the control law due to the frequency separation principle. Thus, each component is synthesized in order to comply with one or a set of quality demand(s), the latter being grouped by contradictory pairs. This is possible because the WECS dynamic properties involved in the imposed quality requirements are reflected in disjoint frequency ranges.

In Chapter 6 particular attention has been paid to frequency separation optimal control in relation to two requirements: 1) the low-frequency energy optimization by keeping the static operating point on the optimal regimes characteristic (ORC) and 2) the LQG optimization according to a desired energy-reliability trade-off. For the first demand the low-frequency WECS model is important, which corresponds to the long- and medium-term wind speed component, whereas the LQG problem is defined in the turbulence frequency range. Even in this case, when the two frequency ranges cannot be clearly separated, the proposed approach is of interest for the following reasons:

- the first component of the control input is implemented by a classical (already present) PI or PID loop; to this component the autonomously synthesized LQG control input is added;
- supposing the operating point is kept on the ORC by the low-frequency control action, then the variation of the linearized model with the wind speed is sensibly reduced; this renders the LQG problem an invariant one.

Frequency separation is easier when the control goals focus on the mechanical oscillatory phenomena or on fluctuant components of the active power. In these cases, design of different control input components is usually achieved through this approach.

In the case of low-power WECS, the performance demands are generally few and aim at limiting power at rated and at energy optimizing, possibly imposing an optimization criterion of energy-reliability type. Many wind plants belong to hybrid renewable energy conversion systems, which have specific control problems. In order to obtain well-performing solutions concerning the structure and control strategy of such systems, the experimental investigation under reproducible conditions is necessary. In this context, the development of laboratory test rigs based upon electromechanical WECS simulators is of clear interest. This issue has been approached in this book within a general design methodology.

In contrast, numerous performance requirements are imposed on multi-megawatt WECS; they mainly envisage the global mechanical loads being reduced. The control problem complexity depends on the WECS structure and operation mode. For example, in the case of fixed-speed WECS (*e.g.*, implementing the "Danish concept"), operation on the natural characteristic of the generator makes that the drive train torsional oscillation to have a good damping factor. Under variable-speed operation, the generator torque control induces weakly damped drive train modes; therefore, one must impose the torsional oscillation damping by control action. Even if supplementary demands arise – the one mentioned is only an example – the variable-speed operation is coherent with the general evolution trend of WECS technology. Due to the continuously increasing number and complexity of performance requirements, many aspects of variable-speed WECS are still in the phase of developing technical solutions, as none of the WECS control techniques has become classical as to be widely used by wind turbine integrators.

A

Features of WECS Used in Case Studies

In this appendix the parameters and features of WECS used for case studies are listed. For the reader's guidance, Table A.1 shows which feature set (table) corresponds to each reported case study.

Table A.1. WECS feature set used for each case study

Case study	1	2	3	4	5	6	7	8	9	10	11	12
Feature set (table)	A.6	A.2	A.7	A.2	A.2	A.2	A.3	A.3	A.4	A.2	A.4	A.5

The aerodynamic features for the WECS listed in the following as well as the modelling elements employed are common. Thus, the rated wind speed $v_n = 10.5$ m/s and the air density is considered constant, at $\rho = 1.25$ kg/m^3. The torque coefficient is computed according to Relation 3.25 by using the following polynomial coefficients $a_6 \div a_0$:

$$\left[-4.54 \cdot 10^{-7} \quad 1.3027 \cdot 10^{-5} \quad -6.5416 \cdot 10^{-5} \quad -9.7477 \cdot 10^{-4} \quad 0.0081 \quad -0.0013 \quad 0.0061 \right].$$

The power coefficient characteristic has the following optimal parameters: $\lambda_{opt} = 7$, $C_{p\,max} = 0.476$.

Table A.2. Parameters of the low-power (6-kW) rigid-drive-train SCIG-based WECS used for MPPT, PI power, on-off, sliding-mode, LQG (rigid-drive-train case) control laws assessment

Turbine rotor	Drive train	SCIG
Blade length: R=2.5 m	Multiplier ratio: i=6.25 LSS inertia: $J_l = 3.6$ kg·m^2 Efficiency: $\eta = 0.95$	p=2, $R_s = 1.265$ Ω, $R_r = 1.43$ mΩ $L_m = 0.1397$ H, $L_s = 0.1452$ H, $L_r = 0.1452$ H, $\omega_S = 100\pi$ rad/s $\Gamma_{G\,max} = 40$ Nm, $V_S = 220$ V

Table A.3. Parameters of the low-power (3-kW) rigid-drive-train PMSG-based WECS used for feedback linearization and QFT control case studies

Turbine rotor	Drive train	PMSG
Blade length: $R=2.5$ m	Multiplier ratio: $i=7$ HSS inertia: $J_h = 0.5042$ kg·m^2 Efficiency: $\eta = 1$	$p=3$, $R_s = 3.3$ Ω, $\Phi_m = 0.4382$ Wb $L_d = 41.56$ mH, $L_q = 41.56$ mH, $R_{ln} = 80$ Ω, $V_S = 380$ V

Table A.4. Parameters of the low-power (6-kW) flexible-drive-train SCIG-based WECS

Turbine rotor	Drive train	SCIG
Blade length: $R=2.5$ m	Multiplier ratio: $i=6$ LSS inertia: $J_l = 3.6$ kg·m^2 Efficiency: $\eta = 1$ Stiffness: $K_S = 75$ Nm/rad Damping: $B_S = 0.5$ kgm^2/s — Case Study 11 $B_S = 2$ kgm^2/s — Case Study 9	$R_s = 1.265$ Ω, $R_r = 1.43$ mΩ, $L_m = 0.1397$ H, $p=2$, $L_s = 0.1452$ H, $L_r = 0.1452$ H, $\omega_S = 100\pi$ rad/s, $\Gamma_{G\max} = 40$ Nm, $V_S = 220$ V

Table A.5. Parameters of the low-power (6-kW) rigid-drive-train DFIG-based WECS used for building the HIL simulator

Turbine rotor	Drive train	DFIG
Blade length: $R=2.5$ m	Multiplier ratio: $i=6.75$ HSS inertia: $J_h = 0.1$ kg·m^2 Efficiency: $\eta = 1$	$p=2$, $R_s = 75.1$ mΩ, $R_r = 0.13$ Ω $L_m = 13.2$ mH, $L_s = 14.3$ mH, $L_r = 14.3$ mH, $\omega_S = 100\pi$ rad/s $\Gamma_{G\max} = 45$ Nm, $V_S = 380$ V

Table A.6. Parameters of the high-power (2-MW) rigid-drive-train SCIG-based WECS used for reduced-order modelling and linearization case study

Turbine rotor	Drive train	SCIG
Blade length: $R=45$ m	Multiplier ratio: $i=100$ HSS inertia: $J_h = 990$ kg·m^2 Efficiency: $\eta = 0.95$	$p=2$, $R_s = 1.1$ mΩ, $R_r = 1.3$ mΩ $L_m = 2.9936$ mH, $L_s = 3.0636$ mH, $L_r = 3.0686$ mH, $\omega_S = 2\pi \cdot 50$ rad/s $\Gamma_{G\max} = 17e5$ Nm, $V_S = 960$ V

Table A.7. Parameters of the high-power (2-MW) rigid-drive-train SCIG-based WECS used for PI speed control law assessment

Turbine rotor	Drive train	SCIG
Blade length: $R=45$ m Rated wind speed: $v_n = 10.5$ m/s	Multiplier ratio: $i=100$ HSS inertia: $J_h = 990$ kg·m^2 Efficiency: $\eta = 0.95$	$p=2$, $R_s = 4$ mΩ, $R_r = 4$ mΩ $L_m = 5.09$ mH, $L_s = 5.25$ mH, $L_r = 5.25$ mH, $\omega_S = 100\pi$ rad/s $\Gamma_{G\max} = 17e5$ Nm, $V_S = 960$ V

B
Elements of Theoretical Background and Development

B.1 Sliding-mode Control

Computation of the sliding surface and of the sliding-mode control law starts from the state equations in the form

$$\dot{\mathbf{x}} = f(\mathbf{x},t) + \mathbf{B}(\mathbf{x},t) \cdot u,$$

where $\mathbf{x} = [\Omega_h \ \Gamma_G]^T$, $f(\mathbf{x},t) = [\Gamma_{wt}(i \cdot \Omega_h, v)/(J_t \cdot i) - \Gamma_G/J_t \ \ -\Gamma_G/T_G]^T$, $\mathbf{B}(\mathbf{x},t) = [0 \ \ 1/T_G]^T$ and $u = \Gamma_G^*$.

Sliding Surface
Let $\sigma(\mathbf{x},u,t)$ denote the sliding surface. Equation 5.25 and condition $\dfrac{\partial \sigma}{\partial \mathbf{x}} \cdot \mathbf{B}(\mathbf{x},t) \neq 0$ (existence of the equivalent control input – DeCarlo et al. 1996; Young et al. 1999), require that $\dfrac{\partial \sigma}{\partial \Gamma_G} \neq 0$; let $\dfrac{\partial \sigma}{\partial \Gamma_G} = a_3 \neq 0$ by notation. From imposing that the sliding-mode dynamics, *i.e.*, on the sliding surface, be equivalent to some linear ones as to Equation 5.26, a first form of the switching surface may be written

$$\sigma(\Omega_h, \Gamma_G) = \sigma_1(\Omega_h) + a_3 \cdot \Gamma_G \qquad (B.1)$$

Equality $\sigma(\Omega_h, \Gamma_G) = 0$ implies that $\sigma_1(\Omega_h) + a_3 \cdot \Gamma_G = 0$ and, finally

$$\sigma_1(\Omega_h) = -a_3 \cdot \Gamma_G \qquad (B.2)$$

From Equation 5.26 one can obtain

$$\Gamma_G = \frac{1}{1+a_2 \cdot J_h} \cdot \left(\Gamma_{wt}(\Omega_h/i,v)/i - a_1 \cdot J_h \cdot \Omega_h\right) \tag{B.3}$$

For simplicity, let $a_3 = 1 + a_2 \cdot J_h$. Equations B.2 and B.3, combined with Equation B.1, give the expression of the sliding surface (arguments being dropped):

$$\sigma = a_1 \cdot J_h \cdot \Omega_h + \Gamma_G \cdot (1 + a_2 \cdot J_h) - \Gamma_{wt}/i$$

Sliding-mode Control Law
This section is dedicated to computation of the two components of the sliding-mode control law: the equivalent control input, u_{eq}, and the on-off component, u_N (DeCarlo et al. 1996).

To compute the *equivalent control input* the following relation is used:

$$u_{eq} = -\left[\frac{\partial \sigma}{\partial x} \cdot \mathbf{B}\right]^{-1} \cdot \left[\frac{\partial \sigma}{\partial t} + \frac{\partial \sigma}{\partial x} \cdot f(x,t)\right] \tag{B.4}$$

Next, the involved expressions will be computed:

$$\frac{\partial \sigma}{\partial x} \cdot \mathbf{B} = \begin{bmatrix} \frac{\partial \sigma}{\partial \Omega_h} & \frac{\partial \sigma}{\partial \Gamma_G} \end{bmatrix} \cdot \begin{bmatrix} 0 \\ 1/T_G \end{bmatrix} = \frac{1}{T_G} \cdot \frac{\partial \sigma}{\partial \Gamma_G} \tag{B.5}$$

Because the sliding surface does not explicitly depend on time, then $\frac{\partial \sigma}{\partial t} = 0$.
To compute the partial derivative of the sliding surface in relation to the first state variable,

$$\frac{\partial \sigma}{\partial \Omega_h} = a_1 \cdot J_h - \frac{1}{i} \cdot \frac{\partial \Gamma_{wt}}{\partial \Omega_h},$$

one may use the partial derivative of the wind torque in relation to the low-speed shaft rotational speed:

$$\frac{\partial \Gamma_{wt}}{\partial \Omega_h} = 0.5\pi\rho R^3 v^2 \cdot \frac{\partial C_\Gamma(\lambda)}{\partial \Omega_h} = \frac{Kv^2 R}{i} \cdot \frac{\partial C_\Gamma(\lambda)}{\partial \lambda} \cdot \frac{\partial \lambda}{\partial \Omega_l} = \frac{Kv^2 R}{i} \left(\frac{C_p'(\lambda) \cdot \lambda - C_p(\lambda)}{\lambda^2}\right),$$

where $K = 0.5 \cdot \pi \cdot \rho \cdot R^2$ is considered invariant, $C_\Gamma(\lambda) = C_p(\lambda)/\lambda$ and $C_p'(\lambda)$ is the derivative of the power coefficient in relation to λ. One obtains

$$\frac{\partial \sigma}{\partial \Omega_h} = a_1 J_h - \left(KvR^2\right)/i^2 \cdot \left(C_p'(\lambda) \cdot \lambda - C_p(\lambda)\right)/\lambda^2 = a_1 J_h - A(\lambda,v), \tag{B.6}$$

where $A(\lambda,v) = K \cdot v \cdot R^2 \cdot \delta(\lambda)/i^2$, with

$$\delta(\lambda) = \left(C_p'(\lambda)\cdot\lambda - C_p(\lambda)\right)\big/\lambda^2 \qquad (B.7)$$

Using that $\dfrac{\partial\sigma}{\partial\Gamma_G} = 1 + a_2\cdot J_h$ in Equation B.5 and then in Equation B.4 of the equivalent control input, and then replacing Equation B.6 in Equation B.4, one obtains after some algebra

$$u_{eq} = -\frac{T_G}{1+a_2\cdot J_h}\cdot\left(\frac{\Gamma_{wt}}{i} - \Gamma_G\right)\cdot\left(a_1 - A(\lambda,v)\right) - \Gamma_G$$

Provided that the dynamics on the sliding surface are

$$J_h\cdot\dot{\Omega}_h = \Gamma_{wt}\big/i - \Gamma_G \equiv a_1\cdot J_h\cdot\Omega_h + a_2\cdot J_h\cdot\Gamma_G,$$

then the expression of the equivalent control input becomes

$$u_{eq} = \Gamma_G - \frac{T_G}{1+a_2\cdot J_h}\left(a_1\cdot J_h\cdot\Omega_h + a_2\cdot J_h\cdot\Gamma_G\right)\left(a_1 - A(\lambda,v)\right)$$

B.2 Feedback Linearization Control

Feedback linearization is based on differential geometry. The most important notions, needed when feedback linearization control technique is applied are presented in the following (Isidori 1989).

Definition 1. The derivative of a *function f* along the *n*-dimension vector *g* or the *Lie derivative* is a vector obtained by multiplying $\partial f/\partial \mathbf{x}$ with $g(\mathbf{x})$:

$$L_g f(\mathbf{x}) = \frac{\partial f}{\partial \mathbf{x}} g(\mathbf{x})$$

The elements of the Lie derivative are

$$\sum_{i=1}^{n}\frac{\partial f_i}{\partial x_i} g_i(\mathbf{x})$$

If f is the k-th derivative along vector g, the notation $L_g^k f$ is used and it satisfies a recursive relation:

$$L_g^k f(\mathbf{x}) = \frac{\partial\left(L_g^{k-1} f(\mathbf{x})\right)}{\partial \mathbf{x}} g(\mathbf{x})$$

with $L_g^0 f(\mathbf{x}) = f(\mathbf{x})$.

Definition 2. A variable transformation defined by

$$z = \Phi(x)$$

with $\Phi(x) \in \mathbf{R}^n$ is a function of n variables:

$$\Phi(x) = \begin{bmatrix} \phi_1(x) \\ \phi_2(x) \\ \vdots \\ \phi_n(x) \end{bmatrix} = \begin{bmatrix} \phi_1(x_1,...,x_n) \\ \phi_2(x_1,...,x_n) \\ \vdots \\ \phi_n(x_1,...,x_n) \end{bmatrix}$$

with the properties:
- $\Phi(x)$ is invertible, i.e., there is a function $\Phi^{-1}(x)$:

$$\Phi^{-1}(\Phi(x)) = x, \forall x \in \mathbf{R}^n,$$

- $\Phi(x)$ and $\Phi^{-1}(x)$ are smooth functions, having partial derivatives of any order,

is called a *diffeomorphism*.

Using these definitions, the feedback linearization technique can be defined as follows. Considering the nonlinear systems defined by

$$\begin{cases} \dot{x} = f(x) + g(x)u \\ y = h(x) \end{cases} \quad (B.8)$$

where $x \in \mathbf{R}^n$ is the state vector, u is the input, y is the output and f and g are nonlinear smooth function, we are looking for an integer r and a feedback:

$$u = \alpha(x) + \beta(x) \cdot u_v$$

with $\alpha(x)$ and $\beta(x)$ being smooth functions defined in the neighbourhood of a point $x_0 \in \mathbf{R}^n$, where $\beta(x_0) \neq 0$, and $u_v(t)$ is the control input, such that the system

$$\begin{cases} \dot{x} = f(x) + g(x)\alpha(x) + g(x)\beta(x)u_v \\ y = h(x) \end{cases}$$

shown in Figure B.1, has the property that the r-th order derivative of the output is

$$y^{(r)}(t) = u_v(t) \quad (B.9)$$

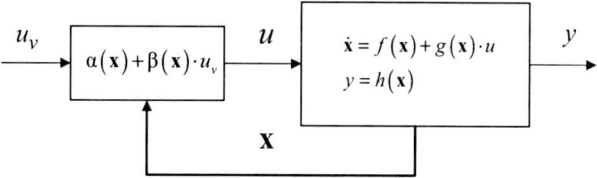

Figure B.1. Feedback linearization control block diagram

Equation B.9 is equivalent to a chain of r integrators corresponding to a linear system with the output $y(t)$ and the control input $u_v(t)$, as shown in Figure B.2.

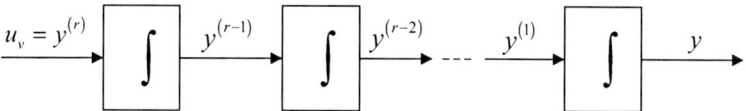

Figure B.2. The linearized system

The transformation of the nonlinear system at Equation B.8 in the linear form shown in Figure B.2 is known as *input-output feedback linearization* (Isidori 1989).

Finding the integer r that fulfils Equation B.9 is also known as finding the relative degree of the system and is defined as follows:

Proposition (Isidori 1989). The nonlinear system at Equation B.8 has *relative degree r* if

$$L_g L_f^{k-1} h(x) = 0 \text{ and } L_g L_f^k h(x) \neq 0$$

for every $k < r-1$ and every $\mathbf{x}$ in the neighbourhood of $\mathbf{x}_0$.

Knowing the relative degree, r, of the system, the linearizing control input

$$u(t) = \frac{1}{L_g L_f^{r-1} h(\mathbf{x})} \left(-L_f^r h(\mathbf{x}) + u_v(t) \right)$$

brings the nonlinear system to the linear system at Equation B.9.

After the relative degree of the system is established, a coordinate transformation $\mathbf{z} = \Phi(\mathbf{x})$, with $\Phi(\mathbf{x}): \mathbf{R}^n \rightarrow \mathbf{R}^n$ being a diffeomorphism, is sought. This transformation leads to expressing the nonlinear system at Equation B.8 as a linear system in $\mathbf{z}$.

If the relative degree of the system is $r = n$, then the system in the new coordinates

$$z_i = \phi_i(\mathbf{x}) = L_f^{i-1} h(\mathbf{x}), \ i = \overline{1, n}$$

is

$$\begin{cases} \dot{z}_1 = z_2 \\ \ldots \\ \dot{z}_{n-1} = z_n \\ \dot{z}_n = b(\mathbf{z}) + a(\mathbf{z})u \end{cases}$$

For this system, the nonlinear feedback

$$u = \frac{1}{a(\mathbf{z})}\left(-b(\mathbf{z}) + u_v\right)$$

leads to a linear and controllable system

$$\begin{cases} \dot{z}_1 = z_2 \\ \ldots \\ \dot{z}_{n-1} = z_n \\ \dot{z}_n = u_v \\ y = z_1 \end{cases}$$

shown in Figure B.3.

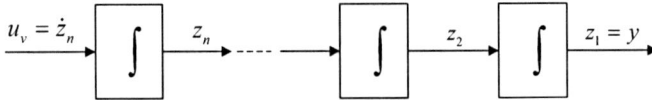

Figure B.3. Input-output linearized system

The control input, written in the old coordinates, is

$$u(t) = \frac{1}{L_g L_f^{n-1} h(\mathbf{x})}\left(-L_f^n h(\mathbf{x}) + u_v\right)$$

If the relative degree of the system is $r < n$, then the system can be brought in the so-called *canonical form*.

Proposition. Consider that the nonlinear system at Equation B.8 has a relative degree $r \leq n$ in $\mathbf{x}_0$. Using the notation

$$\begin{cases} \phi_1(\mathbf{x}) = h(\mathbf{x}) \\ \phi_2(\mathbf{x}) = L_f h(\mathbf{x}) \\ \ldots \\ \phi_n(\mathbf{x}) = L_f^{n-1} h(\mathbf{x}) \end{cases},$$

if $r < n$ then $n-r$ functions cab be found, such that

$$\Phi(\mathbf{x}) = \begin{bmatrix} \phi_1(\mathbf{x}) \\ \phi_2(\mathbf{x}) \\ ... \\ \phi_n(\mathbf{x}) \end{bmatrix}$$

has a non-singular Jacobian matrix in $\mathbf{x}_0$. The expression in $\mathbf{x}_0$ of these functions can be chosen randomly or one can choose $\phi_{r+1}(\mathbf{x}),..., \phi_n(\mathbf{x})$ such that $L_g\phi_i(\mathbf{x}) = 0$, for any i with $r+1 \le i \le n$ and for any $\mathbf{x}$ in the neighbourhood of $\mathbf{x}_0$. The system in the new coordinates is:

- for $z_1,..., z_{r-1}$:

$$\begin{cases} \dot{z}_1(t) = L_f h(\mathbf{x}(t)) = \phi_2(\mathbf{x}(t)) = z_2(t) \\ ... \\ \dot{z}_{r-1}(t) = L_f^{r-1} h(\mathbf{x}(t)) = \phi_r(\mathbf{x}(t)) = z_r(t) \end{cases}$$

- for z_r:

$$\dot{z}_r(t) = L_f^r h(\mathbf{x}(t)) + L_g L_f^{r-1} h(\mathbf{x}(t)) \cdot u(t) \qquad (B.10)$$

Using the notation

$$\begin{cases} a(\mathbf{z}) = L_g L_f^{r-1} h(\mathbf{x}(t)) \\ b(\mathbf{z}) = L_f^r h(\mathbf{x}(t)) \end{cases},$$

Equation B.10 becomes:

$$\dot{z}_r(t) = b(\mathbf{z}(t)) + a(\mathbf{z}(t)) \cdot u(t)$$

with the remark that, by definition, $\mathbf{z}_0 = \Phi(\mathbf{x}_0)$ and $a(\mathbf{z}_0) \ne 0$, leading to $a(\mathbf{z}) \ne 0$, for any $\mathbf{z}$ in the neighbourhood of $\mathbf{z}_0$. As regards the remaining $n-r$ coordinates, they can be chosen such that $L_g\phi_i \mathbf{x} = 0$, leading to

$$\dot{z}_i = \frac{\partial \phi_i}{\partial \mathbf{x}}(f(\mathbf{x}(t)) + g(\mathbf{x}(t))u(t)) = L_g\phi_i(\mathbf{x}(t)) + L_g\phi_i(\mathbf{x}(t))u(t) = L_f\phi_i(\mathbf{x}(t))$$

Using the notation

$$q_i(\mathbf{z}) = L_f\phi_i(\Phi^{-1}(\mathbf{z})), \quad r+1 \le i \le n$$

the last $n-r$ equations can be written as $\dot{z}_i = q_i(\mathbf{z}(t))$.

Under these conditions, the state-space model of the system in the new coordinates is

$$\begin{cases} \dot{z}_1 = z_2 \\ \dot{z}_2 = z_3 \\ \cdots \\ \dot{z}_{r-1} = z_r \\ \dot{z}_r = b(\mathbf{z}) + a(\mathbf{z}) \cdot u \\ \dot{z}_{r+1} = q_{r+1}(\mathbf{z}) \\ \cdots \\ \dot{z}_n = q_n(\mathbf{z}) \\ y = z_1 \end{cases}$$

The input u appears only in z_r, the one that is equal to the relative degree of the system. Such a system, in the new coordinates, where the input u appears only in the expression of z_r, is said to be in the canonical form.

In conclusion, if the system has a relative degree $r < n$, one can obtain a linear subsystem of dimension r, responsible for the input-output mapping, and a subsystem of dimension $n-r$ including the internal dynamics that do not affect the input-output mapping (Figure B.4).

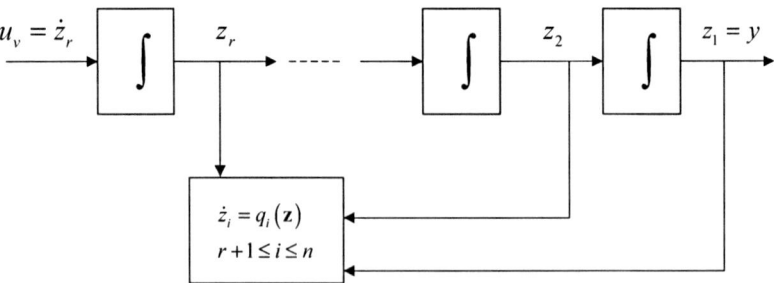

Figure B.4. The system in canonical form

B.3 QFT Robust Control

The design of robust controllers using the QFT method supposes a step-by-step procedure being applied. A general description of this procedure would be difficult to present, that is why here its direct application on a concrete case has been preferred. Thus, the low-power PMSG-based WECS given in Appendix A and Table A.3 has been chosen as an illustrating example.

1. *System formulation.* The PMSG-based WECS was linearized and three operating points on the ORC, defined by the wind speed and the load resistance, were selected. The operating range in focus here is from 3 to 11 m/s and the transfer function is

$$H(s) = \frac{k}{T^2 s^2 + 2\xi T s + 1} \quad (B.11)$$

The values of the parameters, for the three chosen operating points are given in Table B.1.

Table B.1. Operating points chosen on the ORC of a low-power PMSG-based WECS in order to illustrate the QFT design

Operating points		Linearized model parameters		
Wind speed [m/s]	Load resistance [Ω]	K [rad/s/Ω]	T [s]	ζ
4	14,6	3,88	0,0751	0,6147
7	25,9	4,52	0,0456	0,2440
10	38	4,69	0,0326	0,1512

2. *Design specifications.* The tracking specification defines the acceptable variations domain of the closed-loop tracking response due to parameter uncertainty and disturbances. For the PMSG-based WECS, the closed-loop tracking response is imposed to be a second-order system:

$$H(s) = \frac{\omega_0^2}{s^2 + 2\xi\omega_0 s + \omega_0^2},$$

with cut-off frequency $\omega_0 = 20$ rad/s and damping factor $\xi = 0.9$. The limits of the closed-loop tracking response variations are defined by

$$H_{TU}(s) = \frac{\frac{\omega_0^2}{a}(s+a)}{s^2 + 2\zeta\omega_0 + \omega_0^2} \quad H_{TL}(s) = \frac{a_1 \cdot a_2 \cdot a_3}{(s+a_1)(s+a_2)(s+a_3)},$$

where $a = 1.2\omega_0$, $a_1 = 0.5\omega_0$, $a_2 = 1.5\omega_0$ and $a_3 = 2\omega_0$.

The tracking bounds in the time-domain (step response) are presented in Figure B.5.

3. *Choosing the frequency array.* The trial frequency array used in the QFT design method is $\omega = [1 \quad 5 \quad 10 \quad 20 \quad 40 \quad 100]$ rad/s.

256 B Elements of Theoretical Background and Development

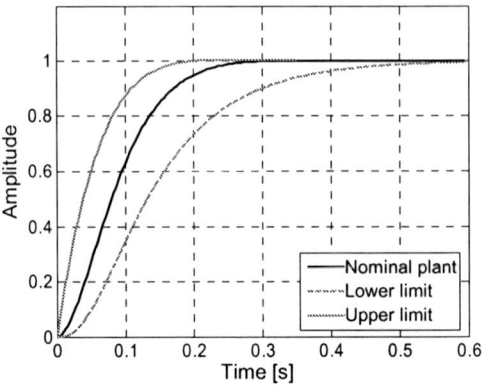

Figure B.5. Tracking bounds in the step response

4. *Plant uncertainty and template computation.* Since the QFT designs are undertaken using a Nichols chart and because a whole set of plants – instead of a single plant – is considered, the magnitude and phase of the plants, at each frequency, yield a set of points on the Nichols chart, instead of a single point. Thus, at each selected frequency a connected region, or so-called *template*, is constructed, which encloses this set of points. In order to be able to construct the templates for the considered set of plants, the LTI model at Equation B.11 is approximated by a similar transfer function, but with only two varying parameters, k and T. Parameters k and T are chosen in such way that their Bode characteristics bound all Bode characteristics of the LTI system models. In Figure B.6 the bounding of the Bode characteristics of all possible combinations of the linearized model parameters (Table B.1) by the transfer function with k varying between 6 and 20 and T ranging from 0.015 to 0.07 for $\xi = 0.8$ is presented. The plant templates at these frequency points are computed with the QFT Toolbox in MATLAB® and presented in Figure B.7.

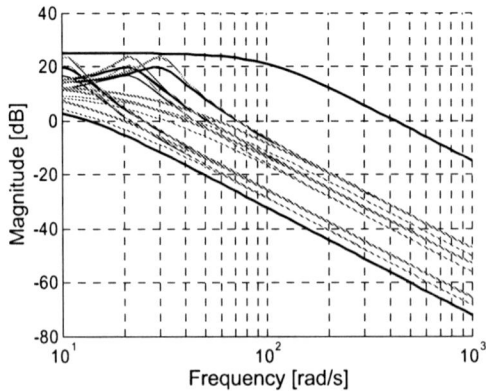

Figure B.6. Bounding of the LTI PMSG-based WECS models Bode characteristic

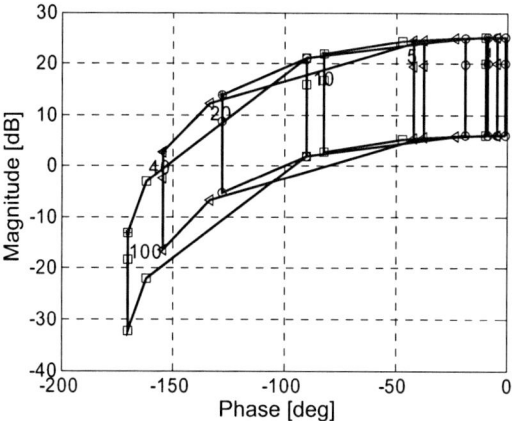

Figure B.7. Plant templates

5. *Selection of a nominal plant model.* The nominal model of the plant, used to compute the bounds, is considered to be

$$H_{nom}(s) = \frac{18}{0.0081s^2 + 0.144s + 1}$$

6. *Generation and integration of bounds.* The two-dimensional magnitude-phase templates are used to define regions (or so-called bounds) in the frequency domain, where the open-loop frequency response must lie in order to satisfy the performance and stability specifications. Stability bounds are calculated using templates and phase margins, while the performance bounds are derived using the templates and upper and lower limits on the frequency-domain response. Both tracking and robust stability margin bounds in the Nichols plane are computed with the QFT Toolbox in MATLAB® and presented in Figure B.8.

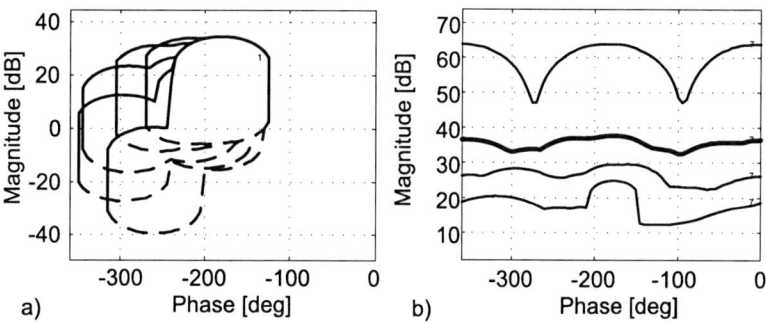

Figure B.8. Stability bounds (**a**) and tracking bounds (**b**) – reprinted (with modifications) from Wind Energy, 9/5, Cutululis NA, Ceangă E, Hansen AD, Sørensen P, Robust multi-model control of an autonomous wind power system, 399-419, Copyright (2006), with permission from John Wiley and Sons Ltd

7. *Loop shaping*. The controller is designed in a very transparent and interactive way, *via* a loop-shaping process in the Nichols plane. The (composite) bounds, for each trail frequency, and the nominal open-loop transfer function are plotted together. The design itself is performed by adding gains, poles and zeros to the nominal plant frequency response in such a way that the boundaries are satisfied at each frequency. The resulted controller is the aggregate of these added gains, poles and zeros. The result of the design is presented in Figure B.9.

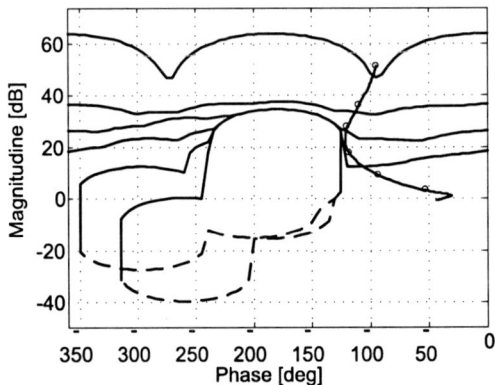

Figure B.9. Open-loop frequency response with controller after the loop-shaping process – reprinted (with modifications) from Wind Energy, 9/5, Cutululis NA, Ceangă E, Hansen AD, Sørensen P, Robust multi-model control of an autonomous wind power system, 399-419, Copyright (2006), with permission from John Wiley and Sons Ltd

The transfer function of the controller, resulting after the QFT design, is

$$C(s) = \frac{6.84(s+62.5)(s+10.06)(s+9.93)}{s(s+1265)(s+5.24)}$$

8. *Prefilter design*. The aim is to ensure that the closed-loop response of the system satisfies the requirements on the closed-loop stability and/or disturbances. The loop shaping is carried out in the frequency domain also, using a Bode diagrams instead of Nichols plots, in the same transparent and interactive manner.

The prefilter transfer function, resulting after the design process, is

$$F(s) = \frac{5478.55}{(s+343.7)(s+15.94)}$$

9. *Analysis and validation*. Finally, the resulting closed-loop system is analysed to make sure that the design specifications are satisfied for frequencies other than those used for computing the bounds. The robust stability of the closed-loop is verified – in a quantitative manner – by plotting the closed-loop Bode characteristic together with the imposed performance, as presented in Figure B.10.

One can note that the robust stability performance is satisfied, since closed-loop Bode characteristic (solid line) is under the imposed one (dotted line) for all frequencies.

The tracking performances are analysed using the closed-loop frequency response of the system, with and without prefilter, as shown in Figure B.11.

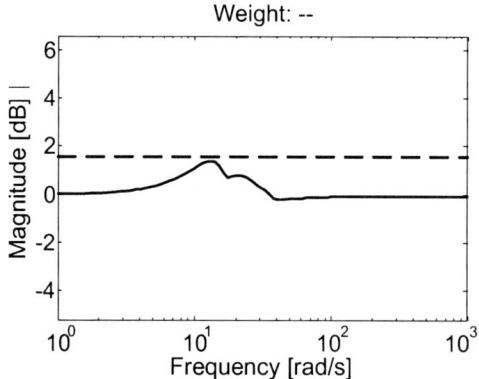

Figure B.10. Robust stability margins – reprinted (with modifications) from Wind Energy, 9/5, Cutululis NA, Ceangă E, Hansen AD, Sørensen P, Robust multi-model control of an autonomous wind power system, 399-419, Copyright (2006), with permission from John Wiley and Sons Ltd

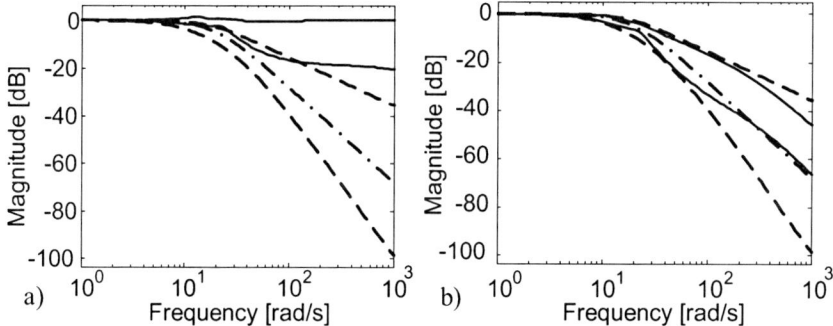

Figure B.11. Closed-loop frequency response without prefilter (**a**) and with prefilter (**b**) – reprinted (with modifications) from Wind Energy, 9/5, Cutululis NA, Ceangă E, Hansen AD, Sørensen P, Robust multi-model control of an autonomous wind power system, 399-419, Copyright (2006), with permission from John Wiley and Sons Ltd

The closed-loop response without prefilter satisfies the stability performances in the frequency domain – the response bandwidth (solid line) is smaller than the bandwidth defined by the upper and lower limit stability specifications (dotted line) – but the robust tracking specification are not satisfied. After inserting the prefilter, the closed-loop frequency response satisfies both conditions, as one can observe in Figure B.11; thus a successful design is achieved.

C

Photos, Diagrams and Real-time Captures

Figure C.1 presents the Simulink® block diagram of the two-loop control structure based upon the frequency separation principle used in Case Study 10 reported in Section 6.4.5.

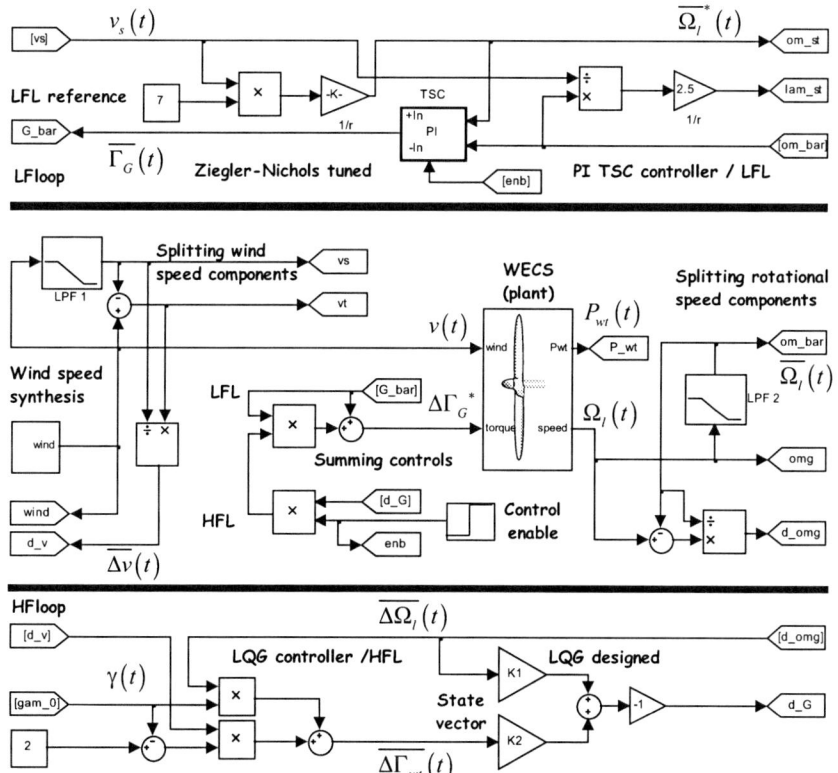

Figure C.1. Simulink® block diagram of the 2LFSP for rigid-drive-train WECS

Figure C.2 shows the structure of the WECS simulator used for real-time assessment of the two-loop control structure based upon the frequency separation principle, as reported in Section 6.4.6.

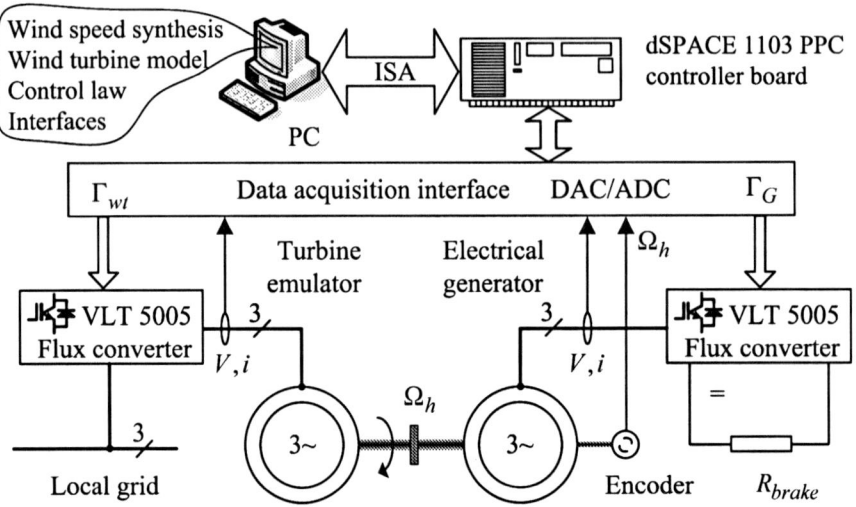

Figure C.2. Hardware structure of the variable-speed rigid-drive-train WECS experimental test rig used for real-time validation of 2LFSP

The simulator consists of two identical rigidly coupled cage induction machines. One of them, denoted by IM (induction motor), emulates the aerodynamic subsystem and the drive train of the WECS; the second one, denoted by IG (induction generator), is a torque controlled generator. The wind speed is computer synthesized and imposes the rotational speed of the coupling.

The two induction machines are controlled *via* power electronics converters, namely VLT 5005 Flux. The computer implements:
- the wind turbine model, utilized to provide the wind torque reference, Γ_{wt} (referred to the high-speed shaft), using the computed wind speed and the measured rotational speed, Ω_h;
- the control law, which provides the generator torque reference, Γ_G^*.

The simulation schemes are built under MATLAB®/Simulink®, using Real Time Interface, and run on the DS1103 PPC controlled board (dSPACE), which is equipped with a Power PC processor for fast floating-point computation at 400 MHz. The bidirectional information flux between the physical part and the computer is supported by a data acquisition interface.

A ControlDesk® panel (Figure C.4) is used to visualize the functional parameters of the simulated WECS and for getting user-supplied data in real time. A more detailed description of this simulator can be found in Cutululis (2005).

Figure C.3 presents the main elements of the experimental rig containing a 1:4.66 HIL WECS simulator used for real-time validations of the optimal control structure developed in Section 6.4.

Figure C.3. Photo of the main elements composing the test rig used for 2LFSP assessment

The simulator contains the following elements:
1. squirrel-cage induction motor from the RTPS (SCIM);
2. squirrel-cage induction generator (SCIG);
3. SCIM–SCIG coupling;
4. power electronics for driving SCIM (VLT 5005 Flux);
5. power electronics for driving SCIG (VLT 5005 Flux);
6. personal computer (PC) implementing user interfaces;
7. Simulink® and ControlDesk® interfaces on PC;
8. DSP DS1103 (dSPACE®) implementing RTPS, WECS and SCIG control;
9. computer I/O connections;
10. rotary encoder;
11. wiring and transducers.

Figure C.4 shows ControlDesk® interface used for real-time assessment of the 2LFSP method. Real-time evolutions of some main variables are visible, as well as the operating point dispersion in the $\Omega_l - P_{wt}$ plane. More real-time captures concerning this method can be found in Figures 6.17–6.21 and also in Munteanu (2006).

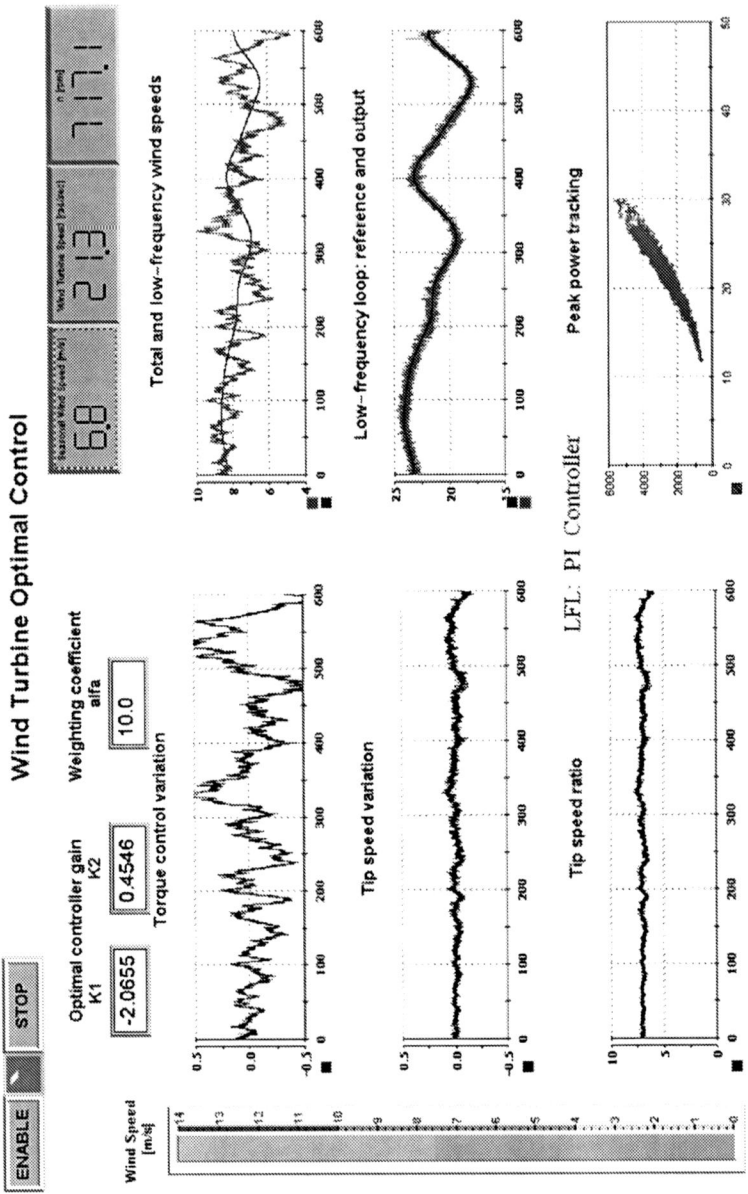

Figure C.4. ControlDesk® panel of the 2LFSP test rig – PI-controller-based LFL – reprinted (with modifications) from Control Engineering Practice, 13, Munteanu I, Cutululis NA, Bratcu AI, Ceangă E, Optimization of variable speed wind power systems based on a LQG approach, 903-912, Copyright (2005), with permission from Elsevier

Figure C.5 presents the real-time physical simulator components from Section 7.3 (Case Study 12). Figure C.5a show the basic software structure of the real-time software simulator, while Figure C.5b depicts the main elements of the effector (electromechanical tracking loop).

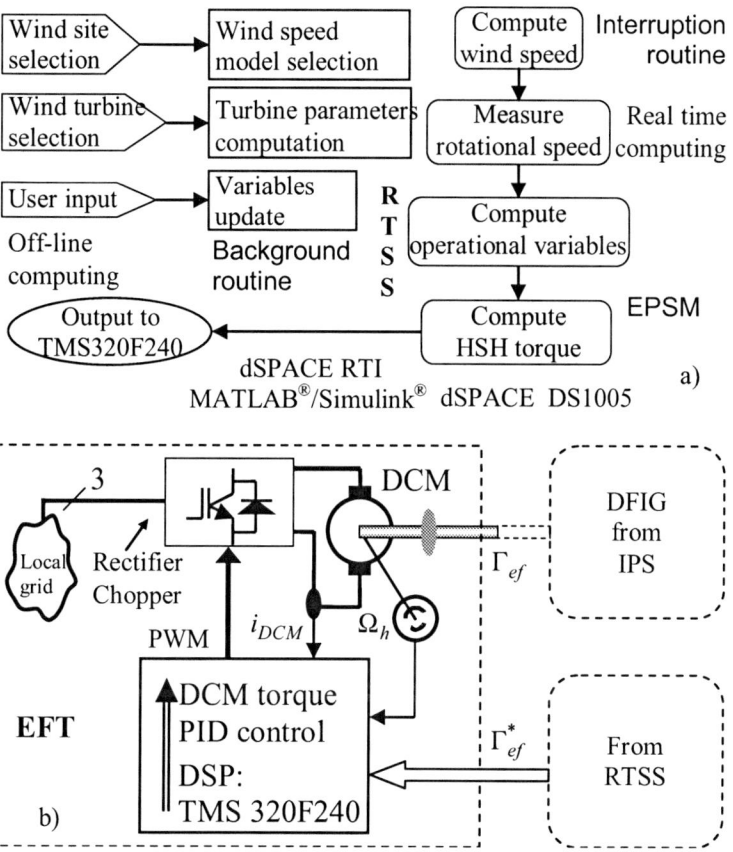

Figure C.5. Software (**a**) and hardware (**b**) components of the real-time physical simulator (RTPS) within the HIL simulator (Case Study 12)

Figure C.6 contains the full structure of the HILS from Case Study 12. In the investigated physical system (IPS, right side of Figure C.6), the control structures realizing the DFIG rotor power transfer to the grid, previously presented in Section 7.3.3, have been built in MATLAB®/Simulink® (on computer PC1) and implemented by means of the Real Time Interface Toolbox (dSPACE® RTI for target processor DS1005). This software component also includes an operating mode controller, responsible for the global behaviour of the WECS. The resulting code runs on the DS1005 Power PC processor together with the real-time software simulator (RTSS). The bidirectional information flux – containing sensor information (A/D), PWM outputs, binary configuring signals for the power system, D/A outputs for visualizing rapid signal variations – is supported by a complex input/output system of the DSP. The interest variables are monitored by means of a ControlDesk® panel. The left side of Figure C.6 contains the structure of the real-time physical simulator (RTPS), previously described in Section 7.3.2, monitored by means of the TestPoint® software on computer PC2. One must note that the RTPS–IPS interaction happens only at the physical level, at the DFIG shaft.

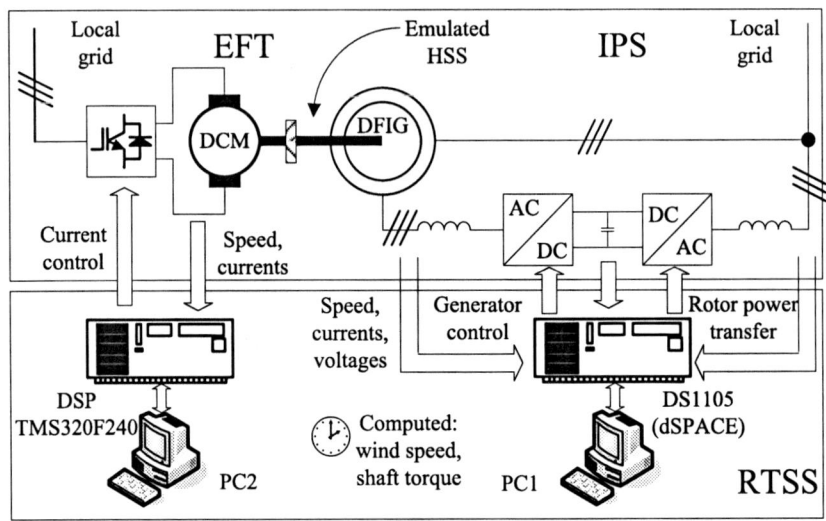

Figure C.6. Functional diagram of the WECS real-time simulator, according to the HILS design methodology (Case Study 12)

Table C.1 and the photo in Figure C.7 shows the components of the HIL WECS simulator described in Case Study 12 (Section 7.3).

Table C.1. Main features of the physical simulator described in Case Study 12

RTPS effector	Simulator components
DC machine Siemens 1G 5 106-0EH-3UV1, with rated power 6.5 kW, rated rotational speed 3850 rpm, maximal torque 20.2 Nm, torque constant 0.87 Nm/A, electromechanical time constant 4.6 ms, inertia 0.02 Kg·m²; stator resistance 78 Ω, rotor resistance 0.78 Ω, rotor inductance 3.6 H, speed constant 0.9 V/rad/s; DC motor speed reduction gear of ratio 1.41; 10 kW rectifier–chopper (IGBT based); real-time DSP TMS320F240, time frame T_e = 200 μs	1. DC motor from RTPS (DCM); 2. doubly-fed induction generator (DFIG); 3. gear box used for DCM–DFIG coupling; 4. power electronics for driving DCM; 5. power electronics for driving DFIG; 6. personal computer PC2 (in Figure C.6); 7. personal computer PC1 (in Figure C.6); 8. TestPoint® interface on PC2; 9. Simulink® and ControlDesk® interface on PC1; 10. DSP TMS320F240 implementing DCM torque control; 11. DSP DS1005 (dSPACE®) implementing WECS and DFIG control; 12. oscilloscope

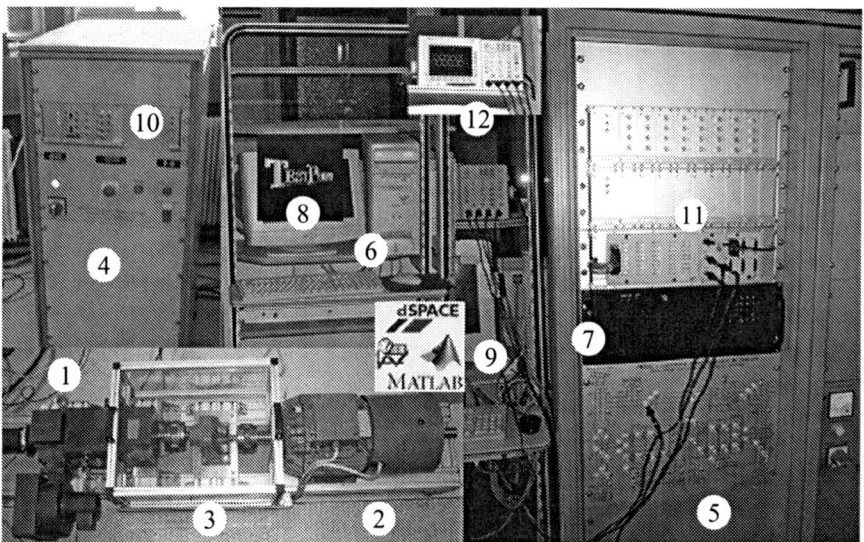

Figure C.7. Components of the HIL WECS simulator (Case Study 12 – courtesy of Grenoble Génie Électrique Laboratory – France)

References

Abo-Khalil AG, Lee D-C, Seok J-K (2004) Variable speed wind power generation system based on fuzzy logic control for maximum output power tracking. In: Proceedings of 2004 IEEE 35th Annual Power Electronics Specialists Conference – PESC 04, vol 3, pp 2039-2043

Ackerman T (ed.) (2005) Wind power in power systems. John Wiley & Sons, Chicester, U.K.

Akhmatov V (2003) Analysis of dynamic behaviour of electric power systems with large amount of wind power. Ph.D. Thesis, Technical University of Denmark, Denmark

Akhmatov V, Knudsen H, Nielsen AH (2000) Advanced simulation of windmills in the electric power supply. International Journal of Electrical Power and Energy Systems 22(6):421-434

Akhrif O, Okou FA, Dessaint LA, Champagne R (1999) Application of a multivariable feedback linearization scheme for rotor angle stability and voltage regulation of power systems. IEEE Transactions on Power Systems 14(2):620-628

Alatalo M (1996) Permanent magnet machines with air gap windings and integrated teeth windings. Technical Report 288, Chalmers University of Technology, Sweden

Amei K, Takayasu Y, Ohji T, Sakui M (2002) A maximum power control of wind generator system using a permanent magnet synchronous generator and a boost chopper circuit. In: Proceedings of the Power Conversion Conference – PCC 2002, vol 3, pp 1447-1452

Apkarian P, Adams R (1998) Advanced gain-scheduling techniques for uncertain systems. IEEE Transactions on Control Systems Technology 6(1):21-32

Apkarian P, Gahinet P, Becker G (1995) Self-scheduled $\mathcal{H}_\infty$ control of linear parameter-varying systems: a design example. Automatica 31(9):1251-1261

Ariyur KB, Krstič M (2003) Real-time optimisation by extremum seeking control. Wiley-Interscience

Åström KJ, Wittenmark B (1995) Adaptive control, 2nd edn. Addison-Wesley

Åström KJ, Hägglund T (1995) PID controllers: theory, design and tuning, 2nd edn. Instrument Society of America

Athans M, Falb P (1966) Optimal control. McGraw-Hill, New-York

Baring-Gould I, Flowers L, Lundsager P, Mott L, Shirazi M, Zimmermann J (2004) Worldwide status of wind-diesel applications. 2004 DOE/AWEA/CanWEA Wind-Diesel Workshop
http://www.eere.energy.gov/windpoweringamerica/pdfs/workshops/2004_wind_diesel

Battaioto PE, Mantz RJ, Puleston PF (1996) A wind turbine emulator based on a dual DSP processor system. Control Engineering Practice 4(9):1261-1266

De Battista H, Mantz RJ (1998) Sliding mode control of torque ripple in wind energy conversion systems with slip power recovery. In: Proceedings of the 24th Annual Conference of the IEEE Industrial Electronics Society – IECON '98, pp 651-656

De Battista H, Mantz RJ (2004) Dynamical variable structure controller for power regulation of wind energy conversion systems. IEEE Transactions on Energy Conversion 19(4):756-763

De Battista H, Puleston PF, Mantz RJ, Christiansen CF (2000a) Sliding mode control of wind systems with DOIG – Power efficiency and torsional dynamics optimization. IEEE Transactions on Power Systems 15(2):728-734

De Battista H, Mantz RJ, Christiansen CF (2000b) Dynamical sliding mode power control of wind driven induction generators. IEEE Transactions on Energy Conversion 15(4):451-457

Betz A (1926) Wind energy and its use by wind-mills (Wind-Energie und ihre Ausnutzung durch Windmühlen). Vandenhoek & Ruprecht, Göttingen

Bhowmik S, Spée R (1998) Wind speed estimation based variable speed wind power generation. In: Proceedings of the Annual IEEE Conference of the Industrial Electronics Society – IECON'98, pp 596-601

Bialasiewicz JT, Muljadi E (2002) Analysis of variable structure diesel/invertor operation control in renewable energy systems. In: Proceedings of American Control Conference, pp 1978-1983

Bianchi FD, Mantz RJ, Christiansen CF (2005) Gain scheduling control of variable-speed wind energy conversion systems using quasi-LPV models. Control Engineering Practice 13(2):247-255

Bianchi F, De Battista H, Mantz RJ (2006) Wind Turbine Control Systems – Principles, Modelling and Gain Scheduling Design. Springer, London

Borghesani C, Chait Y, Yaniv O (2003) The QFT frequency domain control design toolbox for use with MATLAB®. User's Guide, Terasoft, Inc. http://www.terasoft.com/qft/QFTManual.pdf

Borowy BS, Salameh ZM (1997) Dynamic response of a stand-alone wind energy conversion system with battery energy storage to a wind gust. IEEE Transactions on Energy Conversion 12(1):73-78

Bose BK (2001) Modern power electronics and AC drives. Prentice–Hall, Englewood Cliffs, NJ, U.S.A.

Bossanyi EA (2000). The design of closed loop controllers for wind turbines. Wind Energy 3:149-163

Boukhezzar B, Lupu L, Siguerdidjane H, Hand M (2007) Multivariable control strategy for variable speed, variable pitch wind turbines. Renewable Energy 32:1273-1287

Bouscayrol A, Guillaud X, Delarue Ph (2005) Hardware-in-the-loop simulation of a wind energy conversion system using energetic macroscopic representation. In: Proceedings of the 31st Annual Conference of IEEE Industrial Electronics Society – IECON 2005, 6 pages

De Broe AM, Drouilhet S, Gevorgian V (1999) A peak power tracker for small wind turbines in battery charging applications. IEEE Transactions on Energy Conversion 14(4):1630-1635

Burton T, Sharpe D, Jenkins N, Bossanyi E (2001) Wind energy handbook. John Wiley & Sons, New-York

Camblong H, Rodriguez M, Puiggali JR, Abad G (2002a) Comparison of different control strategies to study power quality in a variable speed wind turbine. In: Proceedings of the 1st World Wind Energy Conference (CD-ROM)

Camblong H, Arana J, Rodriguez M, Puiggali JR, Patrouix O (2002b) Wind variations effects on the power quality for different controls of a variable-speed wind turbine. In: Proceedings of the Global Windpower Conference (CD-ROM)

Camblong H, Martinez de Alegria I, Rodriguez M, Abad G (2006) Experimental evaluation of wind turbines maximum power point tracking controllers. Energy Conversion and Management 47(18-19):2846-2858

Cárdenas R, Peña R (2004) Sensorless vector control of induction machines for variable-speed wind energy applications. IEEE Transactions on Energy Conversion 19(1):196-205

Cárdenas R, Asher GM, Ray WF, Peña R (1996) Power limitation in variable speed wind turbines with fixed pitch angle. In: Proceedings of International Conference on Opportunities and Advances in International Electric Power Generation, pp 44-48

Cárdenas R, Peña R, Asher GM, Clare JC (2001) Experimental emulation of wind turbines and flywheels for wind energy applications. In: Proceedings of EPE 2001 (CD-ROM)

Carlin PW, Laxson S, Muljadi EB (2001) The history and state of the art of variable-speed wind turbine technology. Technical Report NREL/TP-500-28607, National Renewable Energy Laboratory, U.S.A.

Ceangă E, Protin L, Nichita C, Cutululis NA (2001) Theory of control systems (Théorie de la commande des systèmes). Technical Publishing House, Bucharest, Romania

Chandrasekharan PC (1996) Robust control of linear dynamical systems. Academic Press

Chapman JA, Ilic MD, King CA, Eng L, Kaufman H (1993) Stabilizing a multimachine power system via decentralized feedback linearizing excitation control. IEEE Transactions on Power Systems 8(3):830-838

Chen Z, Spooner E (1998) Grid interface options for variable-speed, permanent-magnet generators. IEE Proceedings on Electric Power Applications 145(4):273-283

Chen Z, Gomez SA, McCormick M (2000) A fuzzy logic controlled power electronic system for variable speed wind energy conversion systems. In: Proceedings of the 8th Power Electronics and Variable Speed Drives, pp 114-119

Connor B, Leithead WE (1993) Investigation of fundamental trade-off in tracking the Cpmax curve of a variable speed wind turbine. In: Proceedings of the 12th British Wind Energy Conference, pp 313-319

Cutululis NA (2005) Contributions to control strategies synthesis of renewable power systems with hybrid structures (Contribuții privind sinteza strategiilor de conducere automată în sistemele de conversie a energiilor neconvenționale cu structuri hibride). Ph.D. Thesis, "Dunărea de Jos" University of Galați, Romania

Cutululis NA, Ciobotaru M, Ceangă E, Roșu ME (2002) Real time wind turbine simulator based on frequency controlled AC servomotor. Annals of "Dunărea de Jos" University of Galați, III:97-101.
www.ann.ugal.ro/eeai/index.html

Cutululis NA, Bindner H, Munteanu I, Bratcu A, Ceangă E, Sørensen P (2006a) LQ optimal control of wind turbines in hybrid power systems. In: Proceedings of the European Wind Energy Conference – EWEC '06, 6 pages

Cutululis NA, Ceangă E, Hansen AD, Sørensen P (2006b) Robust multi-model control of an autonomous wind power system. Wind Energy 9(5):399-419

Cutululis NA, Sørensen P, Vigueras-Rodriguez A, Jensen L, Hjerrild J, Donovan M, et al. (2007) Models for assessing power fluctuations from large wind farms. In: Proceedings of the 2007 European Wind Energy Conference and Exhibition, 9 pages

Damper RI (1995) Introduction to discrete-time signals and systems. Chapman and Hall

Datta R, Ranganathan VT (2003) A method of tracking the peak power points for a variable speed wind energy conversion system. IEEE Transactions on Energy Conversion 18(1):163-168

DeCarlo RA, Zak SH, Drakunov SV (1996) Variable structure, sliding-mode controller design. In: Levine WS (ed.) The Control Handbook. CRC Press, IEEE Press, pp 941-951

Diop AD (1999) Contributions to development of an electromechanical wind-turbine simulator: simulation and real-time control of a medium-power variable-pitch wind turbine (Contribution au développement d'un simulateur électromécanique d'aérogénérateur : simulation et commande en temps réel d'une turbine éolienne de puissance moyenne à angle de calage variable). Ph.D. Thesis, Université du Havre, France

Diop AD, Nichita C, Belhache JJ, Dakyo B, Ceangă E (1999) Modelling variable pitch HAWT characteristics for a real time wind turbine simulator. Wind Engineering 23(4):225-243

Diop AD, Nichita C, Belhache JJ, Dakyo B, Ceangă E (2000) Error evaluation for models of real time wind turbine simulators. Wind Engineering 24(3):203-221

Dixit A, Suryanarayanan S (2005) Towards pitch-scheduled drive train damping in variable-speed horizontal-axis large wind turbines. In: Proceedings of the 44th IEEE Conference on Decision and Control 2005 European Control Conference – CDC-ECC '05, pp 1295-1300

Dolan DSL, Lehn PW (2005) Real-time wind turbine emulator suitable for power quality and dynamic control studies. In: Proceedings of the International Conference on Power System Transients – IPST '05, 6 pages

Dominguez S, Leithead WE (2006) Size related performance limitations on wind turbine control systems. In: Proceedings of the 2006 European Wind Energy Conference – EWEC '06, 6 pages

DS 742 (2007) Dansk Ingenioerforenings Code of practice for loads and safety of wind turbine constructions. Danish Standards Association

dSPACE (2003) Modular systems based on DS1005. Installation and configuration for Release 3.5. dSPACE Gmbh

Dumitrescu H, Georgescu A, Ceangă V, Popovici J, Ghiţă Gh, Dumitrache A, et al. (1990) Calculation of propellers (in Romanian: Calculul elicei), Romanian Academy Publishing House, Bucharest

Ekelund T (1997) Modeling and linear quadratic optimal control of wind turbines. Ph.D. Thesis, Chalmers University of Göteborg, Sweden

EN (1995) Voltage characteristics of electricity supplied by public distribution systems. EN 50160, www.cenelec.org

van Engelen TG, Schaak P, Lindenburg C (2003) Control for damping the fatigue relevant deformation modes of offshore wind turbines. In: Proceedings of the 2003 European Wind Energy Conference – EWEC '03, 6 pages

Enslin JHR, van Wyk D (1992) A study of a wind power converter with micro-computer based maximal power control utilizing an over-synchronous electronic Scherbius cascade. Renewable Energy 2(6):551-562

EWEA (2005) Large scale integration of wind energy in the European power supply. European Wind Energy Association, Brussels, Belgium

Farret FA, Pfitscher LL, Bernardon DP (2000) An heuristic algorithm for sensorless power maximization applied to small asynchronous wind turbogenerators. In: Proceedings of the IEEE International Symposium on Industrial Electronics – ISIE 2000, vol 1, pp 179-184

Freris LL (1990) Wind energy conversion. Prentice–Hall, Englewood Cliffs, NJ, U.S.A.

Gasch R, Twelve J (2002) Wind power plants: fundamentals, design, construction and operation. James and James, London, and Solarpraxis, Berlin

Le Gourières D (1982) Wind energy – Theory, design and practical calculation of wind energy systems (Énergie éolienne – théorie, conception et calcul pratique des installations). Éditions Eyrolles, Paris

Hanselmann H (1993) Hardware-in-the-loop simulation development and test of electronic control units and mechanical components. In: Proceedings of the Real Time Conference

Hanselmann H (1996) Hardware-in-the-loop simulation testing and its integration into a CACSD toolset. In: Proceedings of the IEEE International Symposium on Computer-Aided Control System Design, 5 pages

Hansen AD, Hansen LH (2007) Wind turbine concept market penetration over 10 Years (1995-2005). Wind Energy 10(1):81-97

Hansen, LH, Helle L, Blaabjerg F, Ritchie E, Munk-Nielsen S, Bindner H et al. (2001) Conceptual survey of generators and power electronics for wind turbines. Technical Report RISØ-R-1205(EN), Roskilde, Denmark

Hansen AD, Jauch C, Sørensen P, Iov F, Blaabjerg F (2003) Dynamic wind turbine models in power system simulation tool DIgSILENT. Technical Report RISØ-R-1400, RISØ National Laboratory, Roskilde, Denmark

Hansen AD, Sørensen P, Iov F, Blaabjerg F (2004) Control of variable speed wind turbines with double-fed induction generators. Wind Engineering 28(4):411-434

Hansen MH, Hansen A, Larsen TJ, Øye S, Sørensen P, Fuglsang P (2005) Control design for a pitch-regulated, variable speed wind turbine. Technical Report RISØ-R-1500, RISØ National Laboratory, Roskilde, Denmark

Hautier JP, Caron JP (1997) Automatic systems. Tome 2: Control systems (Systèmes automatiques. Tome 2 : Commande des processus). Ellipses, Paris

Heier S (2006) Grid integration of wind energy conversion systems, 2nd edn. John Wiley & Sons, Chicester, U.K.

Higuchi Y, Yamamura N, Ishida M, Hori T (2000) An improvement of performance for small-scaled wind power generating system with permanent magnet type synchronous generator. In: Proceedings of the 26th Annual Conference of the IEEE Industrial Electronics Society – IECON 2000, vol 2, pp 1037-1043

Hilloowala RM, Sharaf AF (1994) A utility interactive wind energy conversion scheme with an asynchronous DC link using a supplementary control loop. IEEE Transactions on Energy Conversion 9(3):558-563

Hilloowala RM, Sharaf AM (1996) A rule-based fuzzy logic controller for a PWM inverter in a stand alone wind energy conversion scheme. IEEE Transactions on Industry Applications 32(1):57-65

Hofmann W, Thieme A, Dietrich A, Stoev A (1997) Design and control of a wind power station with double fed induction generator. In: Proceedings of EPE '97, pp 2723-2728

Horowitz IM (1993) Quantitative feedback design theory, vol. 1. QFT Publications, Boulder, Colorado, U.S.A.

IEC (1997) Electromagnetic compatibility (EMC) – Part 4: Testing and measurement techniques – Section 15: Flickermeter – functional and design specifications, IEC 61000-4-15. International Electrotechnical Commision, Geneva, Switzerland

Iov F (2003) Contributions to modelling, analysis and simulation of AC drive systems. Application to large wind turbines. Ph.D. Thesis, "Dunărea de Jos" University of Galați, Galați, Romania

Isidori A (1989) Nonlinear control systems, 2nd edn. Springer-Verlag, Berlin

Jamil M (1990) Wind power statistics and evaluation of wind energy density. Wind Engineering 18(5):227-240

Jeffries WQ, McGowan JG, Manwell JF (1996) Development of a dynamic model for no storage wind/diesel systems. Wind Engineering 20(1):27-38

Kana CL, Thamodharan M, Wolf A (2001) System management of a wind-energy converter. IEEE Transactions on Power Electronics 16(3):375-381

Kanellos FD, Hatziargyriou ND (2002) The effect of variable-speed wind turbines on the operation of weak distribution networks. IEEE Transaction on Energy Conversion 17(4):543-548

Kaura V, Blasko V (1997) Operation of a phase locked loop system under distorted utility conditions. IEEE Transactions on Industry Applications 33(1):58-63

Kiffmeier U (1996) A hardware-in-the-loop testbench for ABS controllers. In: Proceedings of the Conference on Control and Diagnostics in Automotive Applications

Knight AM, Peters GE (2005) Simple wind energy controller for an expanded operating range. IEEE Transaction on Energy Conversion 20(2):459-466

Kojabadi HM, Chang L, Boutot T (2004) Development of a novel wind turbine simulator for wind energy conversion systems using an inverter-controlled induction motor. IEEE Transactions on Energy Conversion 19(3):547-552

Krause PC, Wasynczuk O, Sudhoff SD (2002) Analysis of electric machinery and drive systems, 2nd edition. Wiley-IEEE Press

Krstič M, Wang H-H (2000) Stability of extremum seeking feedback for general nonlinear dynamic systems. Automatica 36:595-601

Kubota Y, Genji K, Miyazato K, Hayashi N, Tokuda H, Fukuyama Y (2002) Verification of cooperative control method for voltage control equipment on distribution networks simulator considering interconnection of wind power generators. In: Proceedings of Transmission and Distribution Conference and Exhibition 2002 Asia Pacific, vol 2, pp 1151-1156

Larsson A (2002) Flicker emission of wind turbines during continuous operation. IEEE Transactions on Energy Conversion 17(1):114-118

Larsson T, Poumarede C (1999) STATCOM, an efficient means for flicker mitigation. In: Proceedings of the IEEE Power Engineering Society 1999 Winter Meeting, pp 1208-1213

Lee DC, Lee GM, Lee KD (2000) DC-bus voltage control of three-phase AC/DC PWM converters using feedback linearization. IEEE Transactions on Industry Applications 36(3):826-833

Leithead WE (1990) Dependence of performance of variable speed wind turbines on the turbulence, dynamics and control. IEE Proceedings 137(6):403-413

Leithead WE, De la Salle S, Reardon D (1991) Role and objectives of control for wind turbines. IEE Proceedings-C 138(2):135-148

Leithead WE, Rogers MCM, Connor B, Pierik JTE, Van Engelen TG, O'Reilly J (1994) Design of a controller for a test-rig for a variable speed wind turbine. In: Proceedings of the Third IEEE Conference on Control Applications, vol 1, pp 239-244

Lenwari W, Sumner M, Zanchetta P, Culea M (2006) A high performance harmonic current control for shunt active filters based on resonant compensators. In: Proceedings of the 2006 IEEE Industrial Electronics Conference – IECON '06, pp 2109-2114

Leonhard W (2001) Control of electrical drives, 3rd edition. Springer, Berlin Heidelberg New-York

Lescher F, Zhao JY, Martinez A (2006) Multiobjective H_2/H_∞ control of a pitch regulated wind turbine for mechanical load reduction. International Conference on Renewable Energies and Power Quality – ICREPQ '06, 6 pages

Long Y, Hanba S, Yamashita K, Miyagi H (1999) Sliding mode controller design via H_∞ theory for windmill power systems. In: Proceedings of the IEEE International Conference on Systems, Man and Cybernetics, vol I, pp 56-61

Lublin L, Athans M (1996) Linear quadratic regulator control. In: Levine WS (ed.) The control handbook. CRC Press, IEEE Press, pp 635-650

Lubosny Z (2003) Wind turbine operation in electric power systems. Springer, Berlin Heidelberg New-York

Lund T, Sørensen P, Eek J (2007) Reactive power capability of a wind turbine with doubly fed induction generator. Wind Energy 10(4):379-394

Manwell JF, McGowan JG, Baring-Gould EI, Jeffries W, Stein W (1993) Hybrid systems modelling: development and validation. Wind Engineering 18(5):241-255

Miller A, Muljadi E, Zinger DS (1997) A variable speed wind turbine power control. IEEE Transactions on Energy Conversion 12(2):451-457

Miller NW, Price WW, Sanchez-Gasca JJ (2003) Dynamic modelling of GE 1.5 and 3.6 wind turbine generators. Technical Report, Power Systems Energy Consulting, General Electric International, Shenectady, U.S.A.

Mirra C (1988) Connection of fluctuating loads. International Union of Electroheat, Paris, France

El Mokadem M, Nichita C, Dakyo B, Koczara W (2003) Control strategy for a variable load supplied by a wind-diesel system. Electromotion, 10:635-640

El Mokadem M, Nichita C, Reghem P, Dakyo B (2005) Experimental system development for maximum power tracking of a wind turbine simulator. In: Proceedings of EPE 2005 (CD-ROM)

El Mokadem M, Nichita C, Reghem P, Dakyo B (2006) Short term energy storage based on reluctance machine control for wind diesel system. In: Proceedings of the EPE-PEMC 2006, pp 1585-1590

Mokhtari M, Marie M (1998) Applications de MATLAB®5 et Simulink®2. Springer, Paris

Molenaar D-P (2003) Cost-effective design and operation of variable speed wind turbines. Ph.D. Thesis, Technical University of Delft, The Netherlands

Muljadi E, Pierce K, Migliore P (2000) A conservative control strategy for variable-speed stall-regulated wind turbines. Technical Report NREL/CP-500-24791, National Renewable Energy Laboratory, Colorado, U.S.A.

Munteanu I (1997) Study and simulation of wind power systems. Application to a direct-current generator based wind power system (Étude et simulation des aérogénérateurs. Application à un aérogénérateur à courant continu). Master Degree Dissertation, Université du Havre, France

Munteanu I (2006) Contributions to the optimal control of wind energy conversion systems. Ph.D. Thesis, "Dunărea de Jos" University of Galaţi, Galaţi, Romania

Munteanu I, Ceangă E, Cutululis NA, Bratcu A (2004a) Linear quadratic optimization of variable speed wind power systems. In: Bars R, Gyurkovics É (eds.) Control Applications of Optimisation 2003 – A Proceedings volume of the 12th IFAC workshop. Elsevier IFAC Publications, pp 157-162

Munteanu I, Bratcu AI, Frangu L (2004b) Nonlinear control for stationary optimization of wind power systems. In: Sgurev V, Dimirovski GM, Hadjiski M (eds.), Preprints of the IFAC Workshop Automatic systems for building the infrastructure in the developing countries – DECOM '04, pp 195-200

Munteanu I, Cutululis NA, Bratcu AI, Ceangă E (2005) Optimization of variable speed wind power systems based on a LQG approach. Control Engineering Practice 13(7):903-912

Munteanu I, Bratcu A, Cutululis NA, Ceangă E (2006a) A two loop optimal control of flexible drive train variable speed wind power systems. In: Zitek P (ed.) Proceedings of the 16th IFAC World Congress (CD_ROM), Elsevier Science

Munteanu I, Guiraud J, Roye D, Bacha S, Bratcu AI (2006b) Sliding mode energy-reliability optimization of a variable speed wind power system. In: Proceedings of the 9th IEEE Workshop on Variable Structure Systems – VSS '06, pp 92-97

Munteanu I, Cutululis NA, Bratcu A, Ceangă E (2006c) Using a nonlinear controller to optimize a variable speed wind power system. Journal of Electrical Engineering 6(4). www.jee.ro

Narendra KS, Balakrishnan J (1997) Adaptive control using multiple models. IEEE Transactions on Automatic Control 42(2):171-187

Nichita C (1995) Study and development of structures and numerical control laws for building up of a 3 kW wind turbine simulator (Étude et développement de structures et lois de commande numériques pour la réalisation d'un simulateur de turbine éolienne de 3 kW). Ph.D. Thesis, Université du Havre, France

Nichita C (1996) Contributions to the synthesis of digital control algorithms for electromechanical simulation systems, with applications at wind energy conversion

(Contribuţii la sinteza algoritmilor de conducere numerică pentru sisteme electrice şi electromecanice de simulare, cu aplicaţii la conversia energiei eoliene). Ph.D. Thesis, "Dunărea de Jos" University of Galaţi, Romania

Nichita C, Ceangă E, Piel A, Belhache JJ, Protin L (1994) Real time servo system for wind turbine simulator. In: Proceedings of the 3rd International Workshop on Advanced Motion Control, pp 1039-1048

Nichita C, Diop AD, Belhache JJ, Dakyo B, Protin L (1998a) Control structures analysis for a real time wind system simulator. Wind Engineering 22(6):275-286

Nichita C, Ceangă E, Bivol I, Munteanu I (1998b) Hardware-in-the-loop techniques for the real time simulation. Annals of "Dunărea de Jos" University of Galaţi III:83-87. http://www.ann.ugal.ro/eeai/index.html

Nichita C, Luca D, Dakyo B, Ceangă E (2002) Large band simulation of the wind speed for real time wind turbine simulators. IEEE Transactions on Energy Conversion 17(4):523-529

Nichita C, El Mokadem M, Dakyo B (2006) Wind turbine simulation procedures. Wind Engineering 30(3):187-200

Niksefat N, Sepehri N (2000) Design and experimental evaluation of a robust force controller for an electro-hydraulic actuator via quantitative feedback theory. Control Engineering Practice 8:1335-1345

Nise N (2000) Control systems engineering, 3rd edn. John Wiley & Sons, New-York

Novak P, Ekelund T (1994) Modelling, identification and control of a variable speed HAWT. In: Proceedings of the European Wind Energy Conference – EWEC '94, pp 441-446

Nuñes AAC, Cortizo PC, Menezes BR (1993) Wind turbine simulator using a DC machine and a power reversible converter. In: Proceedings of the ICEM Conference, pp 536-540

Oppenheim AV, Willsky AS, Nawab SH (1996) Signals and systems, 2nd edn. Prentice Hall

Papadopoulos MP, Papathanassiou SA, Tentzerakis ST, Boulaxis NG (1998) Investigation of the flicker emission by grid connected wind turbines. In: Proceedings of the International Conference on Harmonics and Quality of Power, vol 2, pp 1152-1157

Papathanassiou SA, Papadopoulos MP (2001) Dynamic characteristics of autonomous wind-diesel systems. Renewable Energy 23:293-311

Parfit M, Leen S (2005) Energy of the future (Energia viitorului). National Geographic 8:32-57

Peña R, Clare JC, Asher GM (1996) Doubly fed induction generator using back-to-back PWM converters and its application to variable-speed wind-energy generation. IEE Proceedings on Electric Power Applications 143(3):231-241

Peña RS, Cardenas RJ, Asher GM, Clare JC (2000) Vector controlled induction machines for stand-alone wind energy applications. In: Proceedings of the IEEE Industry Application Annual Meeting – IAS 2000, pp 1409-1415

Pierce K (1999) Control method for improved energy capture below rated power. Technical Report, National Renewable Energy Laboratory, Colorado, U.S.A.

Pike AW, Grimble MJ, Johnson MA, Ordys AW, Shakoor S (1996) Predictive control. In: Levine WS (ed.) The control handbook. CRC Press, IEEE Press, pp 805-814

Pöller MA (2003) Doubly-fed induction machine models for stability assessment of wind farms. In: Proceedings of the 2003 IEEE PowerTech Conference, vol 3, 6 pages

Rabelo B, Hofmann W (2002) DSP-based experimental rig with the doubly-fed induction generator for wind-turbines. In: Proceedings of the 10th International Power Electronics and Motion Control Conference – EPE-PEMC 2002 (CD-ROM)

Rabelo B, Hofmann W (2003) Control of an optimized power flow in wind power plants with doubly-fed induction generators. IEEE 34th Annual Power Electronics Specialist Conference – PESC 2003, 4:1563-1568

Richardson R, Erdman W (1992) U.S. Patent No. 5,083,039

Rodriguez-Amenedo JL, Rodriguez-Garcia F, Burgos JC, Chincilla M, Arnalte S, Veganzones C (1998) Experimental rig to emulate wind turbines. In: Proceedings of the ICEM Conference, vol 3, pp 2033-2038

Ruin S, Carlson O (2000) Wind-hybrid systems with variable speed and DC-link. In: Proceedings of the Wind Power for the 21st Century Conference

Saad-Saoud Z, Jenkins N (1999) Models for predicting flicker induced by large wind turbines. IEEE Transactions on Energy Conversion 14(3):743-748

Saad-Saoud Z, Lisboa ML, Ekanayake JB, Jenkins N, Strbac G (1998) Application of STATCOMs to wind farms. IEE Proceedings – Generation, Transmission and Distribution 145(5):511-516

De La Salle SA, Reardon D, Leithead WE, Grimble MJ (1990) Review of wind turbine control. International Journal of Control 52(6):1295-1310

Savaresi SM (1999) Exact feedback linearization of a fifth-order-model of synchronous generators. IEE Proceedings of Control, Theory and Applications 146(1):53-57

Schiemenz I, Stiebler M (2000) Maximum power point tracker of a wind energy system with a permanent-magnet synchronous generator. In: Proceedings of ICEM 2000, pp 1083-1086

Schiemenz I, Stiebler M (2001) Control of a permanent magnet synchronous generator used in a variable speed wind energy system. In: The IEEE International Electric Machines and Drives Conference – IEDMC 2001, pp 872-877

Seguro JV, Lambert TW (2000) Modern estimation of the parameters of the Weibull wind speed distribution for wind energy analysis. Journal of Wind Engineering and Industrial Aerodynamics 85(1):75-84

Shamma JS (1996) Linearization and gain-scheduling. In: Levine WS (ed.) The control handbook. CRC Press, IEEE Press, pp 388-398

Simoes MG, Bose BK, Spiegel RJ (1997) Fuzzy logic based intelligent control of a variable speed cage machine wind generation system. IEEE Transactions on Power Electronics 12(1):87-95

Skogestad S, Morari M, Doyle JC (1989) Robust control of ill conditioned plants; high purity distillation. IEEE Transactions on Automatic Control 33:1092-1105

Song S-H, Kang S-I, Hahm N-K (2003) Implementation and control of grid connected AC-DC-AC power converter for variable speed wind energy conversion system. Eighteenth Annual IEEE Applied Power Electronics Conference and Exposition 1(9):154-158

Sørensen P, Hansen AD, Rosas PAC (2002) Wind models for simulation of power fluctuations from wind farms. Journal of Wind Engineering and Industrial Aerodynamics 90:1381-1402

Sørensen P, Hansen AD, Iov F, Blaabjerg F, Donovan MH (2005) Wind farm models and control strategies. Technical Report RISØ-R-1464(EN), RISØ National Laboratory, Roskilde, Denmark

Sørensen P, Cutululis NA, Hjerrild J, Jensen L, Donovan M, Cristensen LEA, *et al.* (2006) Power fluctuations from large offshore wind farms. In: Proceedings of the Nordic Wind Power Conference, 5 pages

Spée R, Bhowmik S, Enslin JHR (1994) Adaptive control strategies for variable-speed doubly-fed wind power generation systems. In: Proceedings of the IEEE Industry Applications Society Annual Meeting – IAS 1994, vol 1, pp 545-552

Steurer M, Li H, Woodruff S, Shi K, Zhang D (2004) Development of a unified design, test, and research platform for wind energy systems based on hardware-in-the-loop real time simulation. In: Proceedings of the 35th Annual IEEE Power Electronics Specialists Conference, pp 3604-3608

Sun T (2004) Power quality of grid-connected wind turbines with DFIG and their interaction with the grid. Ph.D. Thesis, Aalborg University, Denmark

Tan K, Islam S (2004) Optimum control strategies in energy conversion of PMSG wind turbine system without mechanical sensors. IEEE Transactions on Energy Conversion 19(2):392-399

Teel AR, Georgiou TT, Praly L, Sontag E (1996) Input-output stability. In: Levine WS (ed.) The control handbook. CRC Press, IEEE Press, pp 895-908

Teodorescu R, Iov F, Blaabjerg F (2003) Flexible development and test system for 11-kW wind turbine. In: Proceedings of the IEEE 34th Annual Power Electronics Specialist Conference – PESC '03, vol 1, pp 67-72

Thiringer T (1996) Periodic pulsation from a three-bladed wind turbine. IEEE Transactions on Energy Conversion 16(3):128-133

Thresher R, Dodge W, Darell M (1998) Trends in the evolution of wind turbines generator configurations and systems. Wind Energy 1(S1):70-85

Tomilson A, Qualcoie J, Gosine R, Hinchey M, Bose N (1998) Application of a static VAR compensator to an autonomous wind/diesel system. Wind Engineering 22(3):131-141

Torres E, Garcia-Sanz M (2004) Experimental results of the variable speed, direct drive multipole synchronous wind turbine TWT1650. Wind Energy 7:109-118

Tsoumas I, Safacas A, Tsimplostefanakis E, Tatakis E (2003) An optimal control strategy of a variable speed wind energy conversion system. In: Proceedings of the 6th International Conference on Electrical Machines and Systems – ICEMS 2003, vol 1, pp 274-277

UpWind (2006) FP6 European research project. www.upwind.eu

Utkin VA (1971) Equations of sliding mode in discontinuous systems. Automation and Remote Control 1(12):1897-1907

Vechiu I, Camblong H, Tapia G, Dakyo B, Nichita C (2004) Dynamic simulation of a hybrid power system: performance analysis. In: Proceedings of the 2004 Wind Energy Conference (CD-ROM)

Veers PS, Butterfield S (2001) Extreme load estimation for wind turbines: issues and opportunities for improved practice. Technical Report AIAA-2001-0044, Sandia National Laboratories, U.S.A.

Vihriälä H (2002) Control of variable speed wind turbines. Ph.D. Thesis, Tampere University of Technology, Finland

Voicu M (1986) Analysis techniques for control system stability (Tehnici de analiză a sistemelor automate). Technical Publishing House, Bucharest, Romania

Wang, Q, Chang L (1999) An independent maximum power extraction strategy for wind energy conversion systems. In: Proceedings of the 1999 IEEE Canadian Conference on Electrical and Computer Engineering, pp 1142-1147

Wang Q, Chang L (2004) An intelligent maximum power extraction algorithm for inverter-based variable speed wind turbine systems. IEEE Transactions on Power Electronics 19(5):1242-1249

Wang Y, Hill DJ, Middleton RH, Gao L (1993) Transient stability enhancement and voltage regulation of power systems. IEEE Transactions on Power Systems 8(2):620-626

Welfonder E, Neifer R, Spanner M (1997) Development and experimental identification of dynamic models for wind turbines. Control Engineering Practice 5(1):63-73

Wilkie J, Leithead WE, Anderson C (1990) Modelling of wind turbines by simple models. Wind Engineering 4:247-274

Williams T, Antsaklis PJ (1996) Decoupling. In: Levine WS (ed.) The control handbook. CRC Press, IEEE Press, pp 795-804

Wu SF, Grimble MJ, Breslin SG (1998) Introduction to quantitative feedback theory for lateral flight control systems design. Control Engineering Practice 6:805-828

Young KD, Utkin VI, Özgüner U (1999) A control engineer's guide to sliding mode control. IEEE Transactions on Control System Technology 7(3):328-342

Zhang X-F, Xu D-P, Liu Y-B (2004) Adaptive optimal fuzzy control for variable speed fixed pitch wind turbines. In: Proceedings of the Fifth World Congress on Intelligent Control and Automation – WCICA 2004, vol 3, pp 2481-2485

Ziegler JG, Nichols NB (1942) Optimum settings for automatic controllers. Transactions of the American Society of Mechanical Engineers 64:759-768

Zmood DN, Donald Grahame Holmes DN, Bode GH (2001) Frequency-domain analysis of three-phase linear current regulators. IEEE Transactions on Industry Applications 37(2):601-611

Index

active stall, 3, 74
actuator disc, 15
aerodynamic efficiency, 18, 72
aerodynamic subsystem, 29
all-pass filter, 212
ARMA model, 182

balance control, 102
basic physical system, 220
basic reproducibility conditions, 222
Betz limit, 16
blade element theory, 37
blade
   edge, 208
   feathering, 84
   flap, 208

centrifugal regulator, 74
chattering, 111
converter
   back-to-back, 22
   grid-side, 22
   rotor-side, 22

damping factor, 208, 215
Danish concept, 20, 103, 218
Delta control, 102
diffeomorphism, 250
Discrete Fourier Transform, 121, 166
doubly-fed induction generator
   modelling, 47
   motion control, 99
drag coefficient, 38, 214
drag force, 38
drive train, 14, 55, 56
driving variable, 221

effector, 221, 232
electromagnetic subsystem, 29
emulated physical system, 221
energy pattern factor, 12
equivalent control input, 112, 136, 145, 248
extremum seeking control, 117

Fast Fourier Transform, 122
flicker, 106
frequency separation principle, 92
full load, 18

gain-scheduling control, 91

high-frequency loop, 177
hill-climbing-like method, 110
HIL simulator
   speed-driven, 224
   torque-driven, 224
horizontal-axis wind turbines, 14
hub, 14
hybrid power system, 5, 23, 151

incidence angle, 38
induction lag, 41, 43
instrumental variables identification
   method, 174
interaction variables, 221
interference factor
   axial flow, 38, 40
   tangential flow, 38, 40
Inverse Fourier Transform, 210
inverter
   turbine-side, 58
   grid-side, 58, 59
investigated physical system, 221

Jacobian matrix, 253

Kaimal's spectrum, 32

Lagrange coefficient, 39
Lie derivative, 249
lift coefficient, 38, 214
lift force, 38
loop shaping, 159, 257
low-frequency loop, 177
LPV control, 93

maximum power point tracking, 76, 117
microgrid, 5
mixed optimization criteria, 92, 186, 207
multi-model control structure, 162

nominal plant, 158, 257
normal form, 153

optimal regimes characteristic, 75

Park Transform, 51
partial load, 18
passive stall, 3, 74
PI power control, 131
PI speed control, 130
pitch
   actuator, 74
   control, 3, 73, 84
pitch-to-feather, 73, 84, 86
pole placement, 113, 130, 155, 201
power coefficient, 16, 79
Prandtl's coefficient, 40
prefilter, 158, 258
probing signal, 117

Rayleigh's distribution, 10, 32
real-time physical simulator, 221, 233
real-time software simulator, 222, 232
reduced order model, 49
reduced-order dynamics, 142
reference chord, 38
relative degree, 153, 251
response variable, 221
Riccati equation, 186, 200
right-half-plane zero, 212
rotational sampling filter, 43
rotationally-sampled wind power spectrum, 44, 210
roughness length, 33
R-S-T controller, 172

self-oscillations, 136
shaft
   high-speed, 14, 46
   low-speed, 14
shaping filter, 34, 183, 198
simulation errors, 226
single-mass model, 143
softstarter, 20
spatial filter, 42
squirrel-cage induction generator
   modelling, 49
   motion control, 96
stall control, 3, 74, 82, 86, 214
state observer, 200
steady-state optimization, 113
switching (sliding) surface, 143, 247
synchronous generator
   permanent-magnet, 51
   wound-rotor, 53

thrust force, 38, 214
tip speed ratio, 17
   optimal, 18
torque coefficient, 18, 41
torque parameter, 63, 91, 178, 198
torsional vibrations, 215
total harmonic distortion, 108
tower
   displacement, 208
   fore-aft, 208
   oscillations, 215
   shadow, 42, 209
   side-to-side, 208
trade-off coefficient, 114
turbulence
   intensity, 32
   length of, 32
   wind speed, 31

van der Hoven's spectrum, 35
von Karman's spectrum, 32, 210

WECS
   fixed-speed, 2, 20, 103
   semi-variable-speed, 103
   slow dynamics of, 177, 182
   turbulent dynamics of, 177, 182
   variable-speed, 2, 21, 105
Weibull distribution, 9
   scale parameter, 10
   shape parameter, 10
weighting parameter, 92

wind power density, 12
wind shear, 42, 209
wind speed
  cut-in, 18
  cut-out, 18
  high-frequency, 31, 177
  low-frequency, 31, 177
  most probable, 13
  predicted, 180
  rated, 18
  relative, 38
wind torque, 40

zero-pole distribution, 65

Other titles published in this series (continued):

Soft Sensors for Monitoring and Control of Industrial Processes
Luigi Fortuna, Salvatore Graziani, Alessandro Rizzo and Maria G. Xibilia

Adaptive Voltage Control in Power Systems
Giuseppe Fusco and Mario Russo

Advanced Control of Industrial Processes
Piotr Tatjewski

Process Control Performance Assessment
Andrzej W. Ordys, Damien Uduehi and Michael A. Johnson (Eds.)

Modelling and Analysis of Hybrid Supervisory Systems
Emilia Villani, Paulo E. Miyagi and Robert Valette

Process Control
Jie Bao and Peter L. Lee

Distributed Embedded Control Systems
Matjaž Colnarič, Domen Verber and Wolfgang A. Halang

Precision Motion Control (2nd Ed.)
Tan Kok Kiong, Lee Tong Heng and Huang Sunan

Model-based Process Supervision
Arun K. Samantaray and
Belkacem Ould Bouamama
Publication due April 2008

Identification of Continuous-time Models from Sampled Data
Hugues Garnier and Liuping Wang (Eds.)
Publication due April 2008

Dry Clutch Control for Automated Manual Transmission Vehicles
Pietro J. Dolcini, Carlos Canudas-de-Wit and Hubert Béchart
Publication due May 2008

Real-time Iterative Learning Control
Xu Jian-Xin, Sanjib K. Panda and Lee Tong Heng
Publication due July 2008

Model Predictive Control Design and Implementation Using MATLAB®
Liuping Wang
Publication due July 2008

Magnetic Control of Tokamak Plasmas
Marco Ariola and Alfredo Pironti
Publication due July 2008

Design of Fault-tolerant Control Systems
Hassan Noura, Didier Theilliol, Jean-Christophe Ponsart and Abbas Chamseddine
Publication due October 2008

Printed in the United States
107230LV00003B/184-210/A